Perspectives in Micro- and Nanotechnology for Biomedical Applications

Perspectives in Micro- and Nanotechnology for Biomedical Applications

Edited by

Juliana M. Chan
Chenjie Xu

Nanyang Technological University, Singapore

Imperial College Press

Published by

Imperial College Press
57 Shelton Street
Covent Garden
London WC2H 9HE

Distributed by

World Scientific Publishing Co. Pte. Ltd.
5 Toh Tuck Link, Singapore 596224
USA office: 27 Warren Street, Suite 401-402, Hackensack, NJ 07601
UK office: 57 Shelton Street, Covent Garden, London WC2H 9HE

Library of Congress Cataloging-in-Publication Data
Names: Chan, Juliana M., editor. | Xu, Chenjie, editor.
Title: Perspectives in micro- and nanotechnology for biomedical applications /
    [edited by] Juliana M. Chan, Chenjie Xu.
Description: Hackensack, New Jersey : Imperial College Press, 2016. |
    Includes bibliographical references and index.
Identifiers: LCCN 2016003861 | ISBN 9781783269600 (hardback : alk. paper)
Subjects: | MESH: Microtechnology | Nanomedicine | Nanostructures--therapeutic use |
    Drug Delivery Systems--trends | Biomedical Engineering
Classification: LCC R857.M3 | NLM QT 36.5 | DDC 610.28--dc23
LC record available at http://lccn.loc.gov/2016003861

British Library Cataloguing-in-Publication Data
A catalogue record for this book is available from the British Library.

Typeset by Stallion Press
Email: enquiries@stallionpress.com

Introduction: Intersection Between and Opportunities for Microtechnologies and Nanotechnologies in Biomedical Applications

Chenjie Xu, Juliana M. Chan and Paul S. Weiss

Microtechnology and nanotechnology manipulate and measure matter on atomic, molecular and supramolecular scales. The key building blocks and functions of biological systems are at the micro- and nanoscales: ranging from proteins, nucleic acids and other macromolecules (1–100 nm) to organelles (100 nm–1 μm), single cells ($\sim$10 μm) and small tissues ($>$100 μm). Likewise, viruses and bacteria sit within these ranges.

Over the past several decades, numerous opportunities have arisen for biomedical researchers to apply micro- and nanotechnologies in healthcare applications. In particular, there are an increasing number of important applications for micro- and nanotechnologies in surgery, cancer diagnosis and therapy, identification of disease markers, molecular imaging, implant technology, tissue engineering, and drug delivery.

Although many medical applications are still in their infancy, an increasing number of products are under clinical translation and some products are already commercially available. These include nanoparticle-based drugs, contrast-enhancing agents for medical imaging, bone-replacement materials, anti-microbial textiles, chips for *in vitro* molecular diagnostics, etc.[1–3] The everyday lives of millions of people have already been improved by innovations in micro- and nanotechnology.

Among the approximately 250 nanomaterial-based medical products that have been approved or are in various stages of clinical study,[4] the first nanoparticle medicines to be FDA approved are liposomes, described by Bangham *et al.* as early as 50 years ago.[5] Liposomes have since been used for the delivery of both small molecules and protein drugs. In the following four decades, research into liposome and lipid nanoparticle drug delivery has led to the development of the first FDA-approved nanomedicine, DOXIL, as well as 12 additional therapeutics.[6]

More recently, liposomes have been used to deliver nucleic acid therapeutics such as small interfering RNAs.[7] They protect nucleic acid agents from degradation by nucleases and also facilitate endosomal escape. One example is ALN-CC5 from Alnylam Pharmaceuticals, which is used to target complement component C5 for the treatment of complement-mediated diseases. In a phase 1/2 trial with 12 healthy volunteers, subcutaneous administration of a single dose of ALN-CC5 resulted in potent, dose-dependent, durable, and statistically significant knockdown of serum C5 of up to 96%.[8]

In addition to organic or polymeric systems, there are substantial roles for inorganic micro- and nanosystems in biomedical applications. For example, superparamagnetic iron oxide nanoparticles act as excellent contrast agents in magnetic resonance imaging (MRI), produce heat as clinical hyperthermia agents, and act as iron supplements for chronic anemia.[6] Aqueous suspensions of silicone-coated iron oxide nanoparticles of approximately 400 nm (Ferumoxsil) have been shown to improve the sensitivity and specificity of iron oxide nanoparticles for the detection of suspected gastrointestinal tract lesions. FDA approval was granted in 1996 to use Ferumoxsil to image the upper gastrointestinal tract.

More recently, German biotechnology company MagForce Nanotechnologies AG directly injected magnetic nanoparticles into recurrent gliomas. The combination of alternating magnetic field heating with radiotherapy resulted in a survival time (from primary diagnosis) of 23.2 months compared to 14.6 months with radiotherapy only.[9] Ferumoxytol (marketed as Feraheme in the USA, Rienso in the EU), which are 1–30 nm carbohydrate (polyglucose sorbitol carboxymethyl ether)-coated iron oxide nanoparticles, are not only an FDA-approved supplement for the treatment of iron deficiency in patients with chronic kidney disease,[10] but also serve as MRI contrast agents. These particles may potentially also be used as imaging enhancers to detect an inflammatory response in patients with acute myocardial infarction.[11]

Motivated by the pioneering work of researchers before them, researchers today are developing an array of new agents and devices for clinical use.[12] To summarize recent developments and understand existing challenges, an international ensemble of leading experts in micro- and nanotechnology have contributed chapters to this book, covering many major topics in the field.

The book starts with a discussion of classical chemical tools for understanding DNA methylation and histone modification by Dr. Kikuchi at Osaka University. Dr. Han from the University of Massachusetts and Dr. Xu from Nanyang Technological University discuss the use of inorganic nanoparticles (lanthanide-doped up-conversion nanoparticles or magnetic nanoparticles) as imaging probes, drug carriers and therapy sensitizers. After a discussion of inorganic nanoparticles, four researchers describe the use of organic or polymeric micro- and nanomaterials for drug delivery. Dr. Hong from the University of Illinois at Chicago covers the development of dendritic nanocarriers for drug delivery applications; Dr. Xu from Tufts University focuses on the development of a combinatorial cationic lipid library for drug delivery; Dr. Green from Johns Hopkins University describes the use of biodegradable polymeric nanoparticles for gene delivery; and Dr. Gu from North Carolina State University focuses on hydrogels. Next, Dr. Machluf from the Israel Institute of Technology summarizes recent developments in the use of cellular membranes called cell ghosts for drug delivery. Finally, Dr. Chan from Nanyang Technological University discusses applications of nanoparticles for the treatment of skin diseases.

This book covers the basic principles and applications of today's most promising micro- and nanotechnologies, and provides an up-to-date reference book for researchers. It is suitable for undergraduate and graduate students in biomedicine and biomedical engineering.

Chenjie Xu
School of Chemical and Biomedical Engineering
Nanyang Technological University

Juliana M. Chan
School of Chemical and Biomedical Engineering and
Lee Kong Chian School of Medicine
Nanyang Technological University

Paul S. Weiss
California NanoSystems Institute and Departments of Chemistry & Biochemistry
and of Materials Science & Engineering
University of California, Los Angeles

References

1. Mitragotri, S. *et al.* Accelerating the translation of nanomaterials in biomedicine. *ACS Nano* 9, 6644–6654 (2015).
2. Farokhzad, O.C., and Langer, R. Impact of nanotechnology on drug delivery. *ACS Nano* 3, 16–20 (2009).
3. Castleberry, S. *et al.* Nanolayered siRNA dressing for sustained localized knockdown. *ACS Nano* 7, 5251–5261 (2013).
4. Etheridge, M.L. *et al.* The big picture on nanomedicine: the state of investigational and approved nanomedicine products. *Nanomedicine* 9, 1–14 (2013).
5. Deamer, D.W. From "Banghasomes" to liposomes: a memoir of Alec Bangham, 1921–2010. *The FASEB Journal* 24, 1308–1310 (2010).
6. Min, Y. *et al.* Clinical translation of nanomedicine. *Chemical Reviews*, DOI: 10.1021/acs.chemrev.5b00116 (2015).
7. Balazs, D.A., and Godbey, W. Liposomes for use in gene delivery. *Journal of Drug Delivery* 2011, 326497 (2011).
8. Alnylam Pharmaceuticals, I. (BUSINESS WIRE, 2015).
9. Huang, H.S., and Hainfeld, J.F. Intravenous magnetic nanoparticle cancer hyperthermia. *International Journal of Nanomedicine* 8, 2521–2532 (2013).
10. Kowalczyk, M. *et al.* Ferumoxytol: A new era of iron deficiency anemia treatment for patients with cronic kidney disease. *Journal of Nephrology* 24, 717–722 (2011).
11. Sosnovik, D.E., and Nahrendorf, M. Cells and iron oxide nanoparticles on the move: magnetic resonance imaging of monocyte homing and myocardial onflammation in patients pith ST-elevation myocardial infarction. *Circulation: Cardiovascular Imaging* 5, 551–554 (2012).
12. Oklu, R. *et al.* Patient-inspired engineering and nanotechnology. *ACS Nano* 9, 7733–7734 (2015).

Contributing Authors

Sarah Altinoglu
Department of Biomedical Engineering
Tufts University
Medford, MA 02155
USA

Jason Bugno
Department of Biopharmaceutical Sciences
College of Pharmacy
University of Illinois
Chicago, IL 60612
USA

Juliana M. Chan
School of Chemical and Biomedical Engineering and
Lee Kong Chian School of Medicine
Nanyang Technological University
Singapore 637457

Hong Y. Cho
Department of Biopharmaceutical Sciences
College of Pharmacy
University of Illinois
Chicago, IL 60612
USA

Jordan J. Green
Department of Biomedical Engineering, Translational Tissue Engineering Center
and Institute for Nanobiotechnology
Johns Hopkins University School of Medicine
733 N Broadway
Baltimore, MD 21205
USA

Departments of Materials Science and Engineering, Ophthalmology,
and Neurosurgery
Johns Hopkins University School of Medicine
733 N Broadway
Baltimore, MD 21205
USA

Zhen Gu
Joint Department of Biomedical Engineering
University of North Carolina at Chapel Hill and North Carolina State University
Raleigh, NC 27695
USA

Pharmaceutics Division
Eshelman School of Pharmacy
University of North Carolina at Chapel Hill
Chapel Hill, NC 27599
USA

Gang Han
Biochemistry and Molecular Pharmacology Department
University of Massachusetts-Medical School
364 Plantation Street, LRB 806
Worcester, MA 01605
USA

Seungpyo Hong
Department of Biopharmaceutical Sciences
College of Pharmacy University of Illinois
Chicago, IL 60612
USA

Kazuya Kikuchi
Graduate School of Engineering, and Immunology Frontier
Research Center (IFReC)
Osaka University
Suita, Osaka, 5650871
Japan

Jayoung Kim
Department of Biomedical Engineering, Translational Tissue Engineering Center,
and Institute for Nanobiotechnology
Johns Hopkins University School of Medicine
733 N Broadway
Baltimore, MD 21205
USA

Kristen L. Kozielski
Department of Biomedical Engineering, Translational Tissue Engineering Center
and Institute for Nanobiotechnology
Johns Hopkins University School of Medicine
733 N Broadway
Baltimore, MD 21205
USA

Sangeetha Krishnamurthy
School of Chemical and Biomedical Engineering
Nanyang Technological University
Singapore 637457

Yue Lu
Joint Department of Biomedical Engineering
University of North Carolina at Chapel Hill
and North Carolina State University
Raleigh, NC 27695
USA

Pharmaceutics Division
Eshelman School of Pharmacy
University of North Carolina at Chapel Hill
Chapel Hill, NC 27599
USA

Marcelle Machluf
The Lab for Cancer Drug Delivery & Cell Based Technologies
Faculty of Biotechnology & Food Engineering
Technion — Israel Institute of Technology
Haifa, 3200003
Israel

Malathi Mathiyazhakan
School of Chemical and Biomedical Engineering
Nanyang Technological University
Singapore 637457

Masafumi Minoshima
Graduate School of Engineering
Osaka University
Suita, Osaka, 5650871
Japan

Beth Schoen
The Lab for Cancer Drug Delivery & Cell Based Technologies
Faculty of Biotechnology & Food Engineering
Technion — Israel Institute of Technology
Haifa, 3200003
Israel

Razina Z. Seeni
School of Chemical and Biomedical Engineering
Nanyang Technological University
Singapore 637457

Paul S. Weiss
California NanoSystems Institute and Departments of Chemistry
& Biochemistry and of Materials Science & Engineering
University of California, Los Angeles
Los Angeles, CA 90095
USA

Yuyan Weng
Center for Soft Condensed Matter Physics and Interdisciplinary Research
Soochow University
Suzhou 215006, China

David R. Wilson
Department of Biomedical Engineering, Translational Tissue Engineering Center,
and Institute for Nanobiotechnology
Johns Hopkins University School of Medicine
733 N Broadway
Baltimore, MD 21205
USA

Chenjie Xu
School of Chemical and Biomedical Engineering
Nanyang Technological University
Singapore 637457

Qiaobing Xu
Department of Biomedical Engineering
Tufts University
Medford, MA 02155
USA

Department of Chemical and Biological Engineering
Tufts University
Medford, MA 02155
USA

School of Medicine, Tufts University
Boston, MA 02111
USA

Program in Cell, Molecular and Developmental Biology
Sackler School of Graduate Biomedical Science
Tufts University
Boston, MA 02111
USA

Yuanwei Zhang
Biochemistry and Molecular Pharmacology Department
University of Massachusetts-Medical School
364 Plantation Street, LRB 806
Worcester, MA 01605
USA

Contents

Chemical Tools for Elucidation of Epigenetic Mechanisms

1

Masafumi Minoshima* and Kazuya Kikuchi†

* Graduate School of Engineering, Osaka University, Suita, Osaka, 5650871, Japan
† Graduate School of Engineering, and Immunology Frontier Research Center (IFReC), Osaka University, Suita, Osaka, 5650871, Japan

Abstract

Epigenetic chemical modifications such as DNA methylation and histone modification play key roles in regulating gene expression without genetic alteration in various physiological events. Aberrant gene expression by altered epigenetic modifications can lead to cancer and other diseases. Epigenetic regulator proteins responsible for writing, erasing and reading epigenetic marks are of interest as novel diagnostic and therapeutic targets. Here, we review the key chemical tools for studying epigenetic processes, including DNA methylation/demethylation, histone acetylation/deacetylation, and histone methylation/demethylation. We describe methods for detecting epigenetic modifications and regulator proteins by using chemical tools in combination with genomic and proteomic analysis. We also discuss the use of chemical probes, such as synthetic histones with site-specific modifications, as well as chemical inhibitors of epigenetic regulators for elucidating the molecular mechanisms of epigenetic gene regulation and its physiological function. These chemical tools are promising strategies for not only uncovering epigenetic roles in biology but also for developing diagnostic or therapeutic drugs applicable to clinical practice.

1.1 Introduction

Epigenetics refers to heritable changes in gene function that occur without the alteration of DNA sequences. In the early 1940s, Waddington used the related term, "epigenesist", to explain that living organisms acquire functions to adapt to their environments during development and differentiation.[1] This acquired phenotype is then passed on without changes in genetic sequence. Later, although researchers had found that DNA methylation and histone acetylation are associated with transcriptional activity,[2,3] the regulatory mechanisms remained unclear. From the late 1980s to the 1990s, the responsible regulatory factors, such as methylated DNA-binding proteins (MDBs), histone acetyltransferases (HATs), and deacetylases (HDACs), were identified; this accelerated our understanding of the mechanisms underlying epigenetic gene regulation. These modifications are regarded as essential signals for controlling nuclear chromatin structures and transcriptional activity. Such gene regulatory mechanisms play key roles in biological phenomena such as X-inactivation and genomic imprinting. In the former example, one of two X chromosomes in females is completely inactivated during development, by the binding of long noncoding RNA, DNA methylation, and histone modifications, without involving any change in sequence information.[4] In the past two decades, DNA methylation and histone modifications have been identified as key players in the regulation of genes involved in a variety of physiological functions and diseases. These changes reflect the status of cells during the cell cycle, development, differentiation, and cell reprogramming, as well as cellular responses to environmental stress. For example, the state of DNA methylation changes dynamically during eukaryotic cell development,[5] contributing to tissue- and lineage-specific gene regulation. Moreover, specific epigenetic marks cause aberrant expression of genes, which are involved in cancer, neurodegenerative disorders, autoimmune diseases, and so on. For instance, aberrantly high DNA methylation in specific genomic regions, the loss of histone acetylation, and the gain of histone methylation are all observed in cancer.[6] Therefore, DNA methylation and histone modifications are potential targets for diagnostic and therapeutic applications. Indeed, inhibitors of DNA methyltransferases and histone deacetylases have already shown efficient anticancer activity in clinical applications.[7]

In this review, we describe the development of chemical tools for elucidating these epigenetic processes. These tools have contributed to advancements in the field of epigenetics. We describe methods, using chemical tools, for detecting and investigating the factors responsible for DNA methylation and histone modification. Although antibodies are also powerful analytical tools in this field, we mainly focus on chemical probes used for genomic and proteomic analysis of epigenetic modifications and associated proteins. We also discuss recent advances in the development of chromatin-regulating tools that employ synthetic chromatins and chemical inhibitors.

1.2   Overview of Epigenetic Chemical Modifications

In eukaryotic cells, DNA is wrapped compactly to fit inside the nucleus. Nuclear DNA is bound to small histone proteins, forming higher-order chromatin structures. The nucleosome, a unit of chromatin, consists of approximately 147 bp of DNA and an octamer comprised of dimers of four histones (H2A, H2B, H3 and H4). The X-ray crystal structure of the DNA–histone octamer complex exhibits a globular core that is separated structurally from a flexible N-terminal tail [Figure 1.1(a)].[8] Such assembled chromatin structures are significantly associated with transcriptional activity. Genes contained in loose chromatin structures (euchromatin) are actively transcribed, whereas genes present in condensed chromatin structures (heterochromatins) are inactivate (silenced). Post-translational modifications of histones regulate the state of the chromatin structures and, therefore, of gene expression. Acetylation and methylation mainly occur at the ε-amino groups of lysines present in the histone tails [Figure 1.1(b)]. Acetylation at the positively-charged lysines alters the charge state of the histone tails and decreases their interaction with negatively-charged DNA; this is thought to result in the loosening of the chromatin structure and enhanced accessibility of transcription factors and RNA polymerases that facilitate gene expression.[9] Other modifications, such as arginine methylation

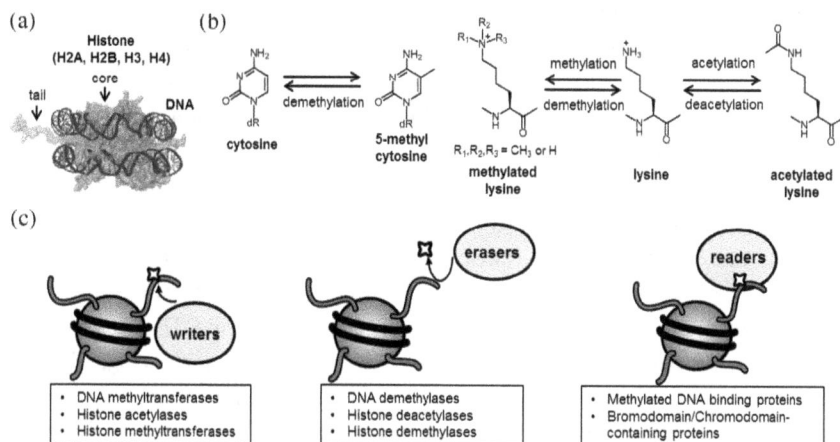

Figure 1.1:   (a) Crystal structure (PDB ID: 1AOI) of a nucleosome. DNA and histones are shown in black and gray, respectively. (b) Epigenetic chemical modifications in DNA methylation and histone acetylation/methylation. (c) Schematic illustration of the roles played by epigenetic writers, erasers and readers.

and deimination, serine/threonine phosphorylation, and lysine ubiquitination of histones also occur.

Eukaryotic DNA is also methylated at the C-5 position of nucleobase cytosines during post-replication processes [Figure 1.1(b)].[10] In the human genome, 5-methyl-cytosines (^{m}Cs) account for approximately 1% of the total bases in the DNA, while approximately 70–80% of cytosines are methylated in 5^t-CpG-3^t sequences. Methylated DNA is mainly located in gene promoters and untranscribed satellite regions that are comprised of highly repeated sequences. DNA methylation is closely associated with heterochromatin formation, which silences gene expression and maintains genomic stability. During DNA replication, the methylated DNA in both strands (full-methylated DNA) is divided into two cells. In these daughter cells, a hybrid between methylated and nascent unmethylated DNA strands (termed hemimethylated DNA) can be converted to fully-methylated DNA by DNA methyltransferases (DNMT1 and DNMT2); in this way, the original DNA methylation status is maintained. Other types of DNMTs (DNMT3a and DNMT3b) attach methyl groups to unmethylated DNA in both strands (termed *de novo* methylation).

Key effectors of epigenetic modifications are classified into three types, viz. writers, erasers and readers, based on their roles [Figure 1.1(c)]. Writers are enzymes that attach epigenetic marks to histones and DNA, e.g., histone acetyltransferases (HATs), histone methyltransferases (KMTs), and DNMTs. HATs catalytically transfer an acetyl group from acetyl-CoA to the histone. In 1996, Brownell *et al.* discovered the acetyltransferase activity of GCN5, which was known as a transcriptional coactivator.[11] This finding directly linked histone acetylation with transcriptional activation. KMTs methylate the ε-amino groups of lysines, by catalyzing the transfer of methyl groups from the cofactor S-adenosylmethionine (SAM).[12] KMTs can produce mono-, di- and trimethylated products. Methylation of lysines has been correlated with both transcriptional activation and silencing, depending on the position of the histones within the DNA and the number of methyl groups. For example, trimethylation of lysine at positions 9 and 27 of histone H3 (H3K9me3 and H3K9me3, respectively) causes transcriptional silencing, whereas trimethylation of K4 of histone H3 (H3K4me3) results in transcriptional activation. Erasers, on the other hand, are enzymes that erase epigenetic marks. Histone deacetylases (HDACs) and histone demethylases (KDMs) are included in this category. HDACs include proteins that belong to the Rpd3/Hda1 family,[13] and hydrolyze an acetamide group from histones in a Zn^{2+}-dependent manner. Deacetylase activity has also been found in yeast;[14] the relevant protein, Sir2, serves as a gene silencer, causing substantial histone deacetylation. However, Sir2 uses a different deacetylation mechanism, as it functions in a nicotinamide dinucleotide (NAD)-dependent manner. Currently, HDACs are classified into four categories: classes I, II, and IV HDACs are Zn^{2+}-dependent, while class III HDACs are NAD-dependent hydrolases (sirtuins). Methylation of histones is reversible, as demonstrated by the discovery of histone demethylases (KDMs). Lysine-specific demethylase 1 (LSD1), the first demethylation enzyme to be discovered, oxidizes methyl groups of methylated lysines with the support of a flavin cofactor.[15] The hydroxymethylated intermediates can then be hydrolyzed, yielding a demethylated product with formaldehyde as a byproduct. However, LSD1 cannot demethylate trimethylated lysines. Researchers searching for novel lysine demethylases have identified lysine demethylation activity in enzymes containing jumonji-C domains.[16] Although these demethylases (JMJDMs) also catalyze the oxidative demethylation of lysines, they use Fe^{2+} and α-ketoglutarate in the oxidation step. JMJDMs are responsible for the removal of methyl groups from trimethylated lysines. While writers and erasers dynamically control the state of DNA and histone modification, readers are proteins that recognize specific epigenetic marks [e.g. methylated DNA-binding proteins (MBPs) and bromodomain-containing proteins]. Readers also serve as transcriptional co-activators or corepressors in the regulation of transcriptional activation or silencing.[17] The collaboration of writers, erasers and readers play a central role in the regulation of genes by epigenetic processes.

Chromatin structure and gene expression are influenced by the modification and position of histones, which reflect the transcriptional state of the relevant gene(s). Furthermore, the combination of epigenetic marks also modifies chromatin states. For example, H3K9me3 and H3K27me3 modifications are found simultaneously in heterochromatin regions. Allis and Jenuwein have suggested that the specific combinations of histone modifications and DNA methylation may determine the regulation of gene expression.[18,19] This "histone code hypothesis" expands the classical "genetic code".

1.3 Chemical Tools for Investigation of DNA Methylation Processes

1.3.1 Detection of DNA Methylation

DNA methylation is a critical epigenetic mark that causes changes in cell development, differentiation, reprogramming, and disease. Thus, selective detection of the amount and genomic location of methylated cytosines (^{m}Cs) is an important tool for evaluating the status of cells. Discrimination of ^{m}Cs from unmethylated cytosines (Cs) is challenging, because methylation at the C-5 position of cytosine does not affect base-pairing with guanines. Thus, detection methods based on DNA hybridization cannot be used to distinguish these nucleobases. Although thin layer chromatography (TLC) and high-performance liquid chromatography (HPLC) can separate ^{m}Cs from Cs, and provide a measure of the extent of methylation, they yield no information on the location of these bases in the genome.

In 1970s, Shapiro *et al.* and Hayatsu *et al.* found that the reactivity of cytosines to sodium bisulfite (NaHSO$_3$) is different from that of ^{m}Cs [Figure 1.2(a)].[20,21] Exposure to NaHSO$_3$ causes deamination of cytosine and cytosine derivatives at C-4 positions in a weak acid solution, yielding uracils. However, the deamination reaction rate of ^{m}Cs is much slower than that of Cs. This difference in reactivity may be due to a decrease in the rate of bisulfite addition to C-6 in C-5-methylated pyrimidines. Two decades after this discovery, these findings were applied to ^{m}C detection.[22] Bisulfite treatment of denatured DNA selectively converts Cs to Us, which base-pairs with As. The resultant DNA fragments can be PCR-amplified, cloned, and sequenced to determine the methylation maps of CpG dinucleotides. Advantages of the bisulfite method include positive detection of methylated CpG sites at a single-base resolution, which can be used to compare tissue-specific levels of methylation in target sequences. A few years after the bisulfite approach was first used to determine methylated sequences, Herman and coworkers developed a more facile and sensitive method, based on PCR, to detect the quality of methylation patterns in the genomic region of interest.[23] PCR amplification was performed on bisulfite-exposed DNA fragments using primers specific for methylated versus unmethylated DNA. Methylated DNA fragments were amplified specifically and detected on a gel, with good sensitivity. Moreover, using real-time PCR, methylation patterns can also be evaluated quantitatively. Such an analysis revealed the aberrant methylation status of the promoter of the p16 gene, a tumor suppressor gene that is down-regulated by methylation in cancer cells.[6] Recent advances in high-throughput DNA sequencing methods allow methylation sites to be detected at single-base resolution in whole genomes.[24] However, one problem with this method is DNA degradation during the long exposure to bisulfite reagents, which interferes with the collection of quantitative and reliable data.[25] Other than the bisulfite method, Tanaka *et al.* have developed a method to chemically lable ^{m}C directly by the formation of a methylcytosine glycol–dioxidoosmium–bipyridine ternary complex.[26,27] Although other detection methods such as ^{m}C-specific reactions using chemical reagents and modified oligonucleotides[28–30] have been reported, the bisulfite method remains the leading protocol for DNA methylation analysis. It is also clinically applicable as a diagnostic tool in cancer.

1.3.2 *Detection of DNA Demethylation*

Since DNA methylation forms a covalent C–C bond, this epigenetic marker is chemically stable and heritable during the replication of somatic cells. However, when somatic cells are reprogrammed during nuclear transfer to eggs, the methylation status is revised, resulting in the removal of methyl groups.[31] Although ^mC-containing oligonucleotides can be removed by the nucleotide excision repair pathway, studies suggest the possibility of an active demethylation pathway that converts ^mCs to Cs at the single-nucleotide level. Candidate enzymes have been reported; these, however, could not explain the mechanism underlying the direct removal of the methyl group from ^mCs.[32,33] In 2009, a novel nucleobase, 5-hydroxymethylcytosine (^{hm}C) was found in the brain, neurons and embryonic stem (ES) cells.[34,35] Approximately 40% of ^mCs in Purkinje neuronal DNA are converted to ^{hm}Cs.[34] In addition, ten–eleven translocation (Tet) family proteins, which are mammalian homologs of the trypanosome proteins JBP1 and JPB2 that belong to the oxygenase superfamily and require Fe^{2+} and 2-oxoglutarate as cofactors, have been demonstrated to be the enzymes responsible for ^{hm}C production.[35] Hydroxylation of ^mCs may affect chromatin structure, because methylated DNA-binding proteins cannot bind to ^{hm}C-containing DNA. Moreover, Tet enzymes further oxidize ^{hm}C into 5-formylcytosine (^fC) and 5-carboxycytosine (^{ca}C).[36] The fully oxidized ^{ca}C can be removed by mammalian thymidine glycosylase (TDG), an enzyme that is involved in base-exclusion repair.[37] This process is one proposed pathway for active DNA demethylation [Figure 1.2(b)]. Although it has been demonstrated that ^{hm}Cs exist in cells as an intermediate of DNA demethylation, its other roles and the reason for the abundance of ^{hm}Cs in specific cells remain unclear.

Detection of ^{hm}C and the related nucleobases are garnering interest in epigenetic studies. Song *et al.* developed a ^{hm}C-specific labeling method using enzymatic and chemical modifications [Figure 1.2(c)].[38] They used viral β-glucosyltransferases, which attach a glucose moiety from uridine diphosphoglucose (UDP-Glu) to the hydroxyl group of ^mC. The glucose moiety employed was chemically modified using click chemistry, allowing this 6-azide-glucose to be labeled with an alkyne-modified biotin. The biotin-tagged ^{hm}C-containing fragment was then enriched and sequenced to determine the distribution of ^{hm}Cs in cell lines and tissues. A similar approach has utilized the oxidation of glycosylated ^{hm}C following labeling with an aldehyde-reactive biotin.[39] Application of high-throughput sequencing analysis revealed that ^{hm}Cs are located at the start sites of genes that have specific histone marks in their promoter regions. Selective chemical oxidation of ^{hm}C to other nucleobase analogs with dinuclear peroxotungstate $[W(=O)(O_2)_2(H_2O)_2(\mu\text{-}O)]$ or

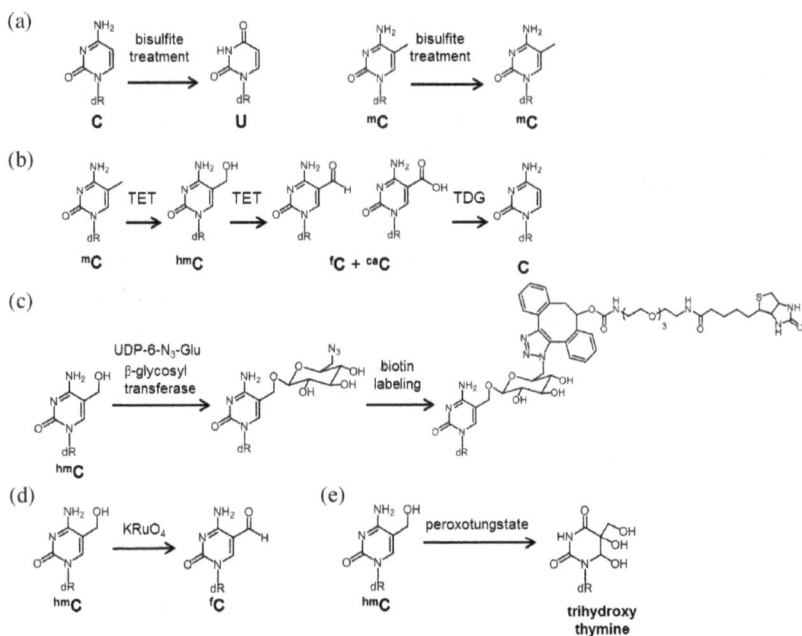

Figure 1.2: (a) Chemoselective conversion of ^{m}C by bisulfite treatment. (b) Proposed pathway of active DNA demethylation. (c) Labeling of ^{hm}C using enzyme-coupled glycosylation and click chemistry. (d) and (e) Chemoselective conversion of ^{hm}C to other detectable nucleobases.

perruthenate (KRuO$_4$) has also been reported [Figures 1.2(d) and (e)].[40,41] KRuO$_4$ selectively converts ^{hm}Cs to further oxidized ^{f}Cs, which can then be deaminated by bisulfite treatment, which, in turn, can then be used to map ^{m}Cs and ^{hm}Cs in genomic DNA quantitatively, at single-base resolution.[41] The recent discovery of ^{hm}Cs suggests a novel possible mechanism for the reversible control of methylated marks. Further biochemical analysis of ^{hm}Cs and related factors will elucidate the unknown roles of this nucleobase and shed further light on the maintenance of DNA methylation.

1.3.3 Regulation of DNA Methylation

DNMTs transfer a methyl group from SAM to the nucleobase. Aberrantly high DNA methylation at the promoter sequences of tumor suppressor genes represses the expression of these genes, which results in cancer progression.[6] Hence, decreasing

DNA methylation by inhibiting DNMTs is a promising approach to cancer therapy. The methylation mechanism involves an attack of the nucleobase by a nucleophilic cysteine residue, at C-6, to form a covalent 5,6-dihydropyrimidine adduct in the catalytic center of DNMTs. Subsequently, the activated C-5 carbon attacks a methyl group of SAM, and the subsequent beta elimination yields a 5-methylcytosine product. Based on this mechanism, several unnatural nucleobases and nucleosides, such as 5-azacytidine (^{aza}C), 5-aza-2^t-deoxycytidine (^{aza}dC), and 5-fluorodeoxycytidine, function as suicide inhibitors of DMNTs.[42] They can inhibit the elimination reaction to form a covalent adduct between a cysteine of DNMT and the C-6 position of a cytosine in DNA. If the inhibitors can be incorporated into cells, DNMT is inactivated, resulting in a decrease in DNA methylation; thus, treatment with ^{aza}C induces expression of tumor suppressor genes and inhibition of proliferation in cancer cells.[43] Kuch et al. have demonstrated the trapping of DNMT and inhibition of cancer cell growth by transfecting a stable dumbbell-shaped oligonucleotide, into which ^{aza}C is incorporated.[44] The modified oligonucleotide inhibited DNMT activity, without incorporating ^{aza}Cs into the genome.

1.4 Chemical Tools for Investigating the Histone Modification Processes

1.4.1 Detection of Histone Modifications

Post-translational histone modifications are also closely associated with gene regulation. As described above, the loss of histone acetylation and the gain of histone methylation are observed in malignant cells, resulting in aberrant gene expression. Thus, detection of histone modifications, such as lysine acetylation and methylation, is essential for understanding the status of cells. Antibodies against specific histone modifications are widely available for application in various biological experiments, such as western blotting, dot blotting, enzyme-linked immunosorbent assay (ELISA) and immunostaining. Moreover, chromatin immunoprecipitation (ChIP) is a powerful tool to investigate the DNA regions that interact with histones with specific modifications [Figure 1.3(a)].[45] Thereafter, genomic analysis of the DNA can be performed using oligonucleotide arrays (e.g. ChIP-chip) or high-throughput DNA sequencing (e.g. ChIP-Seq),[46,47] in order to reveal genomic DNA sequences localized within specific histone marks of interest. However, antibody approaches may present difficulties in terms of validity and reproducibility, due to the quality of the antibodies.[48] Some antibodies show cross-reactivity

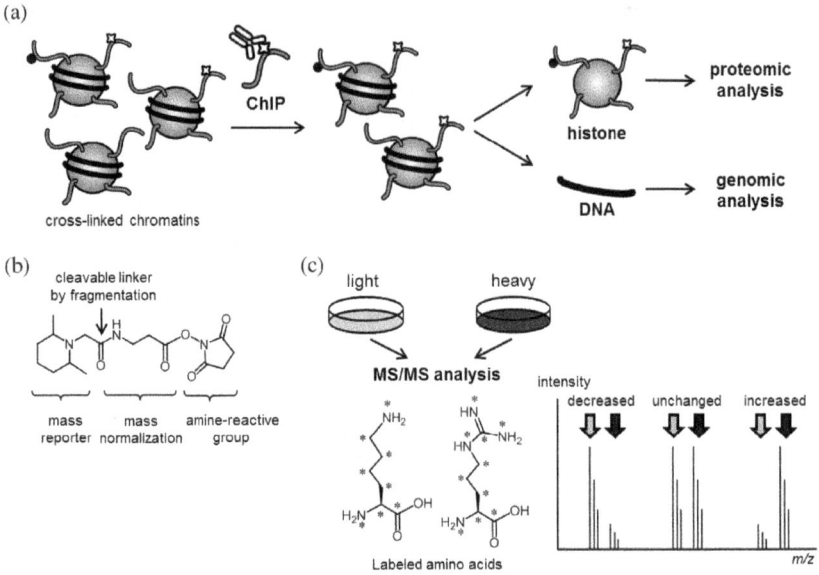

Figure 1.3: (a) Schematic illustration of chromatin immunoprecipitation (ChIP) and sub-sequent analyses. (b) Epigenetic chemical modifications in DNA methylation and histone modifications. (c) Schematic illustration of the roles of epigenetic writers, erasers and readers.

between modifications or sites, or epitope occlusion through interference by adjacent modifications.

Mass spectrometry (MS) is an alternative detection tool for evaluating histone modifications.[49] Although MS methods are limited to providing global proteomic information, they have certain advantages, such as a rapid detection time, accurate assignment of the site and the type of modification, good sensitivity, and the availability of intact, non-labeled histone proteins. In bottom-up approaches, purified histone proteins are enzymatically digested into small peptide fragments and then separated by HPLC. However, treating histones with typical enzymes, such as trypsin, causes problems with the separation and analytical processes, because histones are rich in basic amino acids. Garcia et al. overcame this problem by using propionic anhydride to chemically modify amino groups at the N-terminus as well as at internal unmodified and monomethylated lysines.[50] The fact that trypsin only cleaves at arginine residues facilitates analysis of the mass spectrum of histones. In top-down approaches, the mass spectra of intact proteins are analyzed without enzymatic digestion. The fragmented peptides are analyzed by electrospray ionization mass spectrometry (ESI-MS) or matrix-associated laser desorption ionization mass spectrometry (MALDI-MS). Further fragmented mass spectra (MS/MS) of the peptide can be obtained by the collision-induced dissociation (CID) method. In this approach, the ionized peptide collides with inert gas molecules at low pressure, which determines the sequence of a peptide and its modification state at a single-amino acid level. Other fragmentation techniques, such as electron capture dissociation (ECD) or electron transfer dissociation (ETD), utilize the reaction between the charged peptides and low-energy electrons, followed by rapid bond dissociation via radical generation. ECD and ETD processes can detect MS fragments without the loss of post-translational modifications, and thus can be used in top-down MS analyses of histone proteins. Recent advances in high-resolution instruments such as linear ion traps and Orbitrap MS analyzers can provide accurate, high-resolution data. Phanstiel and coworkers have analyzed and quantified the modification states in intact histone H4 tails from human embryonic stem (ES) cells.[51] They found 74 individual combinatorial codes of histone H4 and dynamic changes in global methylation and acetylation patterns during the differentiation of ES cells.

Incorporation of stable isotopes into peptide samples by chemical or metabolic processes is an accurate method for quantitative analysis because of the low background signals generated. The labeled peptides represent specific MS/MS spectra that can be discriminated from the control samples. Several chemical labeling techniques, such as tandem mass tags (TMT) and isobaric tags for relative and absolute quantification (iTRAQ), have been developed [Figure 1.3(b)].[52,53] In these methods, digested peptides react with amine-reactive tag molecules of identical masses; the fragmented ions show characteristic MS/MS spectra by the labeling positions of each tag. Another useful approach for the metabolic labeling of specific amino acids is stable isotope-labeling by amino acids in cell culture (SILAC).[54] For histone analysis by SILAC, cofactors and amino acids labeled with heavy isotopes can be used to incorporate labels into cellular proteins in cultured cells. The heavy isotope labeling facilitates differentiation between amino acids by mass. Comparative analysis of digested peptide fragments between light (control) and heavy (target) samples allows relative quantification of the modified histone proteins [Figure 1.3(c)]. Vermeulen et al. have reported a quantitative SILAC analysis that uses isotope-labeled arginines and lysines to identify reader proteins relevant to specific histone modifications. After ChIP enrichment of several trimethylated-lysine modified histones, the interacting proteins could be detected and a novel reader protein of trimethylated lysines was identified.[55]

1.4.2 *Chemical Probes for Detection of Effectors in Histone Modifications*

1.4.2.1 *Activity-based probes for proteomic analysis*

Activity-based probes are useful in proteomic analysis. The chemical structures of these probes are based on those of inhibitors, substrates and cofactors of enzymes, allowing them to be directed to the active sites of enzymes. They also possess tags that can be labeled to allow for the detection and capture of enzymes, substrates and associated proteins. An outstanding example of the use of such a probe is the discovery of HDAC by the Schreiber group, which used a chemical probe based on a natural product called trapoxin that inhibits histone deacetylation.[13] Trapoxin has an aliphatic epoxyketone side-chain that mimics an acetylated lysine. The electrophilic epoxyketone reacts with the nucleophilic amino group of lysine to form a covalent bond. Using radiolabeled trapoxin, the factor responsible for histone deacetylation was found to be primarily present in the nuclear fraction of cell extracts. Based on this finding, an affinity agarose matrix that had been modified with trapoxin was used to trap trapoxin-binding proteins [Figure 1.4(a)]. Proteomic analysis of the bound fraction then revealed the first histone deacetylase, HDAC1. Salisbury *et al.* used a chemical probe based on a synthetic HDAC inhibitor, suberoylanilide hydroxamic acid (SAHA), conjugated with a photoreactive benzophenone moiety and an alkyne tag, to analyze HDACs and its associated proteins [Figure 1.4(b)].[56,57] This probe can bind tightly to histone deacetylases with a hydroxamic acid moiety. Photo-irradiation with UV light activates the benzophenone moiety to form a covalent bond with molecules in its proximity. Labeling the alkyne tag with an azide-fluorophore by click chemistry allows for the captured proteins to be visualized in a gel. The captured proteins can also be biotinylated and purified using streptavidin-coated beads. Further analysis using LC-MS/MS techniques can then be done to identify the various associated proteins, such as methylated CpG-binding proteins. Li *et al.* have developed a similar strategy using a trimethylated H3 peptide-based probe to investigate proteins associated with histones containing an H3K4me3 modification.[58,59] Further proteomic analysis, combined with the SILAC approach, allows for quantitative analysis of the interacting proteins.

In addition to inhibitor-based probes, cofactor-based probes have been developed to investigate the substrates of HATs and KMTs. Derivatives of acetyl-CoA can also serve as HAT cofactors for the transfer of functionalized acetyl groups to the nucleophilic lysine amino group of the substrate proteins. Yu *et al.* have demonstrated this concept using CoA modified with a chloroacetyl group.[60] Incubation of the probe with a mixture of histones and yeast acetyltransferase HAT1

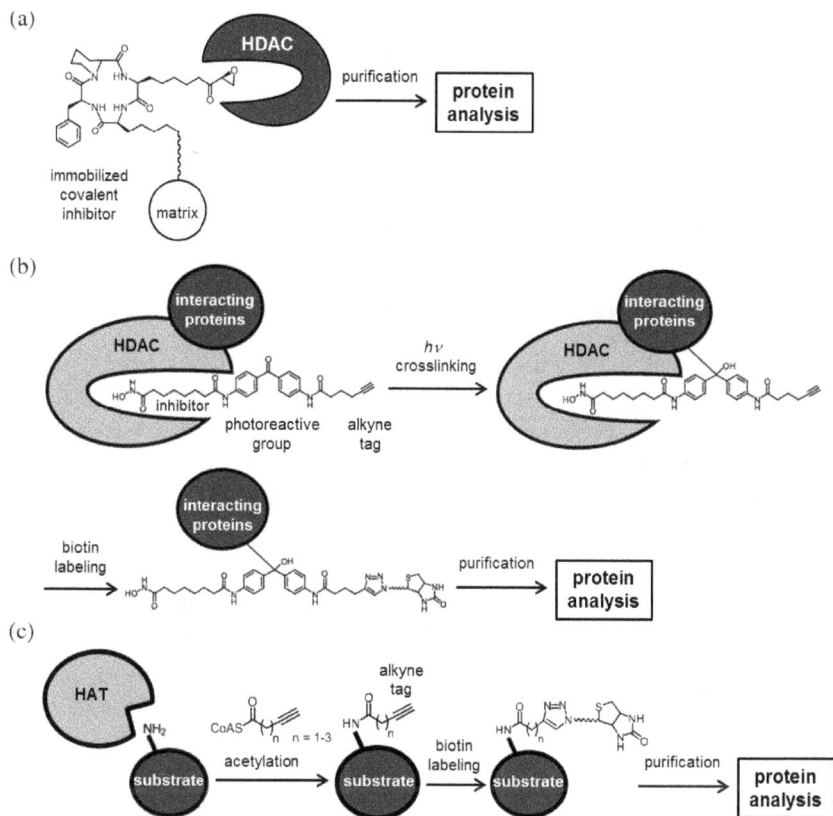

Figure 1.4: Activity-based probes and their applications. (a) Capturing of HDAC using a trapoxin-based probe for detection of HDACs. (b) Labeling of HAT substrates using a cofactor-based probe with an alkyne tag. (c) Capturing of HDAC-interacting proteins using an activity-based probe based on a HDAC inhibitor.

selectively yielded chloroacetylated histone H4 proteins, which could be detected by labeling with a thiol-modified fluorophore. Hwang *et al.* have also conjugated CoA and a biotin-derivative using a reactive thiocarbamate sulfoxide linker.[61] This probe allows for the purification and identification of biotinylated HAT substrates. Yang *et al.* detected and identified acetylated proteins using alkynyl-acetyl-CoA derivatives and fluorescently-labeled them by click chemistry; 4-pentynoyl and 5-hexynoyl-CoA probes efficiently detected the acetylation of histone H3 by p300 HAT [Figure 1.4(c)].[62] The authors also incorporated alkynyl-acetate analogs into living cells and metabolically labeled the proteins expressed in these cells. In the case of KMTs, similar strategies using chemically modified SAM-derivatives have been developed to analyze protein substrates. Peters *et al.* have demonstrated that SAM analogs with a pent-2-en-4-ynyl side-chain, instead of a methyl group, can act as a cofactor of KMT for the detection of histone H3 methylation.[63] Wang *et al.* have also reported methods for labeling the substrates of KMTs using KMT mutants to introduce alkyne-modified SAM-derivatives into the substrates.[64,65] The labeled protein substrates could be detected with fluorescence microscopy and used in the proteomic analysis of cell lysates expressing mutated KMT.

1.4.2.2 *Chemical detection of writer and eraser enzyme activity*

Methods for detecting histone-modifying enzymes are essential to understanding the catalytic activity and substrate specificity of these enzymes. In addition, linked to recent advances in the development of drugs targeting histone-modifying enzymes, a rapid and high-throughput system is required to evaluate the potency and selectivity of candidate compounds. Here, we describe chemical approaches for detecting histone-modifying enzymes.

Initially, radioactivity-based methods were used to detect the acetylation states of histones.[66] Incubation with $[^3H]$ (or $[^{14}C]$)-labeled acetyl-CoA, as a HAT cofactor, allows for the incorporation of a $[^3H]$-labeled acetyl group into proteins or peptides by HATs. The products are then bound to a phosphocellulose filter paper and the unincorporated, labeled cofactors are washed out. The incorporated label is subsequently detected using a scintillation counter. Radiolabeled [Me-^3H] or [Me-^{14}C] SAM, as a cofactor, can be used to measure the activity of KMTs. Advances in peptide array synthesis technology allow for high-throughput analysis of enzyme activity. Rathert *et al.* have prepared a peptide array on a membrane and evaluated the sequence specificity of Dim-5 and G9a methyltransferase activity using radiolabeled SAM.[67] The activity of HDACs and KDMs can also be detected from the amount of $[^3H]$ (or $[^{14}C]$)-acetyl or -methyl groups released, using a similar method. However, radioactive probes are technically problematic and require cumbersome procedures before they can be used. Analysis of these probes also involves multiple, time-consuming steps.

As described above, MS is a useful tool for the detection of acetylated or methylated histones. The acetylation and methylation of the substrates shift the mass spectra by 42.0111 Da or 14.0156 Da, respectively. This difference in mass can easily be detected; Moreover, the number of methyl groups on lysines can be discriminated clearly. Another advantage of this method is the availability of label-free substrates, which allows the intrinsic activity and specificity of the enzymes to be evaluated. Gurard-Levin *et al.* have developed a modified MALDI-MS analysis approach combined with a self-assembled monolayer substrate.[68] In this substrate, peptide libraries containing cysteines at the C-terminus are immobilized on a gold surface. Deacetylation of the peptides can be analyzed directly using MALDI-MS spectrometry. The authors used this technique to perform high throughput analysis of HDAC substrate specificity. Using the TMT chemical-labeling method, Bantscheff *et al.* combined affinity capture and quantitative MS analysis to evaluate the potency and selectivity of HDAC inhibitors for cellular HDAC complexes.[69] They prepared a series of HDAC inhibitors immobilized on a sepharose matrix to capture HDACs and their associated complexes from cell extracts. LC-MS/MS analysis of the captured protein complexes revealed the profile of HDAC inhibitors in cellular HDAC complexes, yielding greater insight into the selectivity of the inhibitors as compared to purified HDAC systems.

Spectrophotometric methods such as absorptiometry, fluorometry and luminometry allow for sensitive and rapid detection of enzyme activity. In addition, by noting changes in absorbance and fluorescence, enzymatic reactions can be continuously monitored so that kinetic data can be collected. Essentially, substrate analogs containing reporter dyes, which change the spectra upon enzymatic action, are used as probes for detection. A number of probes based on fluorescent resonance energy transfer (FRET) or photo-induced electron transfer (PET) have been developed.[70,71] However, it is challenging to develop small molecular probes for the detection of histone-modifying enzymes, because the above mechanisms are often not applicable to modify aliphatic amines on lysine residues. For this reason, enzyme-coupled methods have been developed that allow the conversion of products into detectable molecules. In the case of HATs, the amount of the byproduct CoASH can be measured by subsequent enzymatic treatment with α-ketoglutarate dehydrogenase or pyruvate dehydrogenase and the relevant substrates. CoASH-dependent oxidation of the substrates reduces NAD to NADH, which can be measured from an increase in absorption at 340 nm.[72] On the other hand, direct detection of the products using a fluorogenic probe is much simpler and allows for high-throughput analysis. Trievel *et al.* have developed a HAT activity assay by detecting the byproduct CoASH using a thiol-reactive fluorogenic probe.[73] Wegener *et al.* have developed an enzyme-coupled HDAC detection assay using a peptide-based probe.[74] This probe is comprised of a 7-aminocoumarine dye and an acetylated lysine-containing peptide [Figure 1.5(a)].

The 7-amino group of the coumarin dye is acylated with the peptide at the +1 position of the acetylated lysine, which decreases the fluorescence. Once the probe is deacetylated, the peptide bond adjacent to the basic lysine can be cleaved by trypsin. Thus, the product can be quantified by measuring the recovery of fluorescence. This system is widely applied to measure HDAC activity and inhibition. An improved substrate using an Nε-trifluoroacetylated probe can be used for all Zn^{2+}-dependent

(a)

(b)

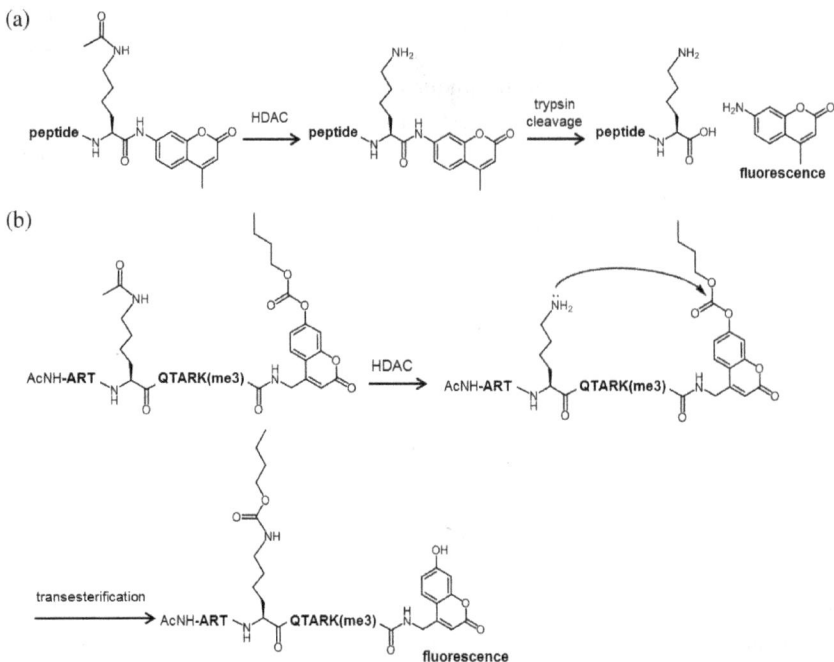

Figure 1.5: Methods for detecting HDAC activity. (a) Enzyme-coupled detection using a peptide-based fluorescent probe. (b) One-step HDAC detection system using a fluorogenic probe based on spontaneous transesterification upon deacetylation.

HDACs, to identify novel inhibitors with unexpected selectivity.[75] However, it has a limitation in that additional protease digestion of the probe is required for fluorescent detection. Additionally, the design of the probe is restricted, because conjugation of the fluorophore with the carboxylate of the terminal lysine is essential for fluorogenic digestion by trypsin. Our group has developed novel fluorogenic probes for HDAC detection that do not require enzyme-coupled procedures. We focused on the nucleophilicity of the amino group in deacetylated lysines and designed a probe comprising of a substrate peptide and a 7-acetylated coumarin dye with an amine-reactive carbonate linker. Deacetylation of the probe spontaneously leads to intramolecular transesterification, producing deacetylated coumarin which greatly enhances the fluorescence [Figure 1.5(b)].[76] A tetraphenylethylene-derived fluorogenic probe, which is based on aggregation-induced emission that is induced by changes in electrostatic interaction, has also been reported.[77] Such direct methods can be used to evaluate enzyme activity in a one-step procedure, which is applicable to the development of drugs targeting histone-modifying enzymes.

Researchers have detected KMT activity by measuring the byproduct, S-adenosylhomocysteine (SAH). Using SAH hydrolase as the coupling enzyme to convert SAH to homocysteine, the amount of SAH can be quantified using a thiol-reactive fluorogenic dye.[78] Detection of KDMs is currently limited to the detection of byproduct formaldehydes, except when using antibodies. More specifically, the formaldehydes produced are oxidized enzymatically and the concomitant reduction of NAD to NADH is detected by measuring absorbance.[15,16]

1.4.3   Synthetic Chromatins with Site-specific Modifications

Site-specific chemical modification of histones provides structural information about proteins at the amino acid level. Moreover, the distinct roles played by combinations of specific histone modifications are of great interest, as they are key to solving the "histone code". To address this issue, reconstruction of histones or nucleosomes containing site-specific modifications is a useful approach. Combinatorial peptide synthesis allows for the construction of short peptide libraries with randomized histone modifications, which can reveal discrimination of combinatorial modifications by chromatin reader proteins.[79] However, it is challenging to make uniform preparations of larger histone proteins containing site-specific modifications from cultured cells or by synthesis of recombinant proteins. Histones isolated from cultured cells are heterogeneous in their modification states, making it impossible to obtain a single pattern of modification by purification. On the other hand, recombinant histones expressed in bacteria are not modified, and attempting to modify them *in vitro* using enzymes, again, does not yield homogeneously-modified histones. Advances in chemical peptide synthesis, protein engineering and *in vitro* nucleosome constitution can yield nucleosomes with specific chemical modifications. Using acetylated or methylated lysine analogs, we can prepare acetylated or methylated histones at arbitrary positions. This section describes the methods for preparing modified histones using chemical modification, incorporation of unnatural amino acids, and protein ligation techniques.

1.4.3.1   Chemical modification

One strategy for site-specific chemical modification in proteins is to conjugate functional molecules onto unique cysteines via thiol-maleimide chemistry. In the case of histones, cysteine residues are replaced with alanine residues at the targeted modification site. Li *et al.* have reconstituted nucleosomes composed of Cy3-labeled DNA and site-specific Cy5-labeled histones and then evaluated the efficiency of

Förster energy transfer (FRET) between Cy3 and Cy5 dyes, which reflects the distance between the two fluorophores in the nucleosome.[80] Changes in FRET efficiency revealed that unwrapping of the nucleosomal DNA depends on the salt concentration and protein binding to the DNA. Simon *et al.* developed a method for site-specific installation of methylated lysines using a chemical modification approach.[81] They allowed cysteine-incorporated histone mutants to react with alkyl halides (2-bromoethyl)-trimethylammonium bromide, yielding trimethylated lysine-mimicking histones [Figure 1.6(a)]. Other methylated (mono- or dimethylated) histones can be prepared in a similar way by using different substrates. Although the reconstructed histones have an S atom in the modified position, this residue can be recognized by antibodies and modifying enzymes.

1.4.3.2 *Incorporation of unnatural amino acids*

Another method involves site-specific incorporation of unnatural amino acids into recombinant histone proteins using a genetically-modified code. This approach of incorporating unnatural amino acids has been developed further by several groups. Representative work by the Schultz group incorporates unnatural amino acids at an UAG (amber) suppressor codon position using a bio-orthogonal tRNA synthetase

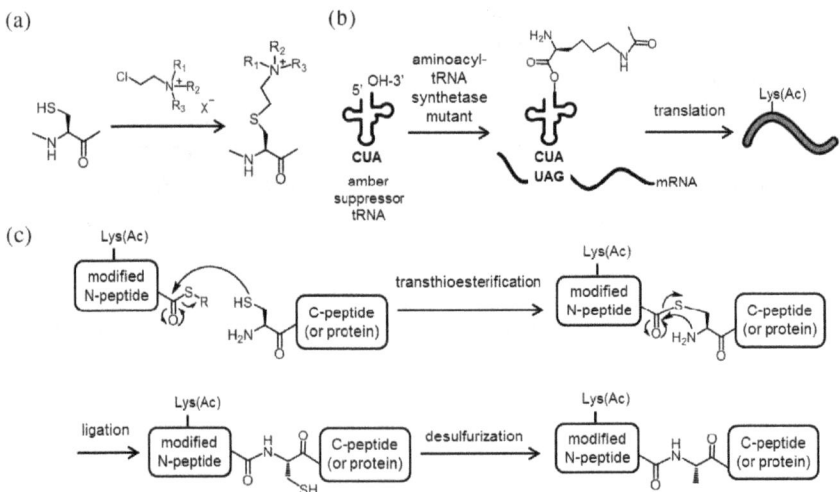

Figure 1.6: Methods used to prepare histones with site-specific modifications. (a) Chemical modification of a thiol in a cysteine residue, used to prepare methylated lysine analogs. (b) Incorporation of an unnatural acetylated lysine using amber suppressor tRNA. (c) Chemical ligation used for fragment condensation of histone proteins.

mutant/tRNA pair.[82] Based on this method, Neumann et al. have developed genetic means to incorporate N-acetyllysine into recombinant proteins via the generation of a bio-orthogonal N-acetyllysyl-tRNA synthetase/tRNA pair [Figure 1.6(b)].[83] The mutant tRNA synthetases that incorporate the acetylated lysine were evolved using a random-mutated library in E. coli. The reconstituted nucleosome, including a histone acetylated at the H3K56 position, showed subtle structural changes upon single-molecule FRET analysis.[84] A different approach incorporating unnatural amino acids into histones, using a reprogrammed genetic code, has also been developed.[85] The codons are reprogrammed and reassigned to incorporate unnatural amino acids, including acetylated or methylated lysines. Using a reprogrammed genetic code and an in vitro translation system, Kang et al. have reported the preparation of histone H3 tail peptides with combinatorial lysine modifications and investigated the effects of these modifications on the binding of reader proteins.

1.4.3.3 Chemical ligation

As histones are small proteins, they can be synthesized by chemical peptide synthesis and fragment condensation. Native chemical ligation (NCL) is an excellent fragment condensation method; in this method, a thioester group at the C-terminus is condensed with a cysteine at the N-terminus via the transesterification of thioesters, followed by an S-to-N acyl shift [Figure 1.6(c)].[86] This reaction efficiently and selectively proceeds under neutral and reductive conditions in an aqueous buffer to yield longer peptides. NCL can also be applied to condensation between a synthetic peptide and an expressed recombinant protein, termed expressed chemical ligation (ECL). As modified histones can be obtained using modified amino acid building blocks, such full or semi-synthetic methods are a powerful means for preparing site-specifically modified histones. The first example of ECL was reported for the ligation of modified N-terminal histone peptides to C-terminal fragments of expressed H3 and H4 histones.[87] The semi-synthesized histones, with multiple acetylation or methylation modifications, can form nucleosomes and be recognized by other histone-modifying enzymes. Subsequently, Shogren-Knaak et al. reported the preparation of a histone with an acetylated lysine at position 16 in histone H4, using a similar ECL protocol.[88] The nucleosomal array, reconstituted with H4K16ac, displayed an effect on chromatin compaction identical to that by adding Mg^{2+}, which demonstrates that acetylation of H4K16 significantly affects higher-order chromatin assembly. The reconstituted nucleosome also inhibits the sliding of ACF, an ATP-dependent chromatin remodeling enzyme. Furthermore, Ottesen and Poirier have investigated the effects on nucleosomal disassembly and DNA unwrapping using different acetylated histones synthesized by NCL and ECL approaches. Site-specific ubiquitinated histone H2B (H2BK120ub) protein has also been prepared by Muir using a semisynthetic approach that included two ligation steps.[89] The ubiquitinated histone H2B affects local and higher-order chromatin structures. Moreover, the K79-specific methyltransferase hDot1L induced methylation at H3K79 in reconstructed ubiquitinated nucleosomes. This result provided direct evidence of the intranucleosomal crosstalk between histone ubiquitination and methylation.

1.4.4 Chemical Probes for Regulation of Histone Modification

Chemical regulators of histone-modifying enzymes are essential tools for investigating the biological effects of specific histone modifications. Moreover, the activities of these enzymes are related to diseases such as cancers, age-related diseases, inflammation, metabolic disorders and neurodegenerative diseases.[90] To date, two HDAC inhibitors have been approved as drugs for cutaneous T cell lymphoma. Therefore, a number of inhibitors have been developed to target the responsible writer and eraser enzymes (HATs, HDACs, KMTs and KDMs) as well as reader proteins (bromodomain-containing proteins) for therapeutic applications in the clinic.[91]

1.4.4.1 HDAC inhibitors

HDACs are the most frequently investigated target enzymes for cancer treatment. Prior to the identification of these enzymes, a few natural compounds [trichostatin A (TSA) and trapoxin B] were known to inhibit histone deacetylation and possess anticancer activity.[92,93] One plausible mechanism underlying their anticancer activity involves the transcriptional activation of tumor-suppressor genes by increased acetylation of histones, resulting in cell cycle arrest or apoptosis of cancer cells.[94] While TSA demonstrates potent inhibition of Zn^{2+}-dependent HDACs, at subnanomolar concentrations, its synthetic analog, suberoylanilide hydroxamic acid [SAHA, Figure 1.7(a)], shows a slightly weaker potency but improved bioavailability. Co-crystallized structures of HDAC and TSA show that the hydroxamic acid group of TSA is plunged into a hydrophobic pocket of the enzyme, tightly binding to the zinc ion in the active site.[95] Short chain fatty acids (butylate and valproate), sulfides (romidepsin) and benzamides (MS-275) also inhibit Zn^{2+}-dependent HDAC activity using an inhibition mechanism similar to that of TSA and SAHA. Recent studies have revealed that each HDAC acts on different protein targets and causes distinct effects. Thus, considerable effort has been focused on the development of potent and selective HDAC inhibitors, to improve the efficacy and reduce the side

(a)

Trichostatin A
(HDAC inhibitor)

SAHA
(vorinostat)
(HDAC inhibitor)

Tubacin
(HDAC6-selective inhibitor)

(b)   (c)

C646
(HAT inhibitor)

Sinefungin
(KMT inhibitor)

UNC0638
(G9a and GLP-selective KMT inhibitor)

(d)   (e)

GSK-J1
(H3K27me3-specific KDM inhibitor)

(+)-JQ1
(acetylated lysine reader inhibitor)

UNC1215
(methylated lysine reader inhibitor)

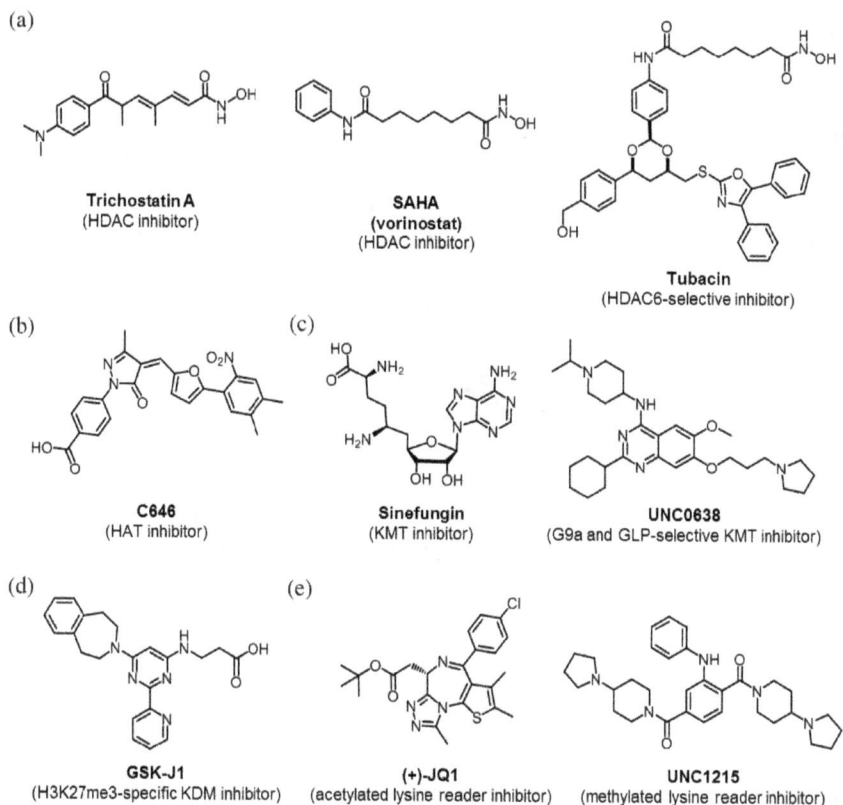

Figure 1.7: Chemical structures of (a) HAT inhibitors, (b) HDAC inhibitors, (c) KMT inhibitors, (d) KDM inhibitors, and (e) inhibitors of the interfaces of the bromodomain and chromodomain.

effects of the drugs. Zn^{2+}-dependent HDAC inhibitors are comprised of a Zn^{2+}-trapping group, a capping aromatic group, and an aliphatic linker that mimics a lysine side-chain. A number of compounds have been developed using chemical derivatization of each of these constitutive parts.[96] For example, HDAC6, belonging to class IIb, can deacetylate tubulins, which are the main component of microtubules. An HDAC6-selective inhibitor, tubacin [Figure 1.7(a)], induces the inhibition of cell motility due to microtubule destabilization.[97] The anti-metastatic activity of tubacin may have potential therapeutic applications in neurodegenerative disorders. Moreover, HDAC inhibitors are also attractive chemical compounds for regenerative therapy using induced pluripotent stem cells (iPSCs). A weak HDAC inhibitor, valproic acid, can improve cell reprogramming efficiency and pluripotent stem cell induction without the need for an oncogenic factor to be introduced.[98]

Class III HDACs (sirtuins) are also potential therapeutic targets for cancer, metabolic diseases and age-related diseases. Since the deacetylation mechanism of these enzymes differs from that of other classes of HDACs, sirtuin inhibitors have distinct structures. EX-527, an indole-based inhibitor developed as a potent and selective sirtuin inhibitor, is being investigated in a clinical trial.[99] Unlike other HDACs, activation of sirtuins is a point of focus in anti-aging research given that the overexpression of Sir2, a homolog of mammalian Sirt1, extends the lifespan of yeast.[100] This finding has inspired research into finding chemical activators of sirtuins. Howitz *et al.* have reported that resveratrol, a polyphenol naturally occurring in red wine, serves as a sirtuin activator and causes lifespan extension in yeast.[101] Several chemical activators of sirtuins have been reported since, and are promising anti-aging drugs.[102] However, other researchers have pointed out that overexpression or activation of sirtuins does not affect lifespan and that the reported compounds do not activate sirtuins;[103,104] thus, these findings are currently controversial.

1.4.4.2 *HAT inhibitors*

Over 50 HATs with modest structural homology have been reported in humans. Given the lack of clear target sites, development of HAT-specific inhibitors remains challenging. However, several natural products, bi-substrate peptide-CoA analogs, and synthetic chemicals screened from libraries [C646, Figure 1.7(b)] have been reported as being HAT inhibitors.[105,106] Analysis of the X-ray crystal structure of a peptide–CoA inhibitor reveals that the binding of CoA induces modest binding of the peptide substrate to the surface of a p300/CBP acetyltransferase.[107] This finding can explain the broad substrate specificity of the p300/CBP enzyme. Studies using a specific HAT inhibitor, C646, have indicated that targeting HATs holds potential for anticancer therapy by regulating gene expression or modulating immune cell responses.[108,109]

1.4.4.3 *Histone methyltransferases inhibitors*

Since histone methylation is prevalent in cancer cells, KMTs are thus promising targets in various cancers, including leukemia. SAM-related analogs such as sinefungin [Figure 1.7(c)] have shown inhibitory activity against a broad range of KMTs.[110] Several SAM-derivatives have also been developed to improve the selectivity of these inhibitors.[111,112] An alternative approach to obtaining selective KMT inhibitors is to screen a chemical library. By screening a small chemical library, Imhof and coworkers identified chaetocin as a SU(VAR)3-9 and G9a methyltransferase inhibitor,[113] and another inhibitor specific for G9a and the related enzyme GLT methyltransferase.[114] Moreover, the structurally-optimized compound UNC0638 [Figure 1.7(c)] specifically inhibited dimethylation at H3K9 and stopped cancer cell proliferation.[115]

1.4.4.4   Histone demethylases inhibitors

KDMs are also potential targets for controlling histone methylation states in disease. KDMs are classified into FAD-dependent LSDs and JMJDM KDMs that use non-heme iron and α-ketoglutarate as cofactors. Polyamine analogs can inhibit the demethylation activity of LSDs, because they also inhibit FAD-dependent polyamine oxidases.[116] Researchers have developed a peptide-based LSD1 inhibitor, which comprises an H3 peptide with an ε-N-propargyl-lysine incorporated at the K4 position.[117] The propargyl moiety is based on a key functional group of inhibitors of monoamine oxidases, which are well-known FAD-dependent oxidative enzymes. Oxidization of the propargyl group can lead to the formation of a covalent adduct with FAD, which inhibits the activity of the enzyme.

On the other hand, bi-substrate inhibitors of JMJDMs have been developed by conjugating iron chelators, such as α-ketoglutarate analogs, with methylated lysine analogs or peptides.[118,119] These inhibitors have shown modest selectivity for several types of JMJDMs. Recently, a selective JMJDM inhibitor against H3K27 demethylation has also been found by screening a compound library [Figure 1.7(d)].[120] This inhibitor induced an increase in methylation levels at H3K27 and proinflammatory cytokine levels in human macrophages, thus revealing a role for the H3K27 methylation mark in inflammatory responses.

1.4.4.5   Inhibitors of the bromodomain/chromodomain interface

Modified histone marks are specifically recognized by several protein motifs, such as bromodomain and chromodomain motifs, which regulate transcription and chromatin structures. The interfaces of these domains and reader proteins are also pharmacological targets. The first HAT to be identified, GCN5, and its homologs have a bromodomain that binds to an acetylated lysine in histones. This domain is known to be a reader of acetylated histone marks that contribute to transcriptional activation. From structural analysis, it is known that the bromodomain has an antiparallel four-helix bundle, with a hydrophobic binding pocket located at one end.[17] This deep and narrow pocket accommodates an acetylated lysine and interacts with a conserved asparagine residue via hydrogen bonding. Several chemical inhibitors mimicking acetylated histones have been developed to target the interface of the bromod-omain and extra terminal (BET) domain family of proteins. A BET domain BRD4-inhibitor, (+)-JQ-1, showed antiproliferative effects in BRD4-dependent malignant cells and in a xenograft model [Figure 1.7(e)].[121] Another inhibitor, I–BET, exerted an anti-inflammatory effect by controlling the activity of macrophages.[122]

Proteins that are members of the Royal superfamily of folds and that contain plant homeodomain (PHD) fingers can serve as readers of methylated lysines. Chro-modomains found in chromatin silencer heterochromatin protein-1 (HP-1) belong to the Royal superfamily, which recognizes highly methylated (tri- and dimethy-lated) lysines. Heterochromatin protein-1 (HP1), a silencer of chromatin structure, contains this domain, which enables it to recognize trimethylated K9 in histone H3. X-ray and NMR structures of the complexes show that the chromodomain forms a β-barrel-like structure and possesses a methylated lysine-binding pocket at one end of the barrel. The tri- and dimethyllysines are accommodated in the pocket, sur-rounded by three aromatic residues that stabilize cation-π interactions. UNC1215 is a recently-developed potent antagonist of the reader protein L3MBTL3, a member of the malignant brain tumor family proteins that recognizes mono- and dimethy-lated lysines [Figure 1.7(e)].[123] This probe was used to clarify a novel interaction between L3MBTL3 and BCLAF1, a protein involved in DNA damage repair and apoptosis.

1.5   Conclusion

In this chapter, we introduced the chemical tools used in epigenetic studies, which is a rapidly developing field in biology. Epigenetic marks (DNA methylation and histone modifications) and their effectors (writers, erasers and readers) control chromatin activity and gene expression in various physiological processes. Moreover, they are involved in diseases such as cancers, age-related diseases, inflammation, metabolic disorders and neurodegenerative disorders. Chemical probes and tools are powerful methods that can support genomic and proteomic analyses in epigenetic research. Chemical treatment with sodium bisulfite is widely used for DNA methylation anal-ysis due to the availability of further genomic analyses by PCR and advanced high-throughput DNA sequencing. The development of methods for detecting histone modifications and identifying the responsible enzymes has provided new insights into the roles of these modifications and the responsible enzymes. Moreover, specific inhibitors of epigenetic effectors serve not only as excellent probes for investigating the effects of specific patterns of histone modifications, but also as candidate drugs for therapeutic applications.

Despite the remarkable advances in epigenetic research in the past two decades, several challenges remain to be addressed to ensure further understanding of epigenetic function and its underlying mechanisms. The first challenge is to probe the epigenetic status of single cells and single molecules. Current genomic and proteomic analysis require samples from a large number of cells. However, the results do not reflect the heterogeneous epigenetic states of the individual cells. Single-cell epigenetic analysis will yield critical information, missed to date, which can contribute to our understanding of the heterogeneity of cells present during development and in tumor tissues. One recent study analyzed the DNA methylation status of early stage embryos in chimeric mice by quantitative PCR in a microfluidics device, after methylation-sensitive restriction enzyme digestion and DNA amplification.[124] This method detected various aberrant DNA methylation patterns in single embryonic cells of chimeric mice and demonstrated epigenetic chimerism during the early pre-implantation phase of development. Other approaches, such as electrical analysis of ion currents through a protein/solid-state nanopore, may be useful for the detection of single translocating biomolecules.[125] Nanopore technology can be applied to single-molecule DNA sequencing, including methylation analysis and assembled chromatin analysis. Although these systems can be improved further, single-molecule methods using advanced nanodevices may potentially help to decipher epigenetic codes in individual cells.

The second challenge involves understanding the dynamics of epigenetic processes in living cells. Marked alterations in epigenetic modifications have been revealed during the cell cycle, circadian cycle, development, differentiation and reprogramming, as well as in response to extracellular stimulation. However, these results were obtained from cell nuclear extracts or fixed cells. Several groups have visualized DNA and histone marks in living cells by expressing fluorescent protein sensors that respond to HDAC activity or by microinjecting fluorescent dye-labeled antibodies that recognize specific epigenetic marks.[126,127] Progress in the development of fluorescent imaging probes may allow cell lineage-specific or tissue-specific dynamics of epigenetic states to be visualized *in vivo*.

The third challenge lies in elucidating the regulation of epigenetic modifications that have a novel mechanism of action. During the identification of epigenetic effector proteins and their structures, new targets, such as the interfaces of writers, erasers and readers, have emerged, which may accelerate the development of selective inhibitors or activators for clinical use. Moreover, recent remarkable advances in the development of gene editing tools, using customized zinc finger, TALE and CRISPR-Cas9 nucleases, have made it possible to target specific gene loci.[128] Using these sequence-specific DNA targeting motifs, combined with epigenetic effectors or drugs, it is possible to achieve epigenetic manipulation of targeted genomic regions. Such approaches may be used to probe the complex epigenetic crosstalk between epigenetic modifications and gene regulation.

Greater understanding of the synchronous regulation of epigenetic information, chromatin structures and gene expression patterns in response to environmental stress will require novel systems for the investigation of epigenetic processes. In combination with new analytical devices, nanotechnology, or new findings in biology, chemical probes can help researchers to elucidate the mechanisms by which living organisms adapt to environmental stress. These findings will contribute not only to progress in epigenetic studies, but also lead to clinical applications in the form of diagnostic tools or therapeutic drugs.

References

1. Waddington, C.H. Canalization of development and the inheritance of acquired characters. *Nature* 150, 563–565 (1942).
2. Allefrey, V.G. *et al.* Acetylation and methylation of histones and their possible role in the regulation of RNA synthesis. *Proc. Natl. Acad. Sci. USA* 51, 786–794 (1964).
3. Bird, A.P. CpG Islands as gene markers in the vertebrate nucleus. *Trends in Genet.* 3, 342–347 (1987).
4. Avner, P., Heard, E. X-chromosome inactivation: Counting, choice and initiation. *Nat. Rev. Genet.* 2, 59–67 (2001).
5. Monk, M., Boubelik, M., Lehnert, S. Temporal and regional changes in DNA methylation in the embryonic, extraembryonic and germ cell lineages during mouse embryo development. *Development* 99, 371–382 (1987).
6. Lo, Y., Wong, I., Zhang, J., Tein, M., Ng, M. Quantitative analysis of aberrant methylation using real-time quantitative methylation-specific polymerase chain reaction. *Cancer Res.* 59, 3899–3903 (1999).
7. Rodrigues-Paredes, M., Esteller, M. Cancer epigenetics reaches mainstream oncology. *Nat. Med.* 17, 330–339 (2011).
8. Luger, K., Mäder, A.W., Richmond, R.K., Sargent, D.F., Richmond, T.J. Crystal structure of the nucleosome core particle at 2.8 Å resolution. *Nature* 389, 251–260 (1997).
9. Lee, D.Y., Hayes, J.J., Pruss, D., Wolffe, A.P. A positive role for histone acetylation in transcription factor access to DNA. *Cell* 72, 73–84 (1993).
10. Bird, A. DNA methylation patterns and epigenetic memory. *Genes Dev.* 16, 6–21 (2002).
11. Brownell, J., Zhou, J., Ranalli, T., Kobayashi, R., Edmondson, D., Roth, S., Allis, C. Tetrahymena histone acetyltransferase A: A homolog to yeast Gcn5p linking histone acetylation to gene activation. *Cell* 84, 843–851 (1996).
12. Rea, S., Eisenhaber, F., O'Carroll, D., Strahl, B., Sun, Z., Schmid, M., Opravil, S., Mechtler, K., Ponting, C., Allis, C., Jenuwein, T. Regulation of chromatin structure by site-specific histone H3 methyltransferases. *Nature* 406, 593–599 (2000).
13. Taunton, J., Hassig, C., Schreiber, S. A mammalian histone deacetylase related to the yeast transcriptional regulator Rpd3p. *Science* 272, 408–411 (1996).

14. Imai, S., Armstrong, C., Kaeberlein, M., Guarente, L. Transcriptional silencing and longevity protein Sir2 is an NAD-dependent histone deacetylase. *Nature* 403, 795–800 (2000).
15. Shi, Y., Lan, F., Matson, C., Mulligan, P., Whetstine, J., Cole, P., Casero, R., Shi, Y. Histone demethylation mediated by the nuclear amine oxidase homolog LSD1. *Cell* 119, 941–953 (2004).
16. Tsukada, Y., Fang, J., Erdjument-Bromage, H., Warren, M., Borchers, C., Tempst, P., Zhang, Y. Histone demethylation by a family of JmjC domain-containing proteins. *Nature* 439, 811–816 (2006).
17. Taverna, S., Li, H., Ruthenburg, A., Allis, C., Patel, D. How chromatin-binding modules interpret histone modifications: Lessons from professional pocket pickers. *Nat. Struct. Mol. Biol.* 14, 1025–1040 (2007).
18. Strahl, B., Allis, C. The Language of Covalent Histone Modifications. *Nature* 403, 41–45 (2000).
19. Jenuwein, T., Allis, C. Translating the histone code. *Science* 293, 1074–1080 (2001).
20. Shapiro, R., Servis, R.E., Welcher, M. Reactions of uracil and cytosine derivatives with sodium bisulfite. A Specific Deamination Method. *J. Am. Chem. Soc.* 92, 422–424 (1970).
21. Hayatsu, H., Wataya, Y., Kai, K., Iida, S. Reaction of sodium bisulfite with uracil, cytosine, and their derivatives. *Biochemistry* 9, 2858–2865 (1970).
22. Frommer, M., McDonald, L., Millar, D., Collis, C., Watt, F., Grigg, G., Molloy, P., Paul, C. A genomic sequencing protocol that yields a positive display of 5-methylcytosine residues in individual DNA strands. *Proc. Natl. Acad. Sci. USA.* 89, 1827–1831 (1992).
23. Herman, J., Graff, J., Myöhänen, S., Nelkin, B., Baylin, S. Methylation-specific PCR: A novel PCR assay for methylation status of CpG islands. *Proc. Natl. Acad. Sci. USA.* 93, 9821–9826 (1996).
24. Cokus, S., Feng, S., Zhang, X., Chen, Z., Merriman, B., Haudenschild, C., Pradhan, S., Nelson, S., Pellegrini, M., Jacobsen, S. Shotgun bisulphite sequencing of the arabidopsis genome reveals DNA methylation patterning. *Nature* 452, 215–219 (2008).
25. Warnecke, P.M., Stirzaker, C., Song, J., Grunau, C., Melki, J.R., Clark, S. J. Identification and resolution of artifacts in bisulfite sequencing. *Methods* 27, 101–107 (2002).
26. Tanaka, K., Tainaka, K., Kamei, T., Okamoto, A. Direct labeling of 5-methylcytosine and its applications. *J. Am. Chem. Soc.* 129, 5612–5620 (2007).
27. Tanaka, K., Tainaka, K., Umemoto, T., Nomura, A., Okamoto, A. An Osmium-DNA interstrand complex: Application to facile DNA methylation analysis. *J. Am. Chem. Soc.* 129, 14511–14517 (2007).
28. Bareyt, S., Carell, T. Selective detection of 5-methylcytosine sites in DNA. *Angew. Chem. Int. Ed.* 47, 181–184 (2008).
29. Yamada, H., Tanabe, K., Nishimoto, S. Fluorometric identification of 5-methylcytosine modification in DNA: Combination of photosensitized oxidation and invasive cleavage. *Bioconjugate Chem.* 19, 20–23 (2008).
30. Ogino, M., Taya, Y., Fujimoto, K. Highly selective detection of 5-methylcytosine using photochemical ligation. *Chem. Commun.* 5996–5998 (2008).

31. Simonsson, S., Gurdon, J. DNA demethylation is necessary for the epigenetic repro-gramming of somatic cell nuclei. *Nat. Cell Biol.* 6, 984–990 (2004).
32. Bhattacharya, S., Ramchandani, S., Cervoni, N., Szyf, M. A mammalian protein with specific demethylase activity for mCpG DNA. *Nature* 397, 579–583 (1999).
33. Barreto, G., Schäfer, A., Marhold, J., Stach, D., Swaminathan, S., Handa, V., Döderlein, G., Maltry, N., Wu, W., Lyko, F., Niehrs, C. Gadd45a promotes epige-netic gene activation by repair-mediated DNA demethylation. *Nature* 445, 671–675 (2007).
34. Kriaucionis, S., Heintz, N. The nuclear DNA base 5-hydroxymethylcytosine is present in purkinje neurons and the brain. *Science* 324, 929–930 (2009).
35. Tahiliani, M., Koh, K., Shen, Y., Pastor, W., Bandukwala, H., Brudno, Y., Agarwal, S., Iyer, L., Liu, D., Aravind, L., Rao, A. Conversion of 5-methylcytosine to 5-hydroxymethylcytosine in mammalian DNA by MLL partner TET1. *Science* 324, 930–935 (2009).
36. Ito, S., Shen, L., Dai, Q., Wu, S., Collins, L., Swenberg, J., He, C., Zhang, Y. Tet proteins can convert 5-methylcytosine to 5-formylcytosine and 5-carboxylcytosine. *Science* 333, 1300–1303 (2011).
37. He, Y.-F., Li, B.-Z., Li, Z., Liu, P., Wang, Y., Tang, Q., Ding, J., Jia, Y., Chen, Z., Li, L., Sun, Y., Li, X., Dai, Q., Song, C.-X., Zhang, K., He, C., Xu, G.-L. Tet-Mediated formation of 5-carboxylcytosine and its excision by TDG in mammalian DNA. *Science* 333, 1303–1307 (2011).
38. Song, C.-X., Szulwach, K., Fu, Y., Dai, Q., Yi, C., Li, X., Li, Y., Chen, C.-H., Zhang, W., Jian, X., Wang, J., Zhang, L., Looney, T., Zhang, B., Godley, L., Hicks, L., Lahn, B., Jin, P., He, C. Selective chemical labeling reveals the genome-wide distribu-tion of 5-hydroxymethylcytosine. *Nat. Biotechnol.* 29, 68–72 (2011).
39. Pastor, W., Pape, U., Huang, Y., Henderson, H., Lister, R., Ko, M., McLoughlin, E., Brudno, Y., Mahapatra, S., Kapranov, P., Tahiliani, M., Daley, G., Liu, X., Ecker, J., Milos, P., Agarwal, S., Rao, A. Genome-wide mapping of 5-hydroxymethylcytosine in embryonic stem cells. *Nature* 473, 394–397 (2011).
40. Okamoto, A., Sugizaki, K., Nakamura, A., Yanagisawa, H., Ikeda, S. 5-hydro xymethylcytosine-selective oxidation with peroxotungstate. *Chem. Commun.* 47, 11231–11233 (2011).
41. Booth, M., Branco, M., Ficz, G., Oxley, D., Krueger, F., Reik, W., Balasubramanian, S. Quantitative sequencing of 5-methylcytosine and 5-hydroxymethylcytosine at single-base resolution. *Science* 336, 934–937 (2012).
42. Jones, P., Taylor, S. Cellular differentiation, cytidine analogs and DNA methylation. *Cell* 20, 85–93 (1980).
43. Laird, P., Jackson-Grusby, L., Fazeli, A., Dickinson, S., Jung, W., Li, E., Weinberg, R., Jaenisch, R. Suppression of intestinal neoplasia by DNA hypomethylation. *Cell* 81, 197–205 (1995).
44. Kuch, D., Schermelleh, L., Manetto, S., Leonhardt, H., Carell, T. Synthesis of DNA dumbbell based inhibitors for the human DNA methyltransferase Dnmt1. *Angew. Chem. Int. Ed.* 47, 1515–1518 (2008).
45. Kuo, M., Allis, C. *In vivo* cross-linking and immunoprecipitation for studying dynamic protein: DNA associations in a chromatin environment. *Methods* 19, 425–433 (1999).

46. Pokholok, D., Harbison, C., Levine, S., Cole, M., Hannett, N., Lee, T., Bell, G., Walker, K., Rolfe, P., Herbolsheimer, E., Zeitlinger, J., Lewitter, F., Gifford, D., Young, R. Genome-wide map of nucleosome acetylation and methylation in yeast. *Cell* 122, 517–527 (2005).

47. Park, P. ChIP-Seq: Advantages and challenges of a maturing technology. *Nat. Rev. Genet.* 10, 669–680 (2009).

48. Egelhofer, T., Minoda, A., Klugman, S., Lee, K., Kolasinska-Zwierz, P., Alekseyenko, A., Cheung, M.-S., Day, D., Gadel, S., Gorchakov, A., Gu, T., Kharchenko, P., Kuan, S., Latorre, I., Linder-Basso, D., Luu, Y., Ngo, Q., Perry, M., Rechtsteiner, A., Riddle, N., Schwartz, Y., Shanower, G., Vielle, A., Ahringer, J., Elgin, S., Kuroda, M., Pirrotta, V., Ren, B., Strome, S., Park, P., Karpen, G., Hawkins, R., Lieb, J. An assessment of histone-modification antibody quality. *Nat. Struct. Mol. Biol.* 18, 91–93 (2011).

49. Eberl, H., Mann, M., Vermeulen, M. Quantitative proteomics for epigenetics. *Chem-biochem* 12, 224–234 (2011).

50. Garcia, B., Mollah, S., Ueberheide, B., Busby, S., Muratore, T., Shabanowitz, J., Hunt, D. Chemical derivatization of histones for facilitated analysis by mass spectrometry. *Nature Protocols* 2, 933–938 (2007).

51. Phanstiel, D., Brumbaugh, J., Berggren, W., Conard, K., Feng, X., Levenstein, M., McAlister, G., Thomson, J., Coon, J. Mass spectrometry identifies and quantifies 74 unique histone H4 isoforms in differentiating human embryonic stem cells. *Proc. Natl. Acad. Sci. USA.* 105, 4093–4098 (2008).

52. Ross, P., Huang, Y., Marchese, J., Williamson, B., Parker, K., Hattan, S., Khainovski, N., Pillai, S., Dey, S., Daniels, S., Purkayastha, S., Juhasz, P., Martin, S., Bartlet-Jones, M., He, F., Jacobson, A., Pappin, D. Multiplexed protein quantitation in *Saccharomyces cerevisiae* using amine-reactive isobaric tagging reagents. *Mol. Cell. Proteomics* 3, 1154–1169 (2004).

53. Dayon, L., Hainard, A., Licker, V., Turck, N., Kuhn, K., Hochstrasser, D., Burkhard, P., Sanchez, J.-C. Relative quantification of proteins in human cerebrospinal fluids by MS/MS using 6-plex isobaric tags. *Anal. Chem.* 80, 2921–2931 (2008).

54. Ong, S.-E., Blagoev, B., Kratchmarova, I., Kristensen, D., Steen, H., Pandey, A., Mann, M. Stable isotope labeling by amino acids in cell culture, SILAC, as a simple and accurate approach to expression proteomics. *Mol. Cell. Proteomics* 1, 376–386 (2002).

55. Vermeulen, M., Eberl, H., Matarese, F., Marks, H., Denissov, S., Butter, F., Lee, K., Olsen, J., Hyman, A., Stunnenberg, H., Mann, M. Quantitative interaction proteomics and genome-wide profiling of epigenetic histone marks and their readers. *Cell* 142, 967–980 (2010).

56. Salisbury, C., Cravatt, B. Activity-based probes for proteomic profiling of histone deacetylase complexes. *Proc. Natl. Acad. Sci. USA.* 104, 1171–1176 (2007).

57. Salisbury, C., Cravatt, B. Optimization of Activity-based probes for proteomic profiling of histone deacetylase complexes. *J. Am. Chem. Soc.* 130, 2184–2194 (2008).

58. Li, X., Kapoor, T. Approach to profile proteins that recognize post-translationally modified histone "Tails". *J. Am. Chem. Soc.* 132, 2504–2505 (2010).

59. Li, X., Foley, E., Molloy, K., Li, Y., Chait, B., Kapoor, T. Quantitative chemical proteomics approach to identify post-translational modification-mediated protein-protein interactions. *J. Am. Chem. Soc.* 134, 1982–1985 (2012).
60. Yu, M., de Carvalho, L., Sun, G., Blanchard, J. Activity-based substrate profiling for Gcn5-related N-acetyltransferases: The use of chloroacetyl-coenzyme A to identify protein substrates. *J. Am. Chem. Soc.* 128, 15356–15357 (2006).
61. Hwang, Y., Thompson, P., Wang, L., Jiang, L., Kelleher, N., Cole, P. A Selective chemical probe for coenzyme A-requiring enzymes. *Angew. Chem. Int. Ed.* 46, 7621–7614 (2007).
62. Yang, Y.-Y., Ascano, J., Hang, H. Bioorthogonal chemical reporters for monitoring protein acetylation. *J. Am. Chem. Soc.* 132, 3640–3641 (2010).
63. Peters, W., Willnow, S., Duisken, M., Kleine, H., Macherey, T., Duncan, K., Litchfield, D., Lüscher, B., Weinhold, E. Enzymatic site-specific functionalization of protein methyltransferase substrates with alkynes for click labeling. *Angew. Chem. Int. Ed.* 49, 5170–5173 (2010).
64. Wang, R., Zheng, W., Yu, H., Deng, H., Luo, M. Labeling substrates of protein arginine methyltransferase with engineered enzymes and matched S-adenosyl-L-methionine analogs. *J. Am. Chem. Soc.* 133, 7648–7651 (2011).
65. Islam, K., Chen, Y., Wu, H., Bothwell, I., Blum, G., Zeng, H., Dong, A., Zheng, W., Min, J., Deng, H., Luo, M. Defining efficient enzyme-cofactor pairs for bioorthogonal profiling of protein methylation. *Proc. Natl. Acad. Sci. USA.* 110, 16778–16783 (2013).
66. Kölle, D., Brosch, G., Lechner, T., Lusser, A., Loidl, P. Biochemical methods for analysis of histone deacetylases. *Methods*, 15, 323–331 (1998).
67. Rathert, P., Zhang, X., Freund, C., Cheng, X., Jeltsch, A. Analysis of the substrate specificity of the Dim-5 histone lysine methyltransferase using peptide arrays. *Chem. Biol.* 15, 5–11 (2008).
68. Gurard-Levin, Z., Kilian, K., Kim, J., Bähr, K., Mrksich, M. Peptide arrays identify isoform-selective substrates for profiling endogenous lysine deacetylase activity. *ACS Chem. Biol.* 5, 863–873 (2010).
69. Bantscheff, M., Hopf, C., Savitski, M., Dittmann, A., Grandi, P., Michon, A.-M., Schlegl, J., Abraham, Y., Becker, I., Bergamini, G., Boesche, M., Delling, M., Dümpelfeld, B., Eberhard, D., Huthmacher, C., Mathieson, T., Poeckel, D., Reader, V., Strunk, K., Sweetman, G., Kruse, U., Neubauer, G., Ramsden, N., Drewes, G. Chemoproteomics profiling of HDAC inhibitors reveals selective targeting of HDAC complexes. *Nat. Biotechnol.* 29, 255–265 (2011).
70. Selvin, P.R. The renaissance of fluorescence resonance energy transfer. *Nat. Struct. Mol. Biol.* 7, 730–734 (2000).
71. de Silva, A.P., H. Q. Gunaratne, H.Q.N., Gunnlaugsson, T., Huxley, A.J.M., McCoy, C.P., Rademacher, J.T., Rice, T.E. Signaling recognition events with fluorescent sensors and switches. *Chem. Rev.* 97, 1515–1566 (1997).
72. Kim, Y., Tanner, K.G., Denu, J.M. A Continuous, Nonradioactive assay for histone acetyltransferases. *Anal. Biochem.* 280, 308–314 (2000).
73. Trievel, R., Li, F.Y., Marmorstein, R. Application of a fluorescent histone acetyltransferase assay to probe the substrate specificity of the human p300/CBP-associated factor. *Anal. Biochem.* 287, 319–328 (2000).

74. Wegener, D., Wirsching, F., Riester, D., Schwienhorst, A. A Fluorogenic histone deacetylase assay well suited for high-throughput activity screening. *Chem. Biol.* 10, 61–68 (2003).
75. Bradner, J., West, N., Grachan, M., Greenberg, E., Haggarty, S., Warnow, T., Mazitschek, R. Chemical phylogenetics of histone deacetylases. *Nat. Chem. Biol.* 6, 238–243 (2010).
76. Baba, R., Hori, Y., Mizukami, S., Kikuchi, K. Development of a fluorogenic probe with a transesterification switch for detection of histone deacetylase activity. *J. Am. Chem. Soc.* 134, 14310–14313 (2012).
77. Dhara, K., Hori, Y., Baba, R., Kikuchi, K. A Fluorescent probe for detection of histone deacetylase activity based on aggregation-induced emission. *Chem. Commun.* 48, 11534–11536 (2012).
78. Collazo, E., Couture, J.-F., Bulfer, S., Trievel, R. A coupled fluorescent assay for histone methyltransferases. *Anal. Biochem.* 342, 86–92 (2005).
79. Garske, A., Oliver, S., Wagner, E., Musselman, C., LeRoy, G., Garcia, B., Kutateladze, T., Denu, J. Combinatorial profiling of chromatin binding modules reveals multisite discrimination. *Nat. Chem. Biol.* 6, 283–290 (2010).
80. Li, G., Widom, J. Nucleosomes facilitate their own invasion. *Nat. Struct. Mol. Biol.* 11, 763–769 (2004).
81. Simon, M., Chu, F., Racki, L., de la Cruz, C., Burlingame, A., Panning, B., Narlikar, G., Shokat, K. The site-specific installation of methyl-lysine analogs into recombinant histones. *Cell* 128, 1003–1012 (2007).
82. Wang, L., Brock, A., Herberich, B., Schultz, P.G. Expanding the genetic code of *Escherichia coli*. *Science* 292, 498–500 (2001).
83. Neumann, H., Peak-Chew, S., Chin, J. Genetically encoding Nε-acetyllysine in recombinant proteins. *Nat. Chem. Biol.* 4, 232–234 (2008).
84. Neumann, H., Hancock, S., Buning, R., Routh, A., Chapman, L., Somers, J., Owen-Hughes, T., van Noort, J., Rhodes, D., Chin, J. A method for genetically installing site-specific acetylation in recombinant histones defines the effects of H3 K56 acetylation. *Mol. Cell* 36, 153–163 (2009).
85. Kang, T., Yuzawa, S., Suga, H. Expression of histone H3 tails with combinatorial lysine modifications under the reprogrammed genetic code for the investigation on epigenetic markers. *Chem. Biol.* 15, 1166–1174 (2008).
86. Dawson, P., Kent, S. Synthesis of native proteins by chemical ligation. *Annu. Rev. Biochem.* 69, 923–960 (2000).
87. He, S., Bauman, D., Davis, J., Loyola, A., Nishioka, K., Gronlund, J., Reinberg, D., Meng, F., Kelleher, N., McCafferty, D. Facile synthesis of site-specifically acetylated and methylated histone proteins: Reagents for evaluation of the histone code hypothesis. *Proc. Natl. Acad. Sci. USA.* 100, 12033–12038 (2003).
88. Shogren-Knaak, M., Ishii, H., Sun, J.-M., Pazin, M., Davie, J., Peterson, C. Histone H4-K16 acetylation controls chromatin structure and protein interactions. *Science* 311, 844–847 (2006).
89. McGinty, R., Kim, J., Chatterjee, C., Roeder, R., Muir, T. Chemically ubiquitylated histone H2B stimulates hDot1L-mediated intranucleosomal methylation. *Nature* 453, 812–816 (2008).

90. Arrowsmith, C., Bountra, C., Fish, P., Lee, K., Schapira, M. Epigenetic protein families: A new frontier for drug discovery. *Nat. Rev. Drug Discov.* 11, 384–400 (2012).

91. Yoshida, M., Kijima, M., Akita, M., Beppu, T. Potent and specific inhibition of mammalian histone deacetylase both *in vivo* and *in vitro* by trichostatin A. *J. Biol. Chem.* 265, 17174–17179 (1990).

92. Kijima, M., Yoshida, M., Sugita, K., Horinouchi, S., Beppu, T. Trapoxin, an antitumor cyclic tetrapeptide, is an irreversible inhibitor of mammalian histone deacetylase. *J. Biol. Chem.* 268, 22429–22435 (1993).

93. Cole, P. Chemical probes for histone-modifying enzymes. *Nat. Chem. Biol.* 4, 590–597 (2008).

94. Marks, P., Rifkind, R., Richon, V., Breslow, R., Miller, T., Kelly, W. Histone deacetylases and cancer: Causes and therapies. *Nat. Rev. Cancer* 1, 194–202 (2001).

95. Finnin, M., Donigian, J., Cohen, A., Richon, V., Rifkind, R., Marks, P., Breslow, R., Pavletich, N. Structures of a histone deacetylase homologue bound to the TSA and SAHA inhibitors. *Nature* 401, 188–193 (1999).

96. Bieliauskas, A., Pflum, M. Isoform-selective histone deacetylase inhibitors. *Chem. Soc. Rev.* 37, 1402–1413 (2008).

97. Haggarty, S., Koeller, K., Wong, J., Grozinger, C., Schreiber, S. Domain-selective small-molecule inhibitor of histone deacetylase 6 (HDAC6)-mediated tubulin deacetylation. *Proc. Natl. Acad. Sci. USA.* 100, 4389–4394 (2003).

98. Huangfu, D., Osafune, K., Maehr, R., Guo, W., Eijkelenboom, A., Chen, S., Muhlestein, W., Melton, D. Induction of pluripotent stem cells from primary human fibroblasts with only Oct4 and Sox2. *Nat. Biotechnol.* 26, 1269–1275 (2008).

99. Napper, A., Hixon, J., McDonagh, T., Keavey, K., Pons, J.-F., Barker, J., Yau, W., Amouzegh, P., Flegg, A., Hamelin, E., Thomas, R., Kates, M., Jones, S., Navia, M., Saunders, J., DiStefano, P., Curtis, R. Discovery of indoles as potent and selective inhibitors of the deacetylase SIRT1. *J. Med. Chem.* 48, 8045–8054 (2005).

100. Kaeberlein, M., McVey, M., Guarente, L. The SIR2/3/4 complex and SIR2 alone promote longevity in *Saccharomyces cerevisiae* by two different mechanisms. *Genes Dev.* 13, 2570–2580 (1999).

101. Howitz, K., Bitterman, K., Cohen, H., Lamming, D., Lavu, S., Wood, J., Zipkin, R., Chung, P., Kisielewski, A., Zhang, L.-L., Scherer, B., Sinclair, D. Small molecule activators of sirtuins extend *Saccharomyces cerevisiae* lifespan. *Nature* 425, 191–196 (2003).

102. Milne, J., Lambert, P., Schenk, S., Carney, D., Smith, J., Gagne, D., Jin, L., Boss, O., Perni, R., Vu, C., Bemis, J., Xie, R., Disch, J., Ng, P., Nunes, J., Lynch, A., Yang, H., Galonek, H., Israelian, K., Choy, W., Iffland, A., Lavu, S., Medvedik, O., Sinclair, D., Olefsky, J., Jirousek, M., Elliott, P., Westphal, C. Small molecule activators of SIRT1 as therapeutics for the treatment of type 2 diabetes. *Nature* 450, 712–716 (2007).

103. Burnett, C., Valentini, S., Cabreiro, F., Goss, M., Somogyvári, M., Piper, M., Hoddinott, M., Sutphin, G., Leko, V., McElwee, J., Vazquez-Manrique, R., Orflla, A.-M., Ackerman, D., Au, C., Vinti, G., Riesen, M., Howard, K., Neri, C., Bedalov, A., Kaeberlein, M., Soti, C., Partridge, L., Gems, D. Absence of effects of Sir2 overexpression on lifespan in *C. elegans* and *Drosophila*. *Nature* 477, 482–485 (2011).

104. Kaeberlein, M., McDonagh, T., Heltweg, B., Hixon, J., Westman, E., Caldwell, S., Napper, A., Curtis, R., DiStefano, P., Fields, S., Bedalov, A., Kennedy, B. Substrate-specific activation of sirtuins by resveratrol. *J. Biol. Chem.* 280, 17038–17045 (2005).

105. Dekker, F., Haisma, H. Histone acetyltransferases as emerging drug targets. *Drug Discov. Today* 14, 942 (2009).

106. Bowers, E., Yan, G., Mukherjee, C., Orry, A., Wang, L., Holbert, M., Crump, N., Hazzalin, C., Liszczak, G., Yuan, H., Larocca, C., Saldanha, S., Abagyan, R., Sun, Y., Meyers, D., Marmorstein, R., Mahadevan, L., Alani, R., Cole, P. Virtual ligand screening of the p300/CBP histone acetyltransferase: Identification of a selective small molecule inhibitor. *Chem. Biol.* 17, 471–482 (2010).

107. Liu, X., Wang, L., Zhao, K., Thompson, P., Hwang, Y., Marmorstein, R., Cole, P. The structural basis of protein acetylation by the p300/CBP transcriptional coactivator. *Nature* 451, 846–850 (2008).

108. Santer, F.R., Höschele, P.P., Oh, S.J., Erb, H.H., Bouchal, J., Cavarretta, I.T., Parson, W., Meyers, D.J., Cole, P.A., Culig, Z. Inhibition of the acetyltransferases p300 and CBP reveals a targetable function for p300 in the survival and invasion pathways of prostate cancer cell lines. *Mol Cancer Ther.* 10, 1644–1655 (2011).

109. Liu, Y., Wang, L., Predina, J., Han, R., Beier, U., Wang, L.-C.S., Kapoor, V., Bhatti, T., Akimova, T., Singhal, S., Brindle, P., Cole, P., Albelda, S., Hancock, W. Inhibition of p300 impairs Foxp3+ T regulatory cell function and promotes antitumor immunity. *Nat. Med.* 19, 1173–1177 (2013).

110. Couture, J.-F., Hauk, G., Thompson, M., Blackburn, G., Trievel, R. Catalytic roles for carbon-oxygen hydrogen bonding in SET domain lysine methyltransferases. *J. Biol. Chem.* 281, 19280–19287 (2006).

111. Yao, Y., Chen, P., Diao, J., Cheng, G., Deng, L., Anglin, J., Prasad, B. V., Song, Y. Selective inhibitors of histone methyltransferase DOT1L: Design, synthesis, and crystallographic studies. *J. Am. Chem. Soc.* 133, 16746–16749 (2011).

112. Zheng, W., Ibáñez, G., Wu, H., Blum, G., Zeng, H., Dong, A., Li, F., Hajian, T., Allali-Hassani, A., Amaya, M., Siarheyeva, A., Yu, W., Brown, P., Schapira, M., Vedadi, M., Min, J., Luo, M. Sinefungin derivatives as inhibitors and structure probes of protein lysine methyltransferase SETD2. *J. Am. Chem. Soc.* 134, 18004–18014 (2012).

113. Greiner, D., Bonaldi, T., Eskeland, R., Roemer, E., Imhof, A. Identification of a specific inhibitor of the histone methyltransferase SU(VAR)3-9. *Nat. Chem. Biol.* 1, 143–145 (2005).

114. Kubicek, S., O'Sullivan, R., August, E., Hickey, E., Zhang, Q., Teodoro, M., Rea, S., Mechtler, K., Kowalski, J., Homon, C., Kelly, T., Jenuwein, T. Reversal of H3K9me2 by a small-molecule inhibitor for the G9a histone methyltransferase. *Mol. Cell* 25, 473–481 (2007).

115. Vedadi, M., Barsyte-Lovejoy, D., Liu, F., Rival-Gervier, S., Allali-Hassani, A., Labrie, V., Wigle, T., Dimaggio, P., Wasney, G., Siarheyeva, A., Dong, A., Tempel, W., Wang, S.-C., Chen, X., Chau, I., Mangano, T., Huang, X.-P., Simpson, C., Pattenden, S., Norris, J., Kireev, D., Tripathy, A., Edwards, A., Roth, B., Janzen, W., Garcia, B., Petronis, A., Ellis, J., Brown, P., Frye, S., Arrowsmith, C., Jin, J. A chemical probe selectively inhibits G9a and GLP methyltransferase activity in cells. *Nat. Chem. Biol.* 7, 566–574 (2011).

116. Huang, Y., Greene, E., Murray Stewart, T., Goodwin, A., Baylin, S., Woster, P., Casero, R. Inhibition of lysine-specific demethylase 1 by polyamine analogs results in reexpression of aberrantly silenced genes. *Proc. Natl. Acad. Sci. USA.* 104, 8023–8028 (2007).

117. Culhane, J., Szewczuk, L., Liu, X., Da, G., Marmorstein, R., Cole, P. A mechanism-based inactivator for histone demethylase LSD1. *J. Am. Chem. Soc.* 128, 4536–4537 (2006).

118. Luo, X., Liu, Y., Kubicek, S., Myllyharju, J., Tumber, A., Ng, S., Che, K., Podoll, J., Heightman, T., Oppermann, U., Schreiber, S., Wang, X. A selective inhibitor and probe of the cellular functions of Jumonji C domain-containing histone demethylases. *J. Am. Chem. Soc.* 133, 9451–9456 (2011).

119. Woon, E., Tumber, A., Kawamura, A., Hillringhaus, L., Ge, W., Rose, N., Ma, J., Chan, M., Walport, L., Che, K., Ng, S., Marsden, B., Oppermann, U., McDonough, M., Schofield, C. Linking of 2-oxoglutarate and substrate binding sites enables potent and highly selective inhibition of JmjC histone demethylases. *Angew. Chem. Int. Ed.* 51, 1631–1634 (2012).

120. Kruidenier, L., Chung, C., Cheng, Z., Liddle, J., Che, K., Joberty, G., Bantscheff, M., Bountra, C., Bridges, A., Diallo, H., Eberhard, D., Hutchinson, S., Jones, E., Katso, R., Leveridge, M., Mander, P., Mosley, J., Ramirez-Molina, C., Rowland, P., Schofield, C., Sheppard, R., Smith, J., Swales, C., Tanner, R., Thomas, P., Tumber, A., Drewes, G., Oppermann, U., Patel, D., Lee, K., Wilson, D. A selective Jumonji H3K27 demethylase inhibitor modulates the proinflammatory macrophage response. *Nature* 488, 404–408 (2012).

121. Filippakopoulos, P., Qi, J., Picaud, S., Shen, Y., Smith, W., Fedorov, O., Morse, E., Keates, T., Hickman, T., Felletar, I., Philpott, M., Munro, S., McKeown, M., Wang, Y., Christie, A., West, N., Cameron, M., Schwartz, B., Heightman, T., La Thangue, N., French, C., Wiest, O., Kung, A., Knapp, S., Bradner, J. Selective inhibition of BET bromodomains. *Nature* 468, 1067–1073 (2010).

122. Nicodeme, E., Jeffrey, K., Schaefer, U., Beinke, S., Dewell, S., Chung, C.-W., Chandwani, R., Marazzi, I., Wilson, P., Coste, H., White, J., Kirilovsky, J., Rice, C., Lora, J., Prinjha, R., Lee, K., Tarakhovsky, A. Suppression of inflammation by a synthetic histone mimic. *Nature* 468, 1119–1123 (2010).

123. James, L., Barsyte-Lovejoy, D., Zhong, N., Krichevsky, L., Korboukh, V., Herold, J., MacNevin, C., Norris, J., Sagum, C., Tempel, W., Marcon, E., Guo, H., Gao, C., Huang, X.-P., Duan, S., Emili, A., Greenblatt, J., Kireev, D., Jin, J., Janzen, W., Brown, P., Bedford, M., Arrowsmith, C., Frye, S. Discovery of a chemical probe for the L3MBTL3 methyllysine reader domain. *Nat. Chem. Biol.* 9, 184–191 (2013).

124. Lorthongpanich, C., Cheow, L., Balu, S., Quake, S., Knowles, B., Burkholder, W., Solter, D., Messerschmidt, D. Single-cell DNA-methylation analysis reveals epigenetic chimerism in preimplantation embryos. *Science* 341, 1110–1112 (2013).

125. Branton, D., Deamer, D., Marziali, A., Bayley, H., Benner, S., Butler, T., Di Ventra, M., Garaj, S., Hibbs, A., Huang, X., Jovanovich, S., Krstic, P., Lindsay, S., Ling, X., Mastrangelo, C., Meller, A., Oliver, J., Pershin, Y., Ramsey, J., Riehn, R., Soni, G., Tabard-Cossa, V., Wanunu, M., Wiggin, M., Schloss, J. The potential and challenges of nanopore sequencing. *Nat. Biotechnol.* 26, 1146–1153 (2008).

126. Sasaki, K., Ito, T., Nishino, N., Khochbin, S., Yoshida, M. Real-Time Imaging of Histone H4 Hyperacetylation in Living Cells. *Proc. Natl. Acad. Sci. USA.* 106, 16257–16262 (2009).
127. Kimura, H., Hayashi-Takanaka, Y., Yamagata, K. Visualization of DNA methylation and histone modifications in living cells. *Curr. Opin. Cell Biol.* 22, 412–418 (2010).
128. Gaj, T., Gersbach, C., Barbas, C. ZFN, TALEN, and CRISPR/Cas-based methods for genome engineering. *Trends in Biotech.* 31, 397–405 (2013).

Lanthanide-doped Upconverting Nanoparticles for Biological Applications

2

Yuanwei Zhang and Gang Han

Biochemistry and Molecular Pharmacology Department,
University of Massachusetts-Medical School,
364 Plantation Street, LRB 806, Worcester, MA 01605, USA

Abstract

Compared with downconverters (organic dyes and quantum dots), upconverting nanoparticles (UCNPs) can switch low energy near-infrared (NIR) photons into high energy UV and/or visible photons with pronounced stability. Due to the fast advancement of colloidal synthesis techniques, lanthanide-doped UCNPs can be fabricated with narrow distribution and modulated physical behaviors. These unique characteristics allow them to outperform conventional bio-imaging materials and offer a versatile platform for biological applications. Here, we discuss several approaches to the synthesis and application of UCNPs as imaging probes, delivery carriers and therapy sensitizers.

2.1 Introduction

Upconversion is a nonlinear optical process in which two or multiple long wavelength (low energy) photons are sequentially absorbed by upconversion materials, before being emitted at a shorter wavelength.[1] Lanthanide-doped upconverting nanoparticles (UCNPs) can be excited with deep tissue penetrating near-infrared (NIR) light and emit light in a broad range from ultraviolet (UV) to NIR with various distinctive characteristics, including a narrow emission band, large anti-Stokes shift, and low light scattering. These characteristics make them unique in contrast to

(a)

(b)

0 s   10 s   60 s   300 s

(c)

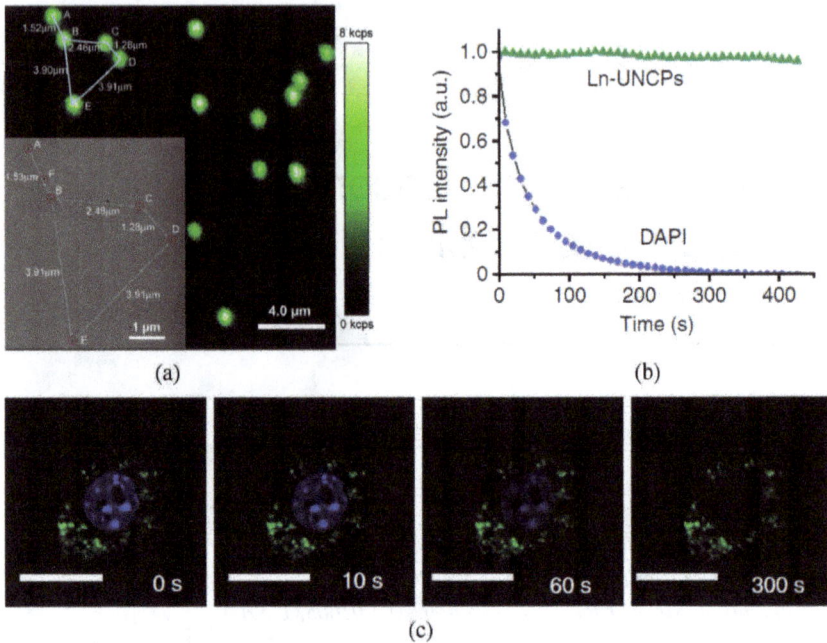

Figure 2.1: (a) Confocal upconverted luminescent image of individual UCNPs on a silicon nitride membrane. (b) Quantitative analysis of the changes in fluorescence intensities of organic dye 4[1],6-diamidino-2-phenylindole (DAPI) and UCNPs in cells. (c) Comparison photobleaching of UCNPs with DAPI by confocal imaging; simultaneous excitations are provided by CW lasers at 405 and 980 nm with powers of ∼1.6 and 19 mW in the focal plane, respectively. The fluorescence signal at 420–480 nm (λ_{ex} = 405 nm) and the upconversion luminescence at 500–600 nm (λ_{ex} = 980 nm) are shown in blue and green, respectively.[5,6]

existing endogenous and exogenous fluorophores, and suitable for various biological studies.[2-4] In addition, these nanoparticles are non-blinking, non-photobleaching, extremely stable, and dodge the endogenous cellular fluorophores spectral window (Figure 2.1).[5,6]

By tuning the emission wavelength of UCNPs from blue to red, multi-color cell imaging has been made possible in organisms such as C. elegans[7] and in tissues.[8] Most importantly, the working window for both NIR excitation and NIR emission (NIR-to-NIR) is particularly interesting for in vivo imaging studies in small animals, because the range is within the biological NIR optical transmission window (700–1000 nm), thus providing greater tissue penetration beyond 3 cm, and lower absorption and scattering of biomolecules (Figure 2.2).[9]

Since the first report of an upconversion phosphor in the early 70s,[10] fabrication technologies have improved rapidly, owing to advancements in colloidal synthesis. For bio-related studies, the wet-chemical route of thermal decomposition and hydrothermal strategies are widely employed, producing nanoparticles within a narrow size distribution and with high dispersibility.[11−13] Moreover, the surface chemistries of various UCNPs have been broadly studied for conjugation to ligands with a wide range of functionalities. The nanoparticles have controllable sizes and shapes, and a large surface area for efficient conjugation to a number of ligands,[14] which provide them the properties of: (1) better cellular uptake, (2) specific cell targeting, and (3) the release of payloads as pro-drugs.[15] UCNPs have promising properties as next generation nanomaterials for a variety of biomedical research applications. For instance, the nanoparticles have been modified for use in cell labeling and tracking,[16,17] disease-targeted imaging,[18] small animal (rat) imaging,[6] drug delivery,[19] as well as photodynamic[20] and photothermal therapy.[21]

All of the above mentioned advantages of UCNPs make them an ideal platform for multimodal applications in imaging and therapy.[22−25] First, they can serve as imaging probes for disease targeting, with many unique benefits including rapid imaging diagnosis and intraoperative feedback,[26] unlike other approaches such as radio-labeled imaging[27] and magnetic resonance imaging.[28] Secondly, they can be adapted as carriers to precisely deliver therapeutics, in which side effects are minimized in nearby tissues and organs. Such multimodal nanoplatforms could be used to elucidate the pharmacokinetics and pharmacodynamics of therapeutic drugs during the course of treatment, thus exhibiting potential as next-generation personalized medicine for treating diseases such as cancer.[29]

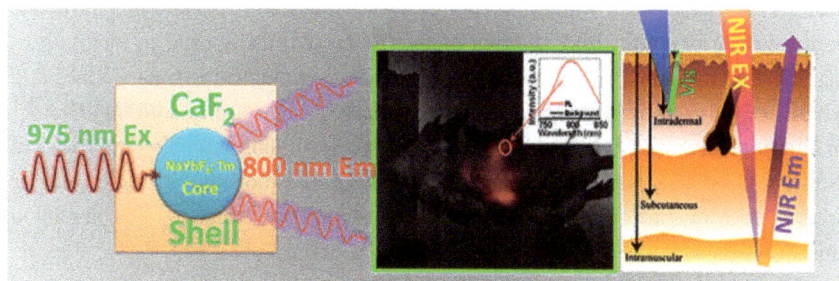

Figure 2.2: (Left) Illustration of lanthanide-doped UCNPs that have core-shell structures, (Middle) and their NIR-to-NIR optical transitions for applications in small animal imaging studies. (Right) An illustration showing the deeper penetration of NIR light into body tissues compared to visible light.[9]

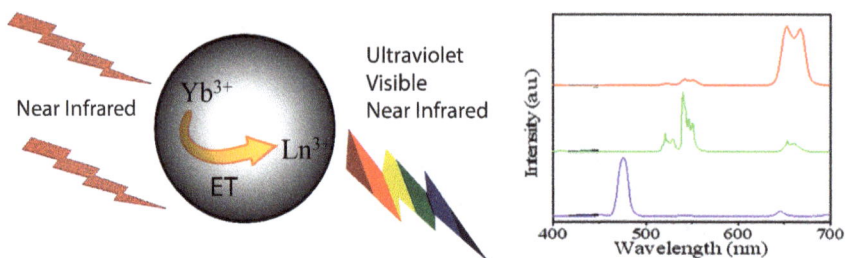

Figure 2.3: Schematic representation of the excitation/emission and interatomic energy transfer profiles of UCNPs, and examples of upconversion emission spectra upon 980 nm excitation.

2.2 Basic Structures of UCNPs

In general, lanthanide-doped UCNPs contain three essential components, namely: host matrix, sensitizer and emitter. The host matrix is normally made of inorganic crystalline and can hold trivalent lanthanides with metastable excited states, among them the sensitizer of higher absorption ability can sequentially absorb two or more excitation photons and leap into excited states, from which the energy is passed down to a neighboring emitter where it consequently emits at a shorter wavelength. For example, the sensitizer of Yb^{3+} can sequentially absorb NIR photons (e.g. 980 nm) and transfer the energy to the emitter of Tm^{3+}/Er^{3+}, and conduct UV to NIR emissions (Figure 2.3).[1]

The choice of host lattice is an important one, because the composition of the host matrix largely determines its luminescence properties.[30] In principle, the host matrix must have low phonon energies, so that lattice stress and non-radiative pathways can be minimized. To this point, the most broadly used inorganic host material for lanthanide-doped UCNPs are the rare earth fluorides: binary REF_3 and $AREF_4$ (RE = rare earth, A = alkali), such as LaF_4, YF_4, $NaYF_4$ and $BaYF_4$.[31,32] In addition, the influence of matrix morphology to emission intensity and ratios of different transitions have been studied. Hexagonal and orthorhombic shapes, among others, are found to outperform other lattice shapes for enhanced upconversion emission.[33,34]

Moreover, the upconversion properties (emission intensity, wavelength, and lifetime) of lanthanide-doped UCNPs can be tuned by the species and doping concentration of both the activator and emitter.[1,35,36] Generally, optimized upconversion efficiency can be attained when energy is sufficiently transferred from activator to emitter. Therefore, two conditions are essential: (1) the activator and emitter must have matched ladder-shape energy level for sufficient energy transfer; (2) the physical distance between activator and emitter shows optimum inter-atomic length.

2.3 Synthesis of Lanthanide-Doped UCNPs

The most well-known synthesis strategies include the hydrothermal method, thermal decomposition, co-precipitation and the sol-gel method.[31,37] Due to the requirement of narrow size distribution and high dispersibility for biological studies, thermal decomposition and hydro-thermal synthesis techniques are the most widely employed strategies. In addition, UCNPs can be fabricated with various shapes under controlled conditions such as temperature, materials ratio, and selected components.[33]

The first solvent dispersible UCNPs was reported by Güdel et al.,[38] in which Er^{3+} and Tm^{3+}-doped $LuPO_4$ and $YbPO_4$ nanoparticles were prepared by co-precipitation in high boiling point solvents. The UCNPs generated can conduct blue, green and red upconversion emissions. However, this co-precipitation approach shows low controllability with respect to nanoparticle size and morphology. To overcome this problem, Mai and Zhang fabricated a variety of high-quality monodispersed cubic and hexagonal $NaYF_4$ based nanocrystals. Generally, the size and shape of nanoparticles can be controlled by decomposing metal trifluoroacetate precursors at high temperatures in oleic acid and oleic amine mixed solvents.[39] This is the most typical method to fabricate UCNPs, and the critical synthesis parameters, such as coordinating solvent compositions,[40,41] decomposition temperature,[42] starting material species and ratio,[43] and core/shell structures, have been thoroughly investigated.[44] For example, to synthesize high luminescent and ultra-small upconversion nanoparticles with biomolecules of comparable size, Chow et al. fabricated ca. 11 nm β-$NaYF_4$:Yb,Er and β-$NaYF_4$:Yb,Tm UCNPs in pure oleylamine solutions, while further upconversion efficiency was improved by coating them with an inert shell structure.[40]

Using a hydro-thermal synthesis strategy, UCNPs with uniform sizes and shapes can also be achieved. Generally, UCNPs are obtained by mixing fluoride salts (e.g. NH_4F) with lanthanide compounds in a solvent with a high boiling point such as ethylene glycol, and having them react under high temperature and pressure. For example, using this method, β-phased $NaYF_4$ UCNPs were fabricated from a corresponding oleate precursor.[12,45] A more sophisticated strategy used a liquid-solid-solution phase-transfer method in water/alcohol/oleic acid mixture to design UCNPs with predictable size, shape and phase.[46,47] For example, polyvinylpyrrolidone-coated α-$NaYF_4$ nanoparticles were co-doped with

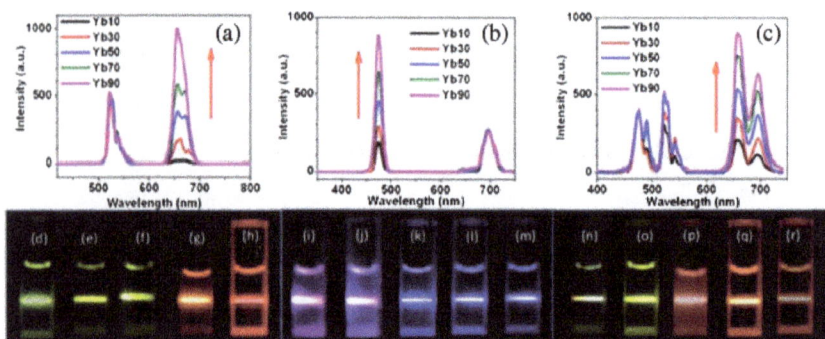

Figure 2.4: The upconversion emission spectra of visible-to-NIR emitting UCNPs, based on YF_3 doped with (a) 0.5% Er^{3+} and 10–90% Yb^{3+}; (b) 2% Tm^{3+} and 10–90% Yb^{3+}; and (c) 0.5% Er^{3+}, 2% Tm^{3+} and 10–90% Yb^{3+}. (d)–(r) shows their corresponding emission imaging graph.[50]

Yb^{3+}/Er^{3+} or Yb^{3+}/Tm^{3+} in ethylene glycol solvents.[48] Furthermore, the emission color and intensity can be precisely controlled using different host-activator systems and dopant concentrations.[49] For example, by studying different dopant concentrations, enhanced multicolor upconversion photoluminescence was achieved in yttrium fluoride nanosystems (Figure 2.4).[50]

2.4 Surface Modification for Upconversion Enhancing and Bio-conjugation

2.4.1 *Core-shell Structures*

As universal bio-probes, UCNPs should have high fluorescent efficiency in aqueous solutions. However, the OH and NH_2 groups surrounding the surface of UCNPs can quench the excited states through multiphoton relaxation procedures. Nevertheless, when nanoparticle sizes shrink, these quenching effects become more pronounced. In recent years, coating the UCNP core structure with a shell is routinely done to enhance luminescence intensity. In this way, lanthanide ions are doped in an interior core that is encapsulated by inert shell layers. Hence, energy loss from the core is significantly reduced and consequently, upconversion efficiency is enhanced. For example, Kong and Zhang studied the core-shell structure of β-$NaYF_4$:Yb^{3+},Er^{3+}@β-$NaYF_4$, and from the kinetic analysis they discovered that the quenching effect of the luminescent centers can be effectively reduced by the homogeneous coating of a $NaYF_4$ shell.[51] Later on, Liu *et al.* showed direct evidence

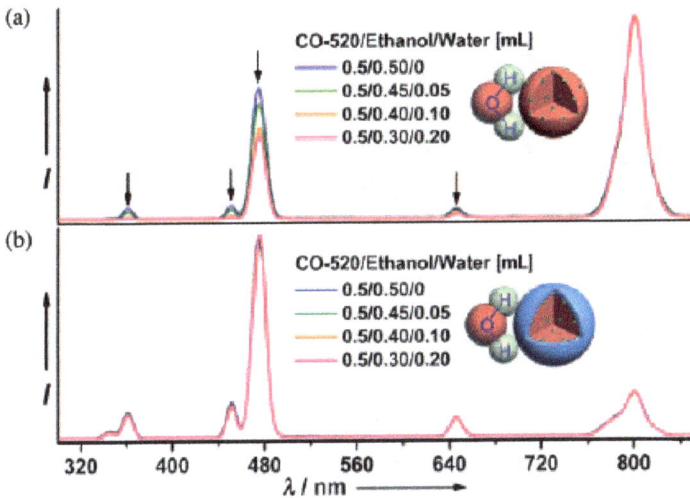

Figure 2.5: Upconversion emission spectra of (a) $NaGdF_4$:25%Yb/0.3%Tm (15 nm) and (b) corresponding core-shell nanoparticles (20 nm) in nonylphenylether/ethanol/water solutions with different water ratio.[52]

for the surface quenching effect related to nanoparticle size. They also intensified the optical integrity and found that surface quenching effects of the nanoparticles can be greatly preserved by coating them with a thin, inert shell (Figure 2.5).[52]

Dopants can be added into the shell layer to synthesize more functional and versatile UCNP systems. For example, Zhao *et al.* introduced Gd^{3+} into the shell layer and synthesized $NaYF_4$:Yb,Er@NaGdF4 nanoparticles.

The components and thickness of the shell layer were studied using cryo-TEM, rigorous EELS, and HAADF techniques at temperatures of 96 K. Greater resistance to the quenching effect of water molecules and an overall enhancement of upconversion were observed.[53] Thus, this strategy of synthesizing high luminescent and stable UCNPs with core-shell structures will facilitate their applications in bio-labeling and bio-imaging. For more practical imaging studies, Zhao and Yao fabricated sandwich-structured rare-earth nanoparticles that have high upconversion emissions when excited at a wavelength of 800 nm. In this system, the nanostructure has been well defined to eliminate potential cross-relaxation pathways in between the activator and the sensitizer. When the interlayer thickness was ~1.45 nm, the emission intensity of the nanoparticles reached a maximum, thus adding new collections of UCNPs for bio-imaging applications (Figure 2.6).[54] Moreover, attaching organic dye antennas onto the nanoparticle surface can achieve broadband excitation, due to energy transfer from the organic dyes to Yb^{3+} in the core.[55] This exciting feature

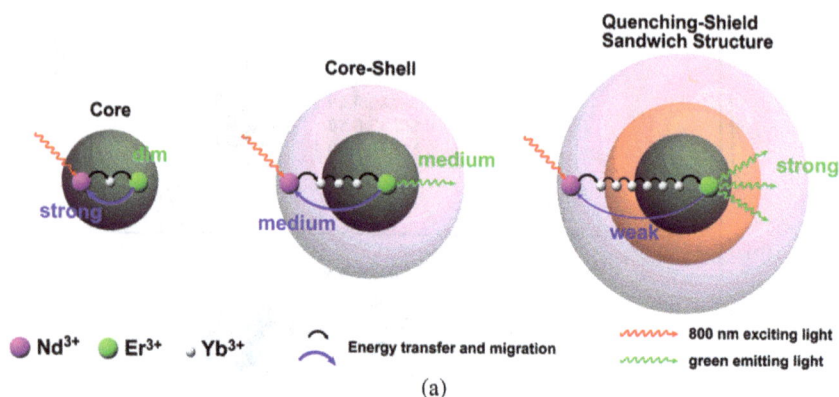

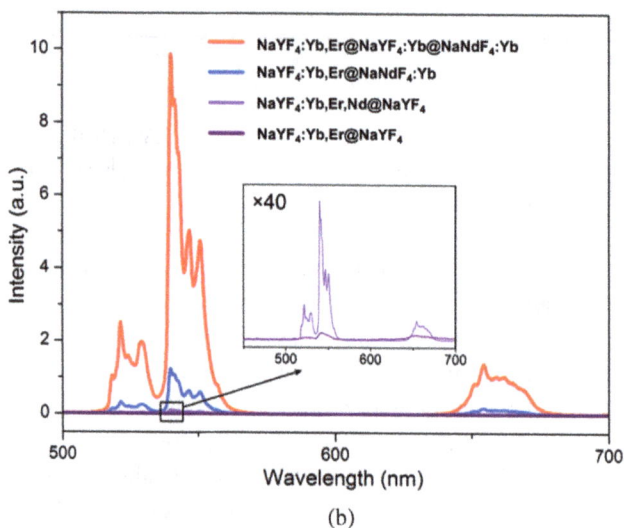

Figure 2.6: (a) Illustration scheme of the energy-transfer and migration processes in the Nd^{3+}-sensitized triple sandwich nanoparticles; (b) UC emission spectra of different core and shell systems under 800 nm excitation.[54]

of dye-sensitized upconversion nanoparticles should lead to important applications in biological imaging.[56]

2.4.2 Surface Modification for Bio-conjugation

Post-modification of UCNP surfaces by charged or polar groups can improve their aqueous solubility and biocompatibility. In this regard, several strategies have been developed to transfer the as-synthesized hydrophobic nanoparticles into aqueous solutions using amphiphilic polymers,[57,58] lipids,[59] silica shells,[60] ligand exchange[61] and ligand oxidation.[62] The hyrdophilic nanoparticle surface can be modified with functional groups to offer the properties of permeability, targeting and delivery. These functional groups can be simple non-specific ligands, or biologically-active molecules. For example, by coating with octylamine/isopropylamine-grafted polyacrylic amphiphilic polymers through hydrophobic van der Waals interactions with surface oleic acid, the UCNPs can be dispersed into an aqueous solution with external carboxyl groups as chemical modification sites.[5,44] Nanoparticle surfaces can also be modified with phospholipids to make them water soluble.[63]

Due to the high chemical stability of lanthanide UCNPs, harsh conditions can be used during synthesis. For example, strong coordinative reagents such as citrate,[5] poly(ethylene glycol)-phosphate,[64] polyacrylic acid,[32] 3-mercaptopropionic acid[65] and nitrosonium tetrafluoroborate[66] can be used to exchange the as-synthesized hydrophobic ligands on the UCNP surface (e.g. oleic acid or oleic amine), generating stabilized UCNPs with emission properties that are inert against acid erosion.[40,67] PEGylated $Yb_2 O_3$:Er nanoparticles were fabricated using a urea-based homogeneous precipitation method. The nanoparticles are facile to construct, and have excellent stability and long blood circulation times *in vivo*.[68] In addition, strong oxidation reagents like ozone and potassium permanganate were also used to react with the carbon-carbon double bond of oleic acid on UCNP surfaces, generating carboxylic acid functional groups for further modification.[69,70]

2.5 UCNPs as Luminescent Imaging Probes

The upconversion property of UCNPs makes them very appealing for bio-imaging studies as they avoid autofluorescence from biological samples. Their near-infrared excitation wavelength also permits deep tissue penetration.[71,72] In addition to their non-blinking and stable properties, UCNPs have improved detection limits with large signal-to-noise ratio compared with classical imaging probes, like organic dyes and quantum dots.[16]

For example, β-NaYF$_4$:Er,Yb nanocrystals were fabricated for luminescence imaging studies. Nanoparticles around 30 nm in diameter were found to have improved signal-to-noise ratio, and to be non-photobleaching and non-blinking. After encapsulating with amphiphilic polymers, the UCNPs can be dispersed in aqueous solutions and taken up by murine fibroblasts by endocytosis.[5] In another example, NaYbF$_4$-based UCNPs have been synthesized with a CaF$_2$ shell coating.

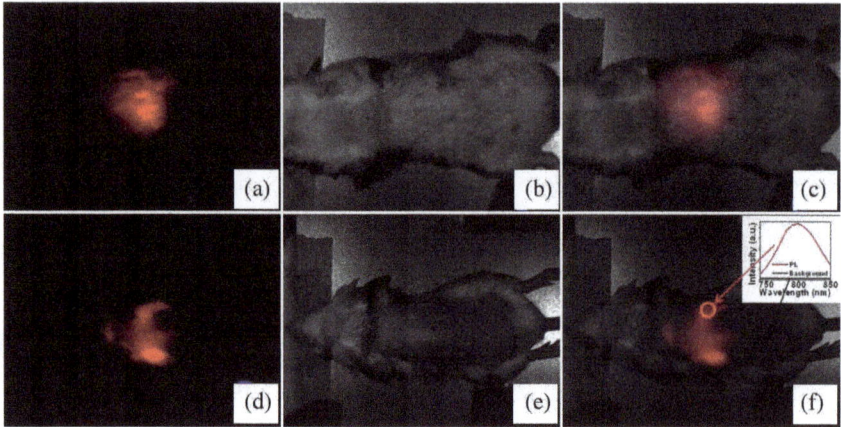

Figure 2.7: Whole animal imaging of mouse after tail vein injection of HA-coated α-(NaYbF$_4$:0.5% Tm^{3+})/CaF$_2$ nanoparticles. (a) and (d) upconversion photoluminescent images; (b) and (e) bright-field images; and (c) and (f) merged images.[9]

The UCNPs were used *in vitro* and *in vivo* for high contrast imaging applications. They were administered into mice by tail vein injection, making them highly practical for imaging studies. (Figure 2.7).[9]

UCNPs with tunable emission colors are suitable for more complex bio-assays, including immune-labeling and cancer cells imaging.[73,74] Their emission color in the visible region can be tuned by changing the lanthanide emitter and/or the doping ratio of Er, Tm and Ho. Also, co-doping with different lanthanide ions in a single NaYbF$_4$ matrix can generate an array of multicolor emissions. Using a multicolor UCNP system followed by further modification of the silica shell with aminopropyltriethyoxysilane, anti-CEA8 antibodies were attached to the nanoparticle surfaces and used for fluorescent imaging in HeLa cells (Figure 2.8).

Because many tumor cells overexpress transmembrane receptors, UCNPs labeled with small peptides are particularly interesting for tumor targeting and tracking within complex biological systems. For example, the peptide motif c(RGD), which shows high binding affinity to the $\alpha_v\beta_3$ integrin receptor, was used to functionalize UCNPs for *ex vivo* and *in vivo* imaging studies. In *in vivo* imaging studies, intense fluorescence at the tumor position was observed 24 h after injection, while luminescence from the liver was significantly reduced (Figure 2.9). Also, due to peptide labeling, the nanosystem was found to have greater affinity to U87MG tumor cell lines compared to MCF-7 tumor cell lines, suggesting a potential application for cancer detection.[75] Another peptide, polypeptide neurotoxin, was shown

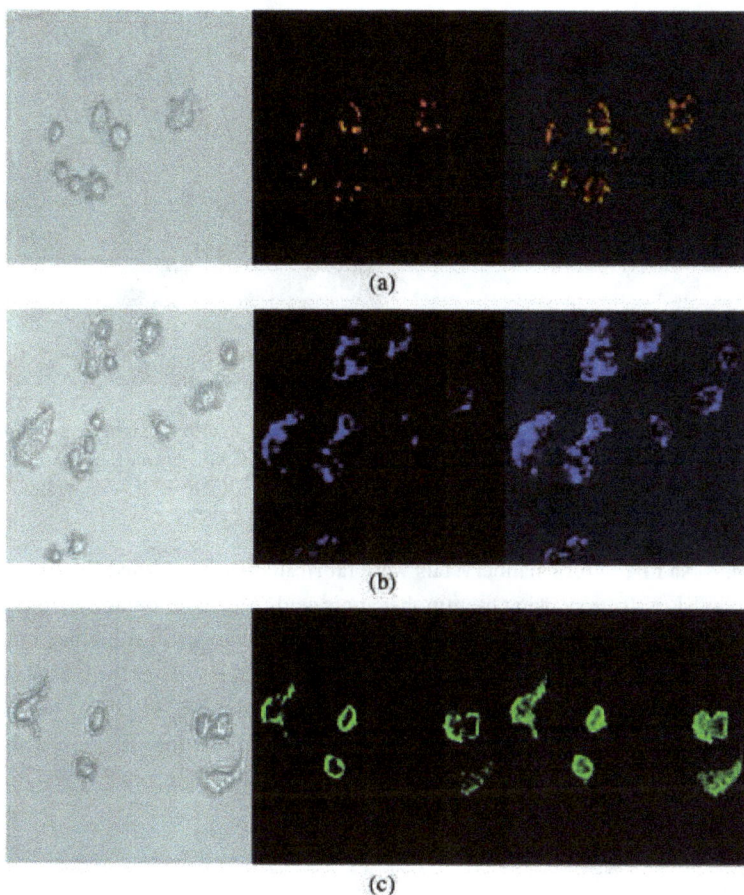

Figure 2.8: Fluorescence imaging of HeLa cells using various UCNPs conjugated to anti-CEA8 antibodies under 980 nm excitation. (a) NaYbF$_4$:Er, (b) NaYbF$_4$:Tm, (c) NaYbF$_4$:Ho.[73]

to effectively bind to many types of cancer cells with high specificity and affinity. For instance, a typical neurotoxin peptide of recombinant chlorotoxin was conjugated to hexagonal-phase NaYF$_4$:Yb,Er/Ce nanoparticles, which were later injected into Balb-C nude mice.[76] Neurotoxin-labeled UCNPs were found to target tumors with sensitivity and specificity when imaged using 980 nm laser excitation.

UCNPs with folic acid (FA) modification were also developed for tumor targeting studies. In one case, FA-coupled NaYF$_4$:Yb,Er UCNPs effectively targeted folate receptor-overexpressing HeLa cells *in vitro* and HeLa tumors *in vivo* and

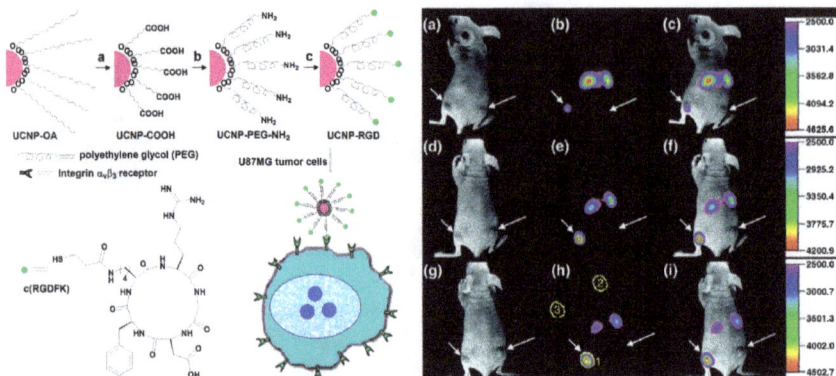

Figure 2.9: (Left) Illustration scheme for UCNP-RGD conjugation. (Right) *In vivo* upconversion luminescence imaging over 24 h of a subcutaneous U87MG tumor (left hind leg) and MCF-7 tumor (right hind leg) after intravenous injection of UCNP-RGD conjugates.[75]

ex vivo.[77] NaYF$_4$:Yb,Er nanocrystals were fabricated by a modified hydrothermal microemulsion strategy using 6-aminohexanoic acid. The exterior amine functional group of the resulting UCNPs not only conferred excellent water solubility, but also enabled further modification by carboxylic acid-activated FA. After the FA-modified UCNPs were injected into HeLa tumor-bearing nude mice for 24 h, an upconverting luminescence signal (around 650 nm) was observed in the tumor regions, while no significant luminescence signal was detected in control mice injected with UCNPs without FA modification (Figure 2.10).

2.6 UCNPs as Therapeutics Delivery Carriers

2.6.1 *UCNPs as Traditional Drug Delivery Tools*

Drug delivery carriers can greatly enhance the solubility, stability and pharmacokinetics of pharmaceutical payloads.[78] A wide variety of materials have been used as drug carriers,[79,80] and more recently, fluorescent quantum dots have been used for real-time optical imaging in live organisms.[81,82] However, quantum dots are highly toxic, limiting their further application in cell culture and animal studies.

Thus, UCNPs with unique optical and biocompatible components have emerged as promising candidates for traceable drug delivery.[83] Classified by the drug loading method, UCNP-based drug delivery systems may contain hydrophobic pockets, mesoporous silica shells, or hollow mesoporous-coated spheres. In the first method, drugs are encapsulated into "hydrophobic pockets" on the surface of UCNPs

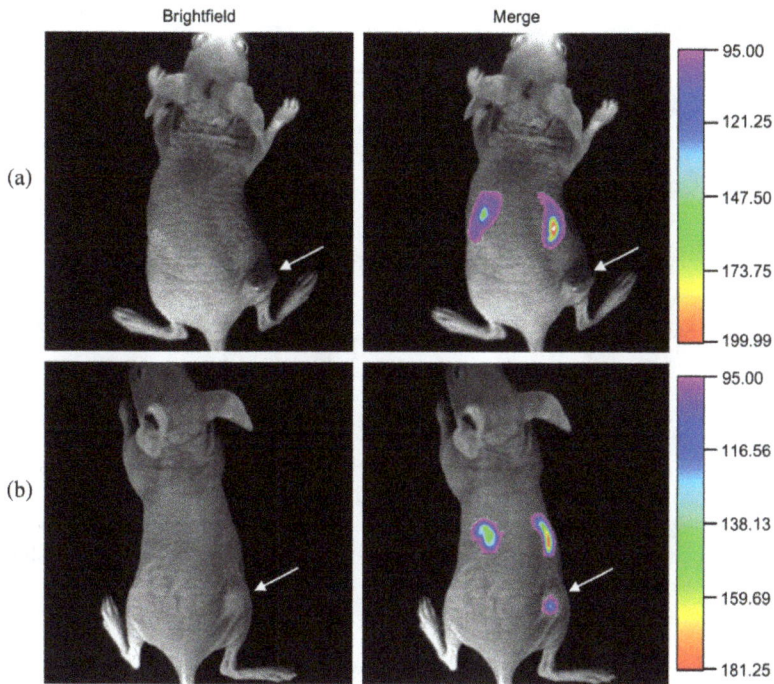

Figure 2.10: *In vivo* upconverting luminescence image of subcutaneous HeLa tumor bearing athymic nude mice (right hind leg, white arrows) after intravenous injection of UCNPs-NH$_2$ (a) and UCNPs-FA (b).[77]

through hydrophobic interactions. For example, by functionalizing the nanoparticle surface with polyethylene glycol-grafted amphiphilic polymers, hydrophobic pockets can be formed between the hydrophobic side chains of the polymers and oleic acid on the UCNP surface, thus enabling anti-cancer drug molecules such as doxorubicin to be encapsulated. The release of doxorubicin is controlled by pH; increased dissociation in acidic conditions favors drug release in tumor cells. The intracellular delivery of doxorubicin was monitored by laser scanning confocal microscopy (Figure 2.11).[19] Later on, this strategy was extended to NaYF4:Yb,Er and iron oxide heterogeneous nanoparticles. Such drug-loaded nanostructures enable concurrent optical imaging and magnetic-targeted drug delivery.[84]

Drugs can also be deposited in the pores of the mesoporous silica shell that is coated onto the surface of UCNPs.[85] For example, ibuprofen, a model drug, was loaded onto mesoporous silica-coated β-NaYF$_4$:Yb,Er UCNPs fibers by an electrospinning process.[86] High levels of drug loading were observed because of the

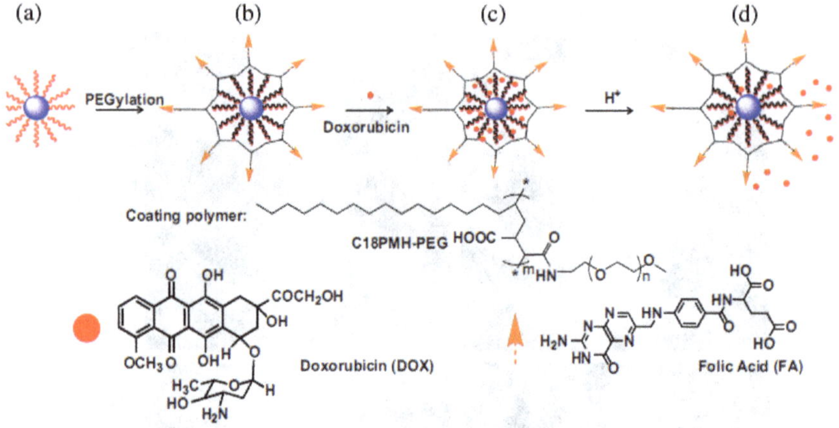

Figure 2.11: Schematic illustration of the UCNP-based drug delivery system. (a) As-synthesized oleic acid capped UCNPs. (b) C18PMH-PEG-FA functionalized UCNPs. (c) Dox loading on UCNPs. Dox molecules are physically adsorbed into the oleic acid layer on the nanoparticle surface by hydrophobic interactions. (d) Release of Dox from UCNPs is triggered by reducing the pH.[19]

high specific surface area and large pore volume of the silica shells. After loading ibuprofen, significant upconverting luminescence quenching was observed, making it possible to determine drug loading by measuring the extent of quenching. More importantly, subsequent drug release can be monitored by the recovery of luminescence intensity.[87]

Drugs can be loaded into the shells of mesoporous hollow spheres as well. The hollow spherical structures allow for significantly higher loading of drugs, without interfering with the upconverting efficiency. For example, a nano-system was recently developed using an ion-exchange strategy.[84] This system is composed of a magnetic nanoparticle in the core of rare-earth-doped $NaYF_4$ shells. By doing so, magnetic guided upconversion imaging was achieved. Doxorubicin was successfully loaded into the hollow volume of the nanorattle via the porous UCNPs layer. *In vivo* experiments showed significant tumor shrinkage with doxorubicin and significant tumor targeting in the presence of an applied magnetic field.[88]

2.6.2 Light Controllable Drug Release Based on UCNPs

To enhance efficacy and minimize side effects during treatment, delivery systems have been designed to release bioactive molecules or drugs at a desired time and location.[89,90] In addition, the ability to deliver multi-payloads in varied doses at different time points would benefit studies on directing cellular processes and disease regulation.[91,92] Among the controlled release methods available, light-triggered molecular cleavage has received growing interest, due to the property of well defined temporal and spatial drug release, compared with other classical stimuli.[93] Using the upconversion properties of UCNPs, the photo-release of imaging probes and drug payloads can be controlled remotely.[94,95] For example, a NIR light-responsive crosslinked mesoporous silica-coated UCNP drug delivery conjugate has been designed as a photocaged nano-carrier. The lanthanide doped UCNPs were first coated with a silica shell, followed by polymerization of 1-(2-nitrophenyl)ethyl photocaged oligo(ethylene) glycol vinyl monomers. The anti-tumor drug doxorubicin (Dox) was encapsulated in the hydrophobic pockets of the aforementioned polymer. Under NIR light irradiation, subsequent emissions from UCNPs can trigger the cleavage of the crosslinked photocaged linker and lead to the precise release of targeted drug at specific locations (Figure 2.12).[96]

Furthermore, drug delivery can also be facilitated by NIR light-mediated photoactivation. For example, a novel NIR light-activated nanoplatform was developed by conjugating a photoactivatable platinum (IV) prodrug and caspase peptide to

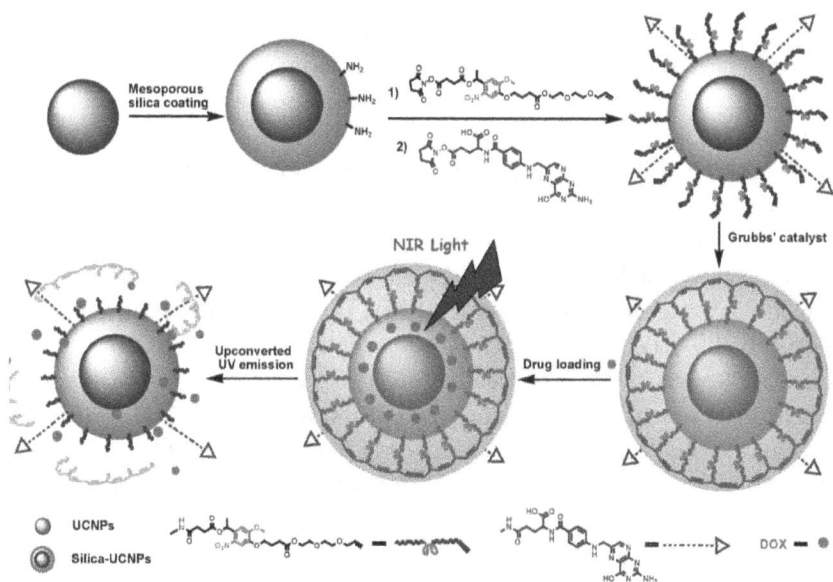

Figure 2.12: Illustration of photo-controlled Dox delivery by photocage mesoporous silica-coated UCNPs.[96]

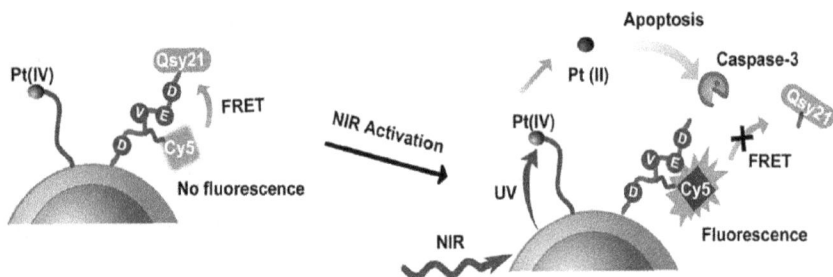

Figure 2.13: Illustration of NIR light activation of Pt (IV) prodrug and intracellular apoptosis imaging using upconversion-luminescent nanoparticles.[97]

silica-coated UCNPs. Upon NIR light excitation, the Pt (IV) prodrug is activated at the surface of the nanoparticle and the activated components selectively released. More importantly, the caspases generated effectively cleave the probe peptide linkers, making direct imaging possible. (Figure 2.13).[97]

2.6.3 *UCNPs for Gene Delivery*

Success in gene therapy depends on the design of effective gene delivery vectors that specifically and controllably deliver gene cargo into living cells and tissues. There is a growing interest in the development of non-viral synthetic nano-carriers for gene delivery.[98] Such carriers require four important properties, namely: (1) safe and non-toxic components, (2) transport traceable by long-term and real-time imaging, (3) controlled release, and (4) temporal and spatial targeting during delivery. UCNPs are promising for the delivery of nucleic acids in gene therapy. For example, silica-coated β-NaYF$_4$:Yb, Er UCNPs were used to track the intracellular uptake and release of siRNA, as well as the biostability of UCNP-bound siRNA, in live cells using the luminescence resonance energy transfer (LRET) system.[99] In this system, cationic silica-coated β-NaYF$_4$:Yb, Er UCNPs act as the donor and siRNA-intercalating dye BOBO-3 acts as the corresponding acceptor. Under 980 nm excitation, the nanoparticles emit upconverted fluorescence at a wavelength of 543 nm, which was consequently absorbed by a BOBO-3-stained siRNA (siRNA-BOBO3) acceptor via an overlapped absorption band. This UCNP was shown to bind to and protect siRNA from RNase cleavage and to effectively deliver siRNA into cells. Under excitation at 980 nm, when the amino-modified UCNP was bound with BOBO-3-stained siRNA, energy transfer from UCNP to BOBO-3 consequently generated characteristic BOBO emissions at 602 nm (Figure 2.14). Once the siRNA was separated from the UCNPs, the LRET process was inhibited. The results showed that

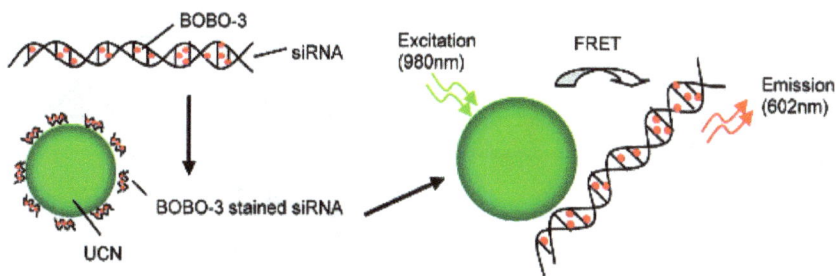

Figure 2.14: Schematic of the FRET-based UCN/siRNA-BOBO3 complex system. siRNA stained with the BOBO-3 dye are attached to the surface of $NaYF_4$:Yb,Er nanoparticles. Upon excitation of the nanoparticles at 980 nm, energy is transferred from the donor (UCN) to the acceptor (BOBO-3).[99]

siRNA was gradually released from the UCNP surface into the cytoplasm over 24 h, correlating quite well with further fluorescent co-localization imaging studies.[99] Subsequently, the same group applied LRET analysis to monitor green fluorescent protein (GFP)-encoded plasmid DNA delivery and release in live cells.[100]

In another example, monodispersed hexagonal $NaGdF_4$-based UCNPs were first conjugated with poly(ethylene glycol) (PEG) to improve their physiological stability and coated with two layers of polyethylenimine (PEI) for gene loading and reduced cytotoxicity. Serum was used to modify the nanoparticle surfaces at the same time, which largely inhibits the transfection activity of PEI. Thus, these nanostructures with well-engineered surfaces were developed for safe and efficient gene delivery *in vitro* as well as for real-time imaging.[101]

2.7 Photodynamic and Photothermal Therapies

2.7.1 *Photodynamic Therapy*

Typical photosensitizers are excited under visible light and generate cytotoxic ROS.[102] Yet, this wavelength of light has rather limited tissue penetration depth. A promising nano-carrier and energy donor for these sensitizers are NIR-excitable UCNPs.[103] For example, a tris(bipyridine)ruthenium(II) ($Ru(bpy)_3^{2+}$)-doped SiO_2 shell was used to encapsulate a $NaYF_4$:Yb,Tm core. Under 980 nm excitation, the blue emission from Tm^{3+} was absorbed by $^1O_2$ generator $Ru(bpy)_3^{2+}$, and the $^1O_2$ generated was detected by chemical methods.[104] To increase photosensitizer loading, β-$NaYF_4$:Yb,Er nanoparticles were fabricated at 100 nm sizes, and a 10 wt% photosensitizer loading ratio was achieved by coating meso-tetraphenylporphine in

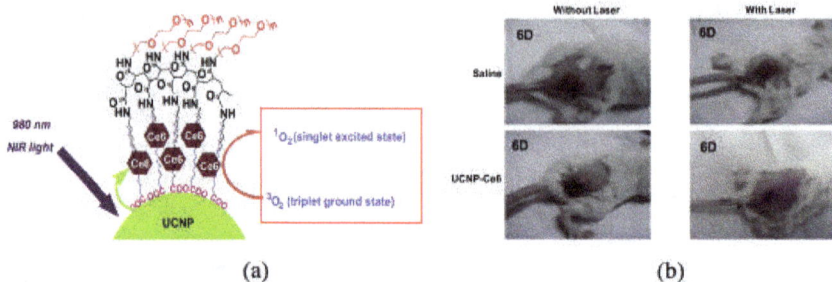

(a)   (b)

Figure 2.15:   (a) Schematic illustration showing NIR-induced PDT using UCNP-Ce6. (b) Representative photos of mice on day 6 after various treatments. Ref. 105.

poly(ethylene glycol-block-(DL)lactic acid).[20] Using this system, efficient cancer cell killing activity was observed upon NIR excitation, and with low cytotoxicity in the absence of NIR light. Moreover, Liu *et al.* designed a UCNP-based photodynamic therapy (PDT) system by incorporating non-covalently photosensitizing porphyrin derivatives into amphiphilic polymer-coated $NaYF_4$:Yb,Er nanoparticles. Excellent tumor regression was observed *in vivo* under 980 nm CW laser excitation, as well as low toxicity and gradual clear out from mouse organs (Figure 2.15).[105]

In another work, mesoporous silica-coated UCNPs were employed as nanoconverters to transfer NIR light (with deep penetration) to visible light and as a carrier of photosensitizers. A mesoporous silica layer was uniformly coated onto the surface of $NaYF_4$ UCNPs, and the visible fluorescence emission from UCNPs at 540 nm and 660 nm matched well with the absorption of two photosensitizers, namely merocyanine 540 and zinc(II) phthalocyanine, respectively. Thus $^1O_2$ can be produced with low-energy NIR light excitation, and this resolved the problem of short wavelength PPT with poor penetration. After conjugating the nanoparticles with folic acid, PDT experiments were carried out both *in vitro* and *in vivo*, demonstrating their cancer cell killing properties (Figure 2.16).[106]

Furthermore, chitosan-wrapped NaYF4:Yb/Er UCNPs co-functionalized with pyropheophorbide a (Ppa) and RGD peptides were developed for targeted targeted NIR photodynamic therapy. In the study, the covalently-linked Ppa photosensitizer was found to be stable and to retain its spectroscopic and functional properties. In U87-MG cells, the nano-system showed strong targeting specificity compared with the corresponding negative group. In addition, this photosensitizer-attached nano-system induced low dark toxicity and high phototoxicity against cancer cells under 980 nm excitation.[107]

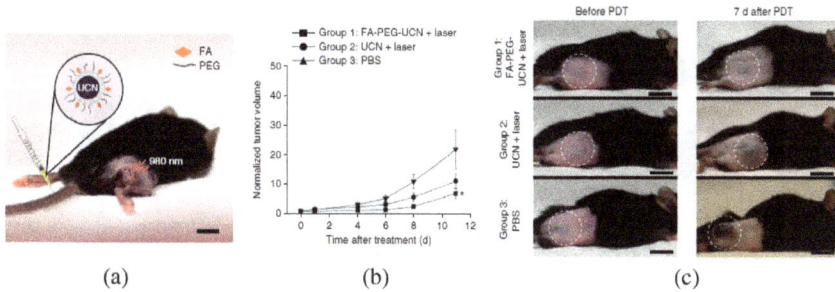

(a)   (b)   (c)

Figure 2.16: Targeted *in vivo* PDT in a subcutaneous tumor model injected with FA-PEG-UCNPs. (a) UCNP-based targeted PDT in a mouse model of melanoma. (b) Changes in tumor size as a function of time after PDT treatment. (c) Representative photos of a mouse from each group, showing the change in tumor size before and 7 days after PDT treatment.[106]

2.7.2 Photothermal Therapy

The basis for photothermal therapy is partly similar to photodynamic therapy, in which vibrational energy (heat) is generated from a photosensitizer for cell killing.[108,109] Multifunctional nanoparticles have been designed to combine upconversion luminescence, magnetism and photothermal properties.[110] Ultra-small Fe_3O_4 nanoparticles were coated onto 200 nm β-$NaYF_4$:Yb,Er upconversion crystals, and then coated with a thin gold shell. The magnetic layer in between the UCNP core and the gold shell can largely reduce the luminescence quenching effects of the gold nanostructure to the UCNP core. Both upconversion imaging and T2-weighted MRI studies were performed *in vitro* and *in vivo*.[111] Instead of gold, β-$NaYF_4$:Yb,Er nanoparticles were coated with a silver shell for stronger plasmonic resonance performance, and PTT was tested *in vitro* in HepG2 hepatic cancer cell lines and BCap-37 cell lines.[112] In addition, UCNPs made of CeO_2:Yb,Er (Tm or Ho) were found to kill lung cancer cells under 980 nm excitation.[113]

To achieve multimodal cancer therapy, multifunctional core/satellite nanotheranostic nanoparticles were developed by attaching ultrasmall CuS nanoparticles onto the surface of silica-coated UCNPs (Figure 2.17). These nano-composites exhibit many advantages, such as: (1) conversion of NIR light into local thermal energy for photothermal therapy agents; (2) large local radiation dose-enhancement by Z elements (Yb, Gd, and Er) around nanoparticles for use as radiosensitizers; and (3) simultaneous UCL/MR/CT trimodal imaging. This development has laid the groundwork for future image-guided therapies.[114]

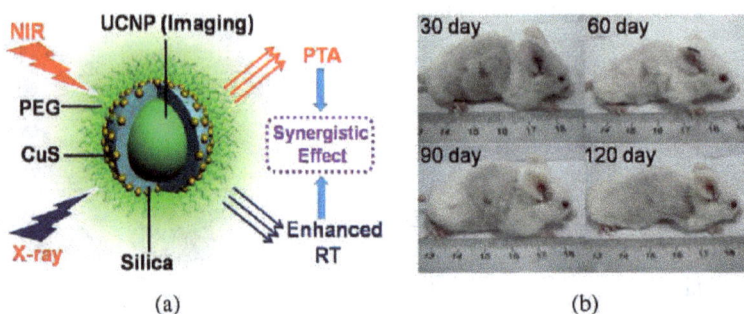

(a)   (b)

Figure 2.17: (a) Illustration of nano-carriers for enhanced photothermal ablation and radiotherapy synergistic therapy. (b) Photographs of mice at 30, 60, 90 and 120 days of treatment, showing complete eradication of the tumor and no visible recurrence in at least 120 days.[114]

2.8 Conclusion and Prospects

The unique optical properties and superior chemical and physical behaviors of UCNPs make them attractive candidates for bio-related applications, especially in bio-imaging and drug delivery. Ultra-small particle sizes are required for the efficient urinary excretion of molecular imaging and drug delivery carriers from the body. However, generating small UCNPs without sacrificing upconversion efficiency is still a big challenge. In addition, to expand the applications of UCNPs, it is critical to develop commercially-available and versatile UCNPs with unique excitation and emission wavelengths.

Of equal importance is the detailed and systematic study of the basic factors that control UCNP luminescent efficiency. This information will help researchers further optimize nanocrystal brightness emission. Meanwhile, further work is needed for a better understanding of the uptake, release and toxicity of UCNPs if these nano-platforms are to be used as *in vivo* imaging tools. Finally, achieving greater specificity in targeting during drug delivery will enhance how diseases such as cancer and HIV/AIDS are treated.

References

1. Auzel, F. Upconversion and anti-stokes processes with f and d ions in solids. *Chem. Rev.* 104, 139–173 (2004).
2. Ong, L.C. *et al.* Upconversion: Road to El Dorado of the fluorescence world. *Luminescence* 25, 290–293 (2010).

3. Feng, W. *et al.* Recent advances in the optimization and functionalization of upcon-version nanomaterials for *in vivo* bioapplications. *NPG Asia Materials* 5, e75; doi:10.1038/am.2013.63 (2013).
4. Chen, G., and Han, G. Theranostic upconversion nanoparticles (I). *Theranostics* 3, 289–291 (2013).
5. Wu, S.W., Han, G., Milliron, D.J., Aloni, S., Altoe, V., Talapin, D.V., Cohen, B.E., and Schuck, P.J. Non-blinking and photostable upconverted luminescence from single lanthanide-doped nanocrystals. *Proc. Natl. Acad. Sci. USA* 106, 10917–10921 (2009).
6. Liu, Q., Feng, W., Yang, T., Yi, T., and Li, F. Upconversion luminescence imaging of cells and small animals. *Nat. Protoc.* 8, 2033–2044 (2013).
7. Zhou, J.C., Yang, Z.L., Dong, W., Tang, R.J., Sun, L.D., and Yan, C.H. Bioimaging and toxicity assessments of near-infrared upconversion luminescent NaYF4:Yb,Tm nanocrystals. *Biomaterials* 32, 9059–9067 (2011).
8. Cheng, L.A., Yang, K., Shao, M.W., Lee, S.T., and Liu, Z.A. Multicolor *in vivo* imaging of upconversion nanopaticles with emissions tuned by luminescence resonance energy transfer. *J. Phys. Chem. C* 115, 2686–2692 (2011).
9. Chen, G., Shen, J., Ohulchanskyy, T.Y., Patel, N.J., Kutikov, A., Li, Z., Song, J., Pandey, R.K., Ågren, H., Prasad, P.N., Han, G. (α-NaYbF4:Tm3+)/CaF2 core/shell nanopar-ticles with efficient near-infrared to near-infrared upconversion for high-contrast deep tissue bioimaging. *ACS Nano*, 6, 8280–8287 (2012).
10. Menyuk, N., Dwight, K., Pierce, J.W. NaYF4: Yb, Er — An efficient upconversion phosphor. *Appl. Phys. Lett.* 21(4), 159 (1972).
11. Yi, H.X., Zhang, Y.W., Si, R., Yan, Z.G., Sun, L.D., You, L.P., Yan, C.H. High-quality sodium rare-earth fluoride nanocrystals: Controlled synthesis and optical properties. *J. Am. Chem. Soc.* 128, 6426–6436 (2006).
12. Qian, H.S., Zhang, Y. Synthesis of hexagonal-phase core-shell NaYF4 nanocrystals with tunable upconversion fluorescence. *Langmuir* 24, 12123–12125 (2008).
13. Sun, L.-D., Wang, Y.-F., and Yan, C.-H. Paradigms and challenges for bioapplication of rare earth upconversion luminescent nanoparticles: Small size and tunable emis-sion/excitation spectra. *Acc. Chem. Res.* 2014, dx.doi.org/10.1021/ar400218t.
14. Verma, A., and Rotello, V.M. Surface recognition of biomacromolecules using nanopar-ticle receptors. *Chem. Commun.* 2005, 303–312.
15. Yang, Y. Upconversion nanophosphors for use in bioimaging, therapy, drug delivery and bioassays. *Microchim. Acta* 181, 263–294 (2014).
16. Mader, H.S., Kele, P., Saleh, S.M., and Wolfbeis, O.S. Upconverting luminescent nanoparticles for use in bioconjugation and bioimaging. *Curr. Opin. Chem. Biol.* 14582–14596 (2010).
17. Nam, S.H., Bae, Y.M., Park, Y.I., Kim, J.H., Kim, H.M., Choi, J.S., Lee, K.T., Hyeon, T., and Suh, Y.D. Long-term real-time tracking of lanthanide ion doped upconverting nanoparticles in living cells. *Angew. Chem. Int. Ed.* 50, 6093–6097 (2011).
18. Naczynski, D.J., Tan, M.C., Zevon, M., Wall, B., Kohl, J., Kulesa, A., Chen, S., Roth, C.M., Riman, R.E., and Moghe, P.V. Rare-earth-doped biological composites as *in vivo* shortwave infrered reporters. *Nat. Comm.* 2013, doi:10.1038/ncomms3199.
19. Wang, C., Cheng, L.A., and Liu, Z.A. Drug delivery with upconversion nanoparti-cles for multi-functional targeted cancer cell imaging and therapy. *Biomaterials* 32, 1110–1120 (2011).

20. Shan, J.N., Budijono, S.J., Hu, G.H., Yao, N., Kang, Y.B., Ju, Y.G., and Prud'homme, R.K. Pegylated composite nanoparticles containing upconverting phosphors and meso-tetraphenyl porphine (TPP) for photodynamic therapy. *Adv. Funct. Mater.* 21, 2488–2495 (2011).

21. Cheng, L., Yang, K., Li, Y.G., Chen, J.H., Wang, C., Shao, M.W., Lee, S.T., and Liu, Z. Facile preparation of multifunctional upconversion nanoprobes for multimodal imaging and dual-targeted photothermal therapy. *Angew. Chem. Int. Ed.* 50, 7385–7390 (2011).

22. Fischer, L.H., Harms, G.S., and Wolfbeis, O.S. Upconverting nanoparticles for nanoscale thermometry. *Angew. Chem. Int. Ed.* 50, 4546–4551 (2011).

23. Wang, G.F., Peng, Q., and Li, Y.D. Lanthanide-doped nanocrystals: Synthesis, optical-magnetic properties, and applications. *Acc. Chem. Res.* 44, 322–332 (2011).

24. Chatterjee, D.K., Gnanasammandhan, M.K., and Zhang, Y. Small upconverting fluorescent nanoparticles for biomedical applications. *Small* 2010, 6, 2781–2795.

25. Zhou, J., Liu, Z., and Li, F. Upconversion nanophosphors for small-animal imaging. *Chem. Soc. Rev.* 41, 1323–1349 (2012).

26. Ntziachristos, V. Fluorescence molecular imaging. *Annu. Rev. Biomed. Eng.* 8, 1–33 (2006).

27. Massoud, T.F., and Gambhir, S.S. Molecular imaging in living subjects: Seeing fundamental biological processes in a new light. *Gene Dev.* 17, 545–580 (2003).

28. Pathak, A.P., Gimi, B., Glunde, K., Ackerstaff, E., Artemov, D., and Bhujwalla, Z.M. Molecular and functional imaging of cancer: Advances in MRI and MRS, *Methods Enzymol.* 386, 3–60 (2004).

29. Shen, J., Zhao, L., and Han, G. Lanthanide-doped upconverting luminescent nanoparticle platforms for optical imaging-guided drug delivery and therapy. *Adv. Drug Deliv. Rev.* 65, 744–755 (2013).

30. Mai, H.X., Zhang, Y.W., Sun, L.D., and Yan, C.H. Highly efficient multicolor upconversion emissions and their mechanisms of monodisperse NaYF4: Yb, Er core and core/shell-structured nanocrystals. *J. Phys. Chem. C* 111, 13721–13729 (2007).

31. Haase, M., and H. Schäfer. Upconverting nanoparticles. *Angew. Chem. Int. Ed.* 50, 5808–5829 (2011).

32. Qiu, H.L., Chen, G.Y., Sun, L., Hao, S.W., Han, G., and Yang, C.H. Ethylenediaminetetraacetic acid (EDTA)-controlled synthesis of multicolor lanthanide doped BaYF(5) upconversion nanocrystals. *J. Mater. Chem.* 21, 17202–17208 (2013).

33. Qiu, P., Zhou, N., Chen, H., Zhang, C., Gao, G., and Cui, D. Recent advances in lanthanide-doped upconversion nanomaterials: Synthesis, nanostructures and surface modification. *Nanoscale* 5, 11512–11525 (2013).

34. Wang, J., Deng, R., MacDonald, M.A., Chen, B., Yuan, J., Wang, F., Chi, D., Hor, T.S.A., Zhang, P., Liu, G., Han, Y., and Liu, X. Enhancing multiphoton upconversion through energy clustering at sublattice level. *Nature Materials* 13, 157–162 (2014).

35. Zhao, J., Jin, D., Schartner, E.P., Lu, Y., Liu, Y., Zvyagin, A.V., Zhang, L., Dawes, J.M., Xi, P., Piper, J.A., Goldys, E.M., and Monro, T.M. Single-nanocrystal sensitivity achieved by enhanced upconversion luminescence. *Nat. Nanotech.* 8, 729–734 (2013).

36. Lu, Y., Zhao, J., Zhang, R., Liu, Y., Liu, D., Goldys, E.M., Yang, X., Xi, P., Sunna, A., Lu, J., Shi, Y., Leif, R.C., Huo, Y., Shen, J., Piper, J.A., Robinson, J.P., and Jin, D.

Tunable lifetime multiplexing using luminescent nanocrystals. *Nat. Photonics* 8, 32–36 (2014).

37. Wang, F., and Liu, X. Recent advances in the chemistry of lanthanide-doped upconversion nanocrystals. *Chem. Soc. Rev.* 38, 976–989 (2009).

38. Heer, S., Lehmann, O., Haase, M., and Güdel, H.U. Blue, green, and red upconversion emission from lanthanide-doped LuPO4 and YbPO4 nanocrystals in a transparent colloidal solution. *Angew. Chem. Int. Ed.* 42, 3179–3182 (2003).

39. Mai, H.X., Zhang, Y.W., Si, R., Yan, Z.G., Sun, L.D., You, L.P., and Yan, C.H. High-quality sodium rare-earth fluoride nanocrystals: Controlled synthesis and optical properties. *J. Am. Chem. Soc.* 128, 6426–6436 (2006).

40. Yi, G.S., and Chow, G.M. Synthesis of hexagonal-phase NaYF4: Yb, Er and NaYF4: Yb, Tm nanocrystals with efficient up-conversion fluorescence. *Adv. Funct. Mater.* 16, 2324–2329 (2006).

41. Chan, E.M., Xu, C., Mao, A.W., Han, G., Owen, J.S., Cohen, B.E., and Milliron, D.J. Reproducible, high-throughput synthesis of colloidal nanocrystals for optimization in multidimensional parameter space. *Nano Lett.* 10, 1874–1885 (2010).

42. Boyer, J.C., Vetrone, F., Cuccia, L.A., and Capobianco, J.A. Synthesis of colloidal upconverting NaYF4 nanocrystals doped with Er3+, Yb3+ and Tm3+, Yb3+ via thermal decomposition of lanthanide trifluoroacetate precursors. *J. Am. Chem. Soc.* 128, 7444–7445 (2006).

43. Boyer, J.C., Cuccia, L.A., and Capobianco, J.A. Synthesis of colloidal upconverting NaYF4:Er3+/Yb3+ and Tm3+/Yb3+ monodisperse nanocrystals. *Nano Lett.* 7, 847–852 (2007).

44. Yi, G.S., and Chow, G.M. Water-soluble NaYF4: Yb, Er(Tm)/NaYF4/polymer core/shell/shell nanoparticles with significant enhancement of upconversion fluorescence. *Chem. Mater.* 19, 341–343 (2007).

45. Zeng, J.H., Su, J., Li, Z.H., Yan, R.X., and Li, Y.D. Synthesis and upconversion luminescence of hexagonal-phase NaYF4: Yb, Er3+, phosphors of controlled size and morphology. *Adv. Mater.* 17, 2119–2123 (2005).

46. Liang, X., Wang, X., Zhuang, J., Peng, Q., and Li, Y.D. Synthesis of NaYF4 nanocrystals with predictable phase and shape. *Adv. Funct. Mater.* 17, 2757–2765 (2007).

47. Wang, X., Zhuang, J., Peng, Q., and Li, Y.D. A general strategy for nanocrystal synthesis, *Nature* 437, 121–124 (2005).

48. Li, Z.Q., and Zhang, Y. Monodisperse silica-coated polyvinylpyrrolidone/NaYF4 nanocrystals with multicolor upconversion fluorescence emission. *Angew. Chem. Int. Ed.* 45, 7732–7735 (2006).

49. Wang, F., and Liu, X.G. Upconversion multicolor fine-tuning: Visible to near-infrared emission from lanthanide-doped NaYF4 nanoparticles. *J. Am. Chem. Soc.* 130, 5642–5643 (2008).

50. Chen, G., Qiu, H., Fan, R., Hao, S., Tan, S., Yang, C., and Han, G. Lanthanidedoped ultrasmall yttrium fluoride nanoparticles with enhanced multicolor upconversion photoluminescence. *J. Mater. Chem.* 22, 20190 (2012).

51. Wang, Y., Tu, L., Zhao, J., Sun, Y., Kong, X., and Zhang, H. Upconversion luminescence of β-NaYF4: Yb3+, Er3+@β-NaYF4 core/shell nanoparticles: excitation power density and surface dependence. *J. Phys. Chem. C* 113, 7164–7169 (2009).

52. Wang, F., Wang, J., and Liu, X. Direct evidence of a surface quenching effect on size-dependent luminescence of upconversion nanoparticles. *Angew. Chem. Int. Ed.* 49, 7456–7460 (2010).

53. Zhang, F., Che, R., Li, X., Yao, C., Yang, J., Shen, D., Hu, P., Li, W., and Zhao, D. Direct imaging the upconversion nanocrystal core/shell structure at the subnanometer level: shell thickness dependent in upconverting optical properties. *Nano Lett.* 12, 2852–2858 (2012).

54. Zhong, Y., Tian, G., Gu, Z., Yang, Y., Gu, L., Zhao, Y., Ma, Y., and Yao, J. Elimination of photon quenching by a transition layer to fabricate a quenching-shield sandwich structure for 800 nm excited upconversion luminescence of Nd3+-sensitized nanoparticles. *Adv. Mater.* 2013, DOI: 10.1002/adma.201304903.

55. Zou, W., Visser, C., Maduro, J.A., Pshenichnikov, M.S., and Hummelen, J.C. Broadband dye-sensitized upconversion of near-infrared light. *Nat. Photonics* 6, 560–564 (2012).

56. Xie, X., and Liu, X. Upconversion goes broadband. *Nat. Materials* 11, 842–843 (2012).

57. Luccardini, C., Tribet, C., Vial, F., Marchi-Artzner, V., and Dahan, M. Size, charge, and interactions with giant lipid vesicles of quantum dots coated with an amphiphilic macromolecule. *Langmuir* 22, 2304–2310 (2006).

58. Zhao, L., Kutikov, A., Shen, J., Duan, C., Song, J., and Han, G. Stem cell labeling using polyethylenimine conjugated (α-NaYbF4:Tm3+)/CaF2 upconversion nanoparticles. *Theranostics* 3, 249–257 (2013).

59. Skajaa, T., Zhao, Y.M., van den Heuvel, D.J., Gerritsen, H.C., Cormode, D.P., Koole, R., van Schooneveld, M. M., Post, J.A., Fisher, E.A., Fayad, Z.A., Donega, C.D., Meijerink, A., and Mulder, W.J.M. Quantum dot and Cy5.5 labeled nanoparticles to investigate lipoprotein biointeractions via forster resonance energy transfer. *Nano Lett.* 10, 5131–5138 (2010).

60. Zhelev, Z., Ohba, H., and Bakalova, R. Single quantum dot-micelles coated with silica shell as potentially non-cytotoxic fluorescent cell tracers. *J. Am. Chem. Soc.* 128, 6324–6325 (2006).

61. Susumu, K., Uyeda, H.T., Medintz, I.L., Pons, T., Delehanty, J.B., and Mattoussi, H. Enhancing the stability and biological functionalities of quantum dots via compact multifunctional ligands. *J. Am. Chem. Soc.* 129, 13987–13996 (2007).

62. Chen, Z.G., Chen, H.L., Hu, H., Yu, M.X., Li, F.Y., Zhang, Q., Zhou, Z.G., Yi, T., and Huang, C.H. Versatile synthesis strategy for carboxylic acid-functionalized upconverting nanophosphors as biological labels. *J. Am. Chem. Soc.* 130, 3023–3029 (2008).

63. Park, Y.I., Kim, J.H., Lee, K.T., Jeon, K.S., Bin Na, H., Yu, J.H., Kim, H.M., Lee, N., Choi, S.H., Baik, S.I., Kim, H., Park, S.P., Park, B.J., Kim, Y.W., Lee, S.H., Yoon, S.Y., Song, I.C., Moon, W.K., Suh, Y.D., and Hyeon, T. Nonblinking and nonbleaching upconverting nanoparticles as an optical imaging nanoprobe and T1 magnetic resonance imaging contrast agent. *Adv. Mater.* 21, 4467–4471 (2009).

64. Boyer, J.C., Manseau, M.P., Murray, J.I., and van Veggel, F.C. Surface modification of upconverting NaYF4 nanoparticles with PEG-phosphate ligands for NIR (800 nm) biolabeling within the biological window. *Langmuir* 26, 1157–1164 (2010).

65. Nyk, M., Kumar, R., Ohulchanskyy, T.Y., Bergey, E.J., and Prasad, P.N. High contrast *in vitro* and *in vivo* photoluminescence bioimaging using near infrared to near infrared up-conversion in Tm3+ and Yb3+ doped fluoride nanophosphors. *Nano Lett.* 8, 3834–3838 (2008).

66. Dong, A., Ye, X., Chen, J., Kang, Y., Gordon, T., Kikkawa, J.M., and Murray, C.B. A generalized ligand-exchange strategy enabling sequential surface functionalization of colloidal nanocrystals. *J. Am. Chem. Soc.* 133, 998–1006 (2011).

67. Naccache, R., Vetrone, F., Mahalingam, V., Cuccia, L.A., and Capobianco, J.A. Controlled synthesis and water dispersibility of hexagonal phase NaGdF(4): Ho(3+)/Yb(3+) nanoparticles. *Chem. Mater.* 21, 717–723 (2009).

68. Liu, Z. *et al.* Long-circulating Er3+-doped Yb2O3 up-conversion nanoparticle as an *in vivo* X-ray imaging contrast agent. *Biomaterials* 33, 6748–6757 (2012).

69. Chen, Z.G., Chen, H.L., Hu, H., Yu, M.X., Li, F.Y., Zhang, Q., Zhou, Z.G., Yi, T., and Huang, C.H. Versatile synthesis strategy for carboxylic acid-functionalized upconverting nanophosphors as biological labels. *J. Am. Chem. Soc.* 130, 3023–3029 (2008).

70. Zhou, H.P., Xu, C.H., Sun, W., and Yan, C.H. Clean and flexible modification strategy for carboxyl/aldehyde-functionalized upconversion nanoparticles and their optical applications. *Adv. Funct. Mater.* 19, 3892–3900 (2009).

71. Xu, C.T., Svensson, N., Axelsson, J., Svenmarker, P., Somesfalean, G., Chen, G., Liang, H., Liu, H., Zhang, Z., and Andersson-Engels, S. Autofluorescence insensitive imaging using upconverting nanocrystals in scattering media. *Appl. Phys. Lett.* 93, 171103 (2008).

72. Chatterjee, D.K., Jalil, R.A., and Zhang, Y. Upconversion fluorescence imaging of cells and small animals using lanthanide doped nanocrystals. *Biomaterials* 29, 937–943 (2008).

73. Wang, M., Mi, C., Zhang, Y., Liu, J., Li, F., Mao, C., and Xu, S. NIR responsive silica-coated NaYbF4: Er/Tm/Ho upconversion fluorescent nanoparticles with tunable emission colors and their applications in immunolabeling and fluorescent imaging. *J. Phys. Chem. C* 113, 19021–19027 (2009).

74. Wang, M., Mi, C., Wang, W., Liu, C., Wu, Y., Xu, Z., Mao, C., and Xu, S. Immonolabeling and NIR-excited fluorescent imaging of HeLa cells by using NaYF4: Yb, Er upconversion nanoparticles. *ACS Nano* 3, 1580–1586 (2009).

75. Xiong, L., Chen, Z., Tian, Q., Cao, T., Xu, C., and Li, F. High contrast upconversion luminescence targeted imaging *in vivo* using peptide-labeled nanophosphors. *Anal. Chem.* 81, 8687–8694 (2009).

76. Yu, X.F., Sun, Z., Li, M., Xiang, Y., Wang, Q.Q., Tang, F., Wu, Y., Cao, Z., and Li, W. Neurotoxin-conjugated upconversion nanoprobes for direct visualization of tumors under near-infrared irradiation. *Biomaterials* 31, 8724–8731 (2010).

77. Xiong, L.Q., Chen, Z.G., Yu, M.X., Li, F.Y., Liu, C., and Huang, C.H. Synthesis, characterization, and *in vivo* targeted imaging of amine-functinalized rare-earth upconverting nanophosphors. *Biomaterials* 30, 5592–5600 (2009).

78. Langer, R., and Tirrell, D.A. Designing materials for biology and medicine. *Nature* 428, 487–492 (2004).

79. Sershen, S., and West, J. Implantable, polymeric systems for modulated drug delivery. *Adv. Drug Del. Rev.* 55, 439–439 (2003).
80. Paleos, C.M., Tsiourvas, D., Sideratou, Z., and Tziveleka, L. Acid- and salt-triggered multifunctional poly(propylene imine) dendrimer as a prospective drug delivery system. *Biomacromolecules* 5, 524–529 (2004).
81. Bagalkot, V., Zhang, L., Levy-Nissenbaum, E., Jon, S., Kantoff, P.W., Langer, R., and Farokhzad, O.C. Quantum dot — aptamer conjugates for synchronous cancer imaging, therapy, and sensing of drug delivery based on bi-fluorescence resonance energy transfer. *Nano Lett.* 7, 3065–3070 (2007).
82. Al-Jamal, W.T., Al-Jamal, K.T., Cakebread, A., Halket, J.M., and Kostarelos, K. Blood circulation and tissue biodistribution of lipid-quantum dot (L-QD) hybrid vesicles intravenously administered in mice. *Bioconj. Chem.* 20, 1696–1702 (2009).
83. Kim, C.S., Tonga, G.Y., Solfiell, D., and Rotello, V.M. Inorganic nanosystems for therapeutic delivery: status and prospects. *Adv. Drug. Deli. Rev.* 65, 93–99 (2013).
84. Xu, H., Cheng, L., Wang, C., Ma, X.X., Li, Y.G., and Liu, Z. Polymer encapsulated upconversion nanoparticle/iron oxide nanocomposites for multimodal imaging and magnetic targeted drug delivery. *Biomaterials* 32, 9364–9373 (2011).
85. Slowing, I.I., Trewyn, B.G., Giri, S., and Lin, V.S.-Y. Mesoporous silica nanoparticles for drug delivery and biosensing applications. *Adv. Funct. Mater.* 17, 1225–1236 (2007).
86. Hou, Z.Y., Li, C.X., Ma, P.A., Li, G.G., Cheng, Z.Y., Peng, C., Yang, D.M., Yang, P.P., and Lin, J. Electrospinning preparation and drug-delivery properties of an upconversion luminescent porous NaYF(4): Yb(3+), Er(3+)@silica fiber nanocomposite. *Adv. Funct. Mater.* 21, 2356–2365 (2011).
87. Kang, X.J., Cheng, Z.Y., Li, C.X., Yang, D.M., Shang, M.M., Ma, P.A., Li, G.G., Liu, N.A., and Lin, J. Core-shell structured up-Conversion luminescent and mesoporous NaYF(4): Yb(3+)/Er(3+)@nSiO(2)@mSiO(2) nanospheres as carriers for drug delivery, *J. Phys. Chem. C* 115, 15801–15811 (2011).
88. Zhang, F., Braun, G.B., Pallaoro, A., Zhang, Y., Shi, Y., Cui, D., Moskovits, M., Zhao, D., and Stuky, G.D. Mesoporous multifunctional upconversion luminescent and magnetic "nanorattle" materials for targeted chemotherapy. *Nano Lett.* 12, 61–67 (2012).
89. Putney, S.D., and Burke, P.A. Improving protein therapeutics with sustainded-release formulations. *Nat. Biotechnol.* 16, 153–157 (1998).
90. Lee, K.Y., Peters, M.C., Anderson, K.W., and Mooney, D.J. Controlled growth factor release from synthetic extracellular matrices. *Nature* 408, 998 (2000).
91. Richardson, T.P., Peters, M.C., Ennett, A.B., and Mooney, D.J. Polymeric system for dual growth factor delivery. *Nat. Biotechnol.* 19, 1029–1034 (2001).
92. Sohier, J., Vlugt, T., J.H., Cabrol, N., Van Blitterswijk, C., de Groot, K., and Bezemer, J.M. Dual release of proteins from porous polymeric. *J. Control. Release* 111, 95–106 (2006).
93. Brieke, C., Rohrbach, F., Gottschalk, A., Mayer, G., and Heckel, A. Light-controlled tools. *Angew. Chem. Int. Ed.* 51, 8446–8476 (2012).

94. Jayakumar, M., K.G., Idris, N.M., and Zhang, Y., Remote activation of biomolecules in deep tissues using near-infrared-to-UV upconversion nanotransducers. *Proc. Natl. Acad. Sci. USA* 109, 8483–8488 (2012).

95. Yang, Y., Liu, F., Liu, X., and Xing, B. NIR light controlled photorelease of siRNA and its targeted intracellular delivery based on upconversion nanoparticles. *Nanoscale* 5, 231–238 (2013).

96. Yang, Y., Velmurugan, B., Liu, X., and Xing, B. NIR photoresponsive crosslinked upconverting nanocarriers toward selective intracellular drug release. *Small* 9, 2937–2944 (2013).

97. Min, Y., Li, J., Liu, F., Yeow, E., K.L., and Xing, B. Near-infrared light-mediated photoactivation of a platinum antitumor prodrug and simultaneous cellular apoptosis imaging by upconversion-luminescent nanoparticles. *Angew. Chem. Int. Ed.* 53, 1012–1016 (2014).

98. Thomas, M., and Klibanov, A.M. Non-viral gene therapy: Polycation-mediated DNA delivery. *Appl. Microbiol. Biot.* 62, 27–34 (2003).

99. Jiang, S., and Zhang, Y. Upconversion nanoparticle-based FRET system for study of siRNA in live cells. *Langmuir* 26, 6689–6694 (2010).

100. Guo, H.C., Idris, N.M., and Zhang, Y. LRET-based biodetection of DNA release in live cells using surface-modified upconverting fluorescent nanoparticles. *Langmuir* 27, 2854–2860 (2011).

101. He, L., Feng, L., Cheng, L., Liu, Y., Li, Z., Peng, R., Li, Y., Guo, L., and Liu, Z. Multilayer dual-polymer-coated upconversion nanoparticles for multimodal imaging and serum-enhanced gene delivery. *ACS Appl. Mater. Interfaces* 5, 10381–10388 (2013).

102. Fisher, A., M.R., Murphree, A.L., and Gomer, C.J. Clinical and preclinical photodynamic therapy. *Laser Surg. Med.* 17, 2–31 (1995).

103. Shan, G., Weissleder, R., and Hilderbrand, S.A. Upconverting organic dye doped core-shell nano-composites for dual-modality NIR imaging and photo-thermal therapy. *Theranostics* 3, 267–274 (2013).

104. Guo, Y.Y., Kumar, M., and Zhang, P. Nanoparticle-based photosensitizers under CW infrared excitation. *Chem. Mater.* 19, 6071–6072 (2007).

105. Wang, C., Tao, H.Q., Cheng, L., and Liu, Z. Near-infrared light induced *in vivo* photodynamic therapy of cancer based on upconversion nanoparticles. *Biomaterials* 32, 6145–6154 (2011).

106. Idris, N.M., Gnanasammandhan, M.K., Zhang, J., Ho, P.C., Mahendran, R., and Zhang, Y. *In vivo* photodynamic therapy using upconversion nanoparticles as remote-controlled nanotransducers. *Nat. Medicine* 18, 1580–1585 (2012).

107. Zhou, A., Wei, Y., Wu, B., Chen, Q., and Xing, D. Pyropheophorbide A and c(RGDyK) comodified chitosan-wrapped upconversion nanoparticle for targeted near-infrared photodynamic therapy. *Mol. Pharmaceutics* 9, 1580–1589 (2012).

108. Zhang, J.Z. Biomedical applications of shape-controlled plasmonic nanostructures: A case study of hollow gold nanospheres for photothermal ablation therapy of cancer. *J. Phys. Chem. Lett.* 1, 686–695 (2010).

109. Saxton, R.E., Paiva, M.B., Lufkin, R.B., and Castro, D.J. Laser photochemotherapy — a less invasive approach for treatment of cancer. *Semin. Surg. Oncol.* 11, 283–289 (1995).

110. Kelkar, S.S., and Reineke, T.M. Theranostics: Combining imaging and therapy, *Bioconj. Chem.* 22, 1879–1903 (2011).

111. Cheng, L., Yang, K., Li, Y.G., Chen, J.H., Wang, C., Shao, M.W., Lee, S.T., and Liu, Z. Facile preparation of multifunctional upconversion nanoprobes for multimodal imaging and dual-targeted photothermal therapy. *Angew. Chem. Int. Ed.* 50, 7385–7390 (2011).

112. Dong, B.A., Xu, S., Sun, J.A., Bi, S., Li, D., Bai, X., Wang, Y., Wang, L.P., and Song, H.W. Multifunctional NaYF(4): Yb(3+), Er(3+)@agcore/shell nanocomposites: integration of upconversion imaging and photothermal therapy. *J. Mater. Chem.* 21, 6193–6200 (2011).

113. Babu, S., Cho, J.H., Dowding, J.M., Heckert, E., Komanski, C., Das, S., Colon, J., Baker, C.H., Bass, M., Self, W.T., and Seal, S. Multicolored redox active upconverter cerium oxide nanoparticle for bio-imaging and therapeutics. *Chem. Commun.* 46, 6915–6917 (2010).

114. Xiao, Q., Zheng, X., Bu, W., Ge, W., Zhang, S., Chen, F., Xing, H., Ren, Q., Fan, W., Zhao, K., Hua, Y., and Shi, J. A core/satellite multifunctional nanoteranostic for *in vivo* imaging and tumor eradication by radiation/photothermal synergistic therapy. *J. Am. Chem. Soc.* 135, 13041–13048 (2013).

Magnetic Nanoparticles for Drug Delivery

3

Malathi Mathiyazhakan and Chenjie Xu

School of Chemical and Biomedical Engineering, Nanyang Technological University, Singapore 637457

Abstract

The pursuit of novel drug delivery systems has continued over decades to maximize therapeutic efficacy and minimize undesirable side effects. Nanotechnology has shown great promise to revolutionize the field of drug delivery by providing better encapsulation and bioavailability, controlled release, and lower toxicity. Magnetic nanoparticles (MNPs), a class of nanomaterials responding to magnetic manipulation, are one of the key areas for innovation. In this chapter, we introduce the principles of MNPs in drug delivery and describe how they are synthesized and functionalized for treating a broad range of diseases.

3.1 Background

Drug delivery researchers aim to administer a pharmaceutical compound (i.e., drug) to achieve a therapeutic effect in humans or animals. Modifying the release profile, absorption, distribution and/or elimination of drugs has led to improved drug efficacy and safety, as well as patient convenience and compliance.

Since its advent in the 1960s, nanoscience has provided new opportunities for researchers to develop nano-sized formulations or devices that control the rate, time and location of drug release in the body in a more effective and safer manner.[1] Drugs are conjugated to or encapsulated within nano-sized carriers, termed as drug delivery systems (DDS). In nanoland, tiny differences in size can add up to huge differences in function. Therefore, medicines composing of nanoscale materials or structures could have dramatically different pharmacokinetic and pharmacodynamic properties. For example, Abraxane®, a paclitaxel albumin-stabilized nanoparticle (NP) formulation (Abraxis BioScience LLC), shows a 43% larger clearance and 53% higher volume of distribution than pure paclitaxel.[2] So far, the marriage between nanoscience and clinical needs has generated many well-known DDSs including lipid-based NPs (liposomes, micelles and solid lipid NPs), polymeric systems, hydrogels, dendrimers and inorganic NPs (silica, gold and iron oxide).[3,4] Some of these DDSs have been successfully applied in the clinic and many more are currently being clinically evaluated. As documented by McCullough *et al.*, over 30 nano-drugs have been commercialized and 90 others are at different stages of clinical trials.[5]

Magnetic NPs (MNPs) are a popular choice among researchers for DDS development. They are usually smaller than the single domain limit of magnetic materials, and exhibit superparamagnetism at room temperature.[6] They maintain colloidal stability and avoid agglomeration in the absence of magnetic field, but accumulate when a magnet is present or generate heat when an alternating magnetic field (AMF) is applied. There are generally two ways to utilize MNPs in drug delivery. The first one is straightforward and simple. When a static magnetic field is placed at a diseased site, drug-containing MNPs administered systematically will accumulate for localized drug delivery.[7] The second strategy is to use the heat-generating ability of MNPs under the AMF.[8] Briefly, drugs are conjugated to MNPs via thermo-labile linker molecules. Upon AMF treatment, the linkers are cleaved to release the drug at sites where the AMF is focused. Moreover, the generated heat can also create nano-sized pores on the lipids or polymeric matrix where the MNPs are entrapped, providing a way to further control drug release. Related research has grown in recent years. As shown in Figure 3.1, there are 2,008 MNP-related research papers from 2005–2015, of which 59 are in preclinical studies. Translation from bench to bedside is also taking place for commercial products such as TargetMAG-doxorubicin NPs, FluidMAG®and MagNaGel®.

3.2 General Principles of Magnetic Nanoparticles

3.2.1 *Physical Principle of Magnetic Nanoparticles*

The origin of magnetic properties in matter lies in the orbital and spin motion of electrons, whose spin and angular momentum are associated with the magnetic moment. The interaction between the magnetic moments of atoms from the same material causes magnetic order below a certain critical temperature. On the basis of these interactions and their influence on the materials behavior in response to

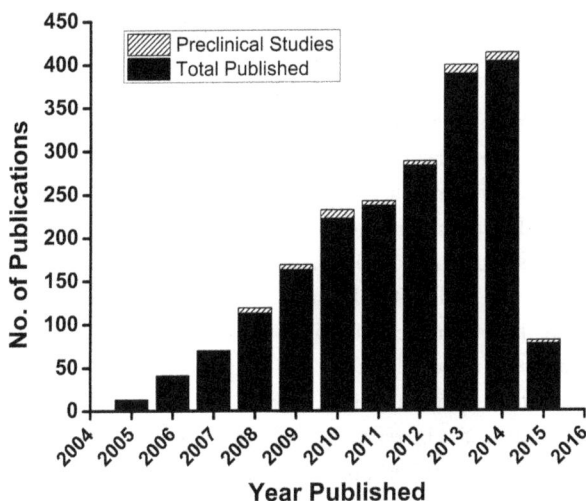

Figure 3.1: Publication analysis in the past 10 years for magnetic nanoparticles in drug delivery. (Search key words: Magnetic nanoparticles, drug delivery, animal study, preclinical study, clinic study. Search engine: Web of Science.)

magnetic fields at different temperatures, materials can be classified as diamagnetic, paramagnetic, ferromagnetic, antiferromagnetic and ferrimagnetic. For bulk ferromagnetic or ferrimagnetic materials, they are composed of regions, called magnetic domains, within which there is an alignment of the magnetic moments. If the volume of the material is reduced, it results in a situation where just one domain is reached and the magnetic properties are no longer similar to those of bulk ferromagnetic or ferrimagnetic materials. The size of this single-domain normally falls between 1 and 1,000 nm.[9] Thus, single-domain ferromagnetic or ferrimagnetic materials can be called MNPs. It should be noted that the exact frontier size between multi-domain and single-domain depends on the crystalline structure as well as the composition of the material.

Single-domain MNPs exhibit superparamagnetism at room temperature,[6] meaning that the thermal energy is enough to change spontaneously the magnetization within each MNP. In other words, the magnetic moment of each MNP will be able to rotate randomly just because of temperature influence. For this reason, in the absence of an electromagnetic field, the net magnetic moment of a system containing MNPs will be zero at high enough temperatures. However, in the presence of a magnetic field, there will be a net statistical alignment of magnetic moments, analogous to what happens to paramagnetic materials, except that now the magnetic moment is not that of a single atom but of the MNPs containing various atoms.

This property, marked by the lack of remnant magnetization after the removal of external fields, enables the MNPs to maintain their colloidal stability and avoid agglomeration.

Besides their unique response to magnetic fields, MNPs can also generate heat when subjected to an external AMF. This is attributed to the re-orientation of magnetic moments in MNPs. In bulk magnetic materials with multidomains, this reorientation is produced through the movement of domain walls. However, in single domain MNPs, this reorientation of magnetic moments can occur due to: 1) the rotation of the moment within MNPs, overcoming their anisotropy energy barrier (Néel loss), or 2) the mechanical rotations of MNPs that create frictional losses with the environment (Brown Loss).[10] For maghemite (γ-Fe$_2$O$_3$) MNPs, below 15 nm, Néel relaxation prevails over Brown relaxation while Brown relaxation is the rotation mechanism for larger sizes and low viscosity media. Interestingly, for MNPs localized within target cells, intracellular organelles such as endosomes may affect particle movement, resulting in heat generation due to Néel's relaxation. As a consequence of magnetization rotation under the AMF, an increase in temperature with time ($.6.T/.6.t$) at a given MNP mass concentration (m) can be measured and used to determine the specific absorption rate (SAR), which is used to compare the heating capacity of different MNPs.

3.2.2 Synthesis of Magnetic Nanoparticles

MNPs can be synthesized by physical, chemical and biological methods, each with their own advantages and disadvantages.[11] Physical methods are used for the ease of synthesis and scale-up, but controlling the particle size, composition and surface chemistry is rather challenging.[12] Chemical methods allow for the efficient and precise control of particle size, composition and surface chemistry, but are detrimental to the environment and human health. Biological methods ensure high yield, reproducibility, scalability as well as environmental friendliness, but the processes are rather time-consuming. Therefore, the ultimate choice of synthetic methods (especially in industry) depends on the specific application and regulatory requirements.

In drug delivery, it is necessary to precisely control the composition, morphology and surface chemistry of MNPs as they determine biodistribution, biocompatibility and therapeutic effects. Thus, researchers have mainly relied on chemical methods to synthesize MNPs. Chemical methods include sol-gel synthesis, oxidation method, co-precipitation, hydrothermal reactions, flow injection synthesis, electrochemical method, aerosol/vapor phase method, thermal decomposition reactions, supercritical fluid method and synthesis using nanoreactors.[11,13] However, taking cost and adaptability into consideration, co-precipitation and thermal decomposition reactions are currently preferred.

Co-precipitation refers to a facile way of synthesizing metal-oxide particles from aqueous metal salt solutions by the addition of a base. The most well-known case is the synthesis of iron oxide MNPs, in which aqueous Fe^{3+} and Fe^{2+} salt solutions are co-precipitated in the presence of ammonium at room temperature to yield iron oxide MNPs.[14] The size, morphology and magnetic properties of MNPs can be controlled by tuning the ratio of Fe^{2+} and Fe^{3+}, the surfactant, the addition of other metal ions, the temperature and pH of the solution, and the stirring rate.[14,15] Besides iron oxide MNPs, spinel ferrite MNPs such as $MnFe_2 O_4$ MNPs have also been successfully prepared by co-precipitation.[16]

Thermal decomposition is a chemical decomposition caused by heat. The decomposition temperature of a substance is the temperature at which the substance chemically decomposes. During MNP synthesis, organometallic precursors and metal salts are used as precursors to prepare monodisperse MNPs.[17] In this method, a burst nucleation occurs when the precursors quickly increase over critical saturation without further formation of nuclei afterwards.[18] The produced nuclei then grow at the same rate, giving monodisperse particles. The addition of stabilizing agents (i.e. surfactant) in the solvent controls the size of the particles as they are formed. This method yields high quality monodisperse, crystalline NPs in bulk quantities as surface defects are often not present due to the high temperature conditions. For example, FeCo NPs were synthesized by reduction of $Fe(acac)_3$ and $Co(acac)_2$ in the presence of oleic acid, oleylamine and 1,2-hexadecanediol under a gas mixture of 93% Ar+7% H_2 at 300°C, or by sodium borohydride reduction of ferrous and cobalt salts.[19] FePt (or FeAu) NPs were synthesized by thermal decomposition of $Fe(CO)_5$ and reduction of $Pt(acac)_2$ (or $AuAc_3$) by 1,2-hexadecanediol in the presence of oleic acid and oleylamine.[20,21]

As mentioned above, chemical methods are detrimental to the environment and human health. Biological methods can be used to achieve precise control over particle size, composition and surface chemistry while being environmental friendly at the same time. Thus the biological synthesis of MNPs might benefit from greater attention. One example is magnetotactic bacteria, which produce highly regular, crystalline MNPs of a specific size and shape.[22] In the bio-mineralization process, the precursor, ferrihydrite ($5Fe_2 O_3 \cdot 9H_2O$), is reduced and oxidized to iron oxide or iron sulphide.[23] The prepared MNPs are single-domain particles ranging from 35 to 120 nm. Their morphology depends on the particular species or strain.[24] Researchers have transfected these genes into other organisms such as *E. coli* to biologically synthesize MNP for biomedical applications such as controlled drug delivery and immunomagneto assays.[25]

3.2.3 Surface Modification of Magnetic Nanoparticles

Without further modification, the as-prepared MNPs are often hydrophobic, prone to aggregation, and may lose their superparamagnetic properties. To maintain MNPs in their colloidal form during storage and to improve their stability and biocompatibility, solid core MNPs are further modified with bulky or charged molecules like polymers, surfactants or silica. These materials provide physical and chemical stability to the magnetic core by electrostatic repulsion or steric hindrance. Furthermore, they provide functional groups for attaching targeting moieties or therapeutics. There are generally two strategies for MNP modification: surfactant addition and surface ligand exchange.[6]

In the first strategy, an additional surfactant is added onto the surface of MNPs without disrupting the original surface structure. The new surfactants could be phospholipids, polymers like poly(ethyl glycol) (PEG) and chitosan, thermos-sensitive polymers such as poly(N-isopropylacrylamide) (pNIPAM), or hydrogels like arabic gum. PEG coating endows water solubility, compatibility and stealth properties to the MNPs.[26,27] In addition, PEG coating helps MNPs evade the reticuloendothelial system[28] and cross the blood brain barrier (BBB).[29] Chitosan coated particles have been used to deliver plasmid DNA and as photo-sensitizing agents for photodynamic therapy.[30–32] Fe_3O_4 MNPs coated with pNIPAM were used for magnetic hyperthermia and delivery of chemotherapeutic agents.[33] In the second strategy, the original coating is substituted by another type of surfactant that can form stronger chemical bonds with MNPs.[34] Thiol-containing molecules (e.g. cystamine, dimercaptosuccinic acid),[35,36] hydroxides (tetramethylammonium hydroxide),[37] and catechol derivatives (e.g. dopamine)[38] are often used to substitute the original molecules on MNPs.

Other biomolecules are then added to the surface-modified MNPs to render additional functionalities for targeting and therapeutic applications. These include peptide molecules such as arginyl-glycyl-aspartic acid (RGD),[39] proteins such as transferrin and annexin,[40,41] aptamers, and antibodies such as Herceptin.[42]

3.2.4 Drug Loading on Magnetic Nanoparticles

There are two ways to formulate drugs with MNPs. The first option is to load drugs onto the surface of MNPs using linker molecules. This usually results in low drug loading efficiency and thus reduced drug efficacy. To improve the efficacy, a high concentration of MNPs has to be injected. This again leads to increased cytotoxicity and reduced biocompatibility.

The second choice is to formulate drugs with hollow MNPs, in which the drug can be loaded onto the surface as well as within the surface.[43–46] For example, cisplatin-loaded hollow Fe_3O_4 MNPs and doxorubicin-loaded hollow Mn_3O_4 MNPs were loaded with drugs within and on the MNPs. The drug loading efficiency was 24.8% and 14% for hollow Fe_3O_4 and Mn_3O_4 MNPs, compared to 4.82% and 4% for solid Fe_3O_4 and Mn_3O_4 MNPs, respectively.[47,48] In the case of Fe_3O_4/SiO_2 — hollow core, double shell MNPs co-loaded with docetaxel and camptothecin — the loading efficiency improved from 1–3% to 14–15%. Thus, hollow MNPs that exhibit improved drug loading efficiency are ideal candidates for drug delivery applications.[49]

3.3 Drug Delivery with MNPs

This section briefly outlines the use of MNPs as drug delivery vehicles for cancer therapy, infectious diseases, and regenerative medicine and tissue engineering. The uses of MNPs in toxin removal and iron replacement therapy are also discussed.

3.3.1 *Cancer Therapy*

Cancer is one of the most devastating diseases known to man and chemotherapy is a common treatment option but results in systemic toxicity. One way to address this challenge is to use MNPs for localized and targeted delivery while monitoring the bio-distribution of drug-MNP complexes via magnetic resonance imaging. In the following four sections, we will review the applications of MNPs in cancer therapy based on their functions.

3.3.1.1 *To circumvent multidrug resistance*

Cancer cells can become resistant to drugs, which is a phenomenon termed multidrug resistance (MDR).[50] Research has shown that MDR is mainly caused by reduced drug influx, activation of detoxifying proteins, activation of cell repair mechanisms and alterations in cell signaling pathways. MDR inhibitors can be used to overcome MDR, but these suffer from unfavorable pharmacokinetic properties and high cytotoxicity.

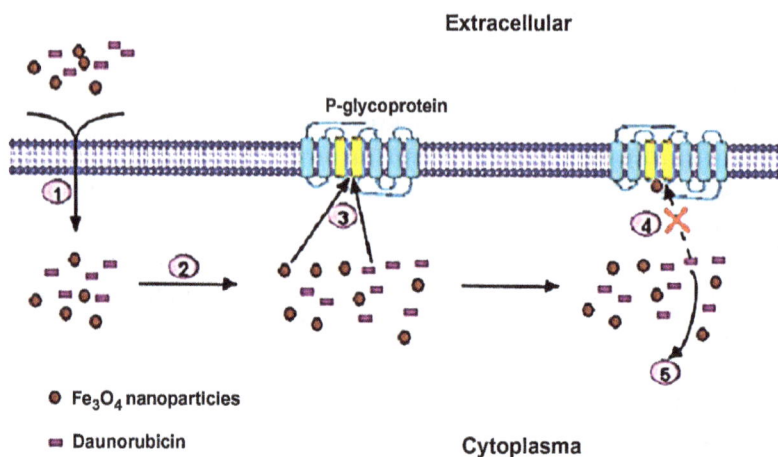

Figure 3.2:   Schematic showing a possible mechanism by which $Fe_3 O_4$ nanoparticles block the Pgp-mediated efflux of daunorubicin from drug resistant leukemia K562 cells. (Reprinted with permission from Wang *et al.*[52])

Interestingly, NPs have been found to circumvent ATP-binding cassette (ABC)-transporters-induced MDR, as the cellular uptake of NPs is mainly through endocytosis.[2,51] P-Glycoprotein (Pgp) is a member of the ABC transporter protein family and resides on the cell membrane. Pgp is responsible for pumping chemotherapeutic agents out of the cytoplasm, leading to MDR in cancer cells. By blocking the activity of Pgp, it is possible to reverse the MDR of cancer cells. Chen *et al.* studied the role of $Fe_3 O_4$ MNPs in preventing daunorubicin-mediated MDR in sensitive and resistant K562 cells (Figure 3.2). Tetraheptylammonium (THA) was coated onto MNPs to improve the interaction between the NPs and the lipid portion of the cell membrane. The study revealed a significant interaction between THA-coated MNPs and the cell membrane and a substantial increase in the uptake of daunorubicin in resistant K652 cells. The control formulation, composed of comparable sizes of THA-coated Ni MNPs, did not have much influence on the uptake of daunorubicin in sensitive and resistant cells, suggesting the exclusivity of THA-capped Fe_3O_4 MNPs in accelerating daunorubicin uptake.[52] Other drugs such as 5-bromotetrandrine[53] and adriamycin (ADM)[54] have also been delivered to leukemia cells using a similar approach.

3.3.1.2   *To improve the drug specificity*

Tumor vasculature is leaky and exhibits poor lymphatic drainage, making it difficult for nanocarriers to reach and accumulate at tumor sites. Non-targeted nanocarriers like liposomes and micelles can improve the local drug concentration at tumor lesions and inflamed tissues, by exploiting the enhanced permeation and retention (EPR) effect of the vasculature. However, not all tumor lesions exhibit the EPR effect and also not all nanocarriers are capable of utilizing the EPR effect. Hence the final tumor drug concentration achieved by passive targeting remains relatively low.[55] Nanocarriers were then modified for active targeting by attaching targeting ligands onto the surface of the carriers. The specificity of targeted nanocarriers is based on the overexpressed surface receptors in tumor tissue. Despite the increased complexity of synthetic procedures, bioaccumulation through active targeting remains challenging.[56] Drug release triggered by an applied magnetic field remains an attractive alternative as it allows for precise spatial and temporal control over drug release.[57] In such cases, magnetic nanocarriers injected intravenously are guided to target lesions by an applied magnetic field.

Maximal bioaccumulation using magnetic targeting depends on the intensity of the applied magnetic field, mode of delivery and the surface and pharmacokinetic properties of the MNPs. Ideally, long circulating MNPs with a size of $>100\,nm$ are usually preferred for effective magnetic targeting. For instance, PEG-modified MNPs were found to be potential candidates for magnetic targeting to brain tumor cells.[58] In yet another work, polymeric micelles encapsulating superparamagnetic iron oxide NPs (SPIONs) and doxorubicin (DOX) with and without folate-targeting ligands were tested in a human oral cavity squamous carcinoma cell line. A significant difference in cytotoxicity was observed between the samples with and without a magnetic field. Also, increased drug accumulation and reduced cell viability was observed in cells treated with micelles containing SPIONs without folate. This work suggests that magnetic targeting alone could be effective in enhancing drug bioavailability at target sites compared to conventional targeting ligands.[59]

In vivo, magnetic targeting is often achieved by placing a small permanent magnet at an external location near the tumor site for superficial tumors, or by non-invasive surgical procedures for deep-tissue tumors.[60,61] However, intravenous injection of MNPs often leads to particle aggregation, which results in rapid clearance. More recently, other possibilities such as intra-arterial delivery, disruption of protective tissue layers, and magnetic resonance navigation (MRN) to maximize drug specificity have been widely explored. MNPs loaded with mitoxantrone were delivered intra-arterially to bone tumors in rabbit model, with the aid of an electromagnet and an external magnetic field. This led to a four-fold increase of drug accumulation at tumor sites compared to that in the liver and kidneys. Magnetic drug targeting combined with intra-arterial delivery evades the first pass clearance and enhances the bioaccumulation of nanocarriers at tumor lesions, resulting in superior therapeutic efficacy (Figure 3.3).[62]

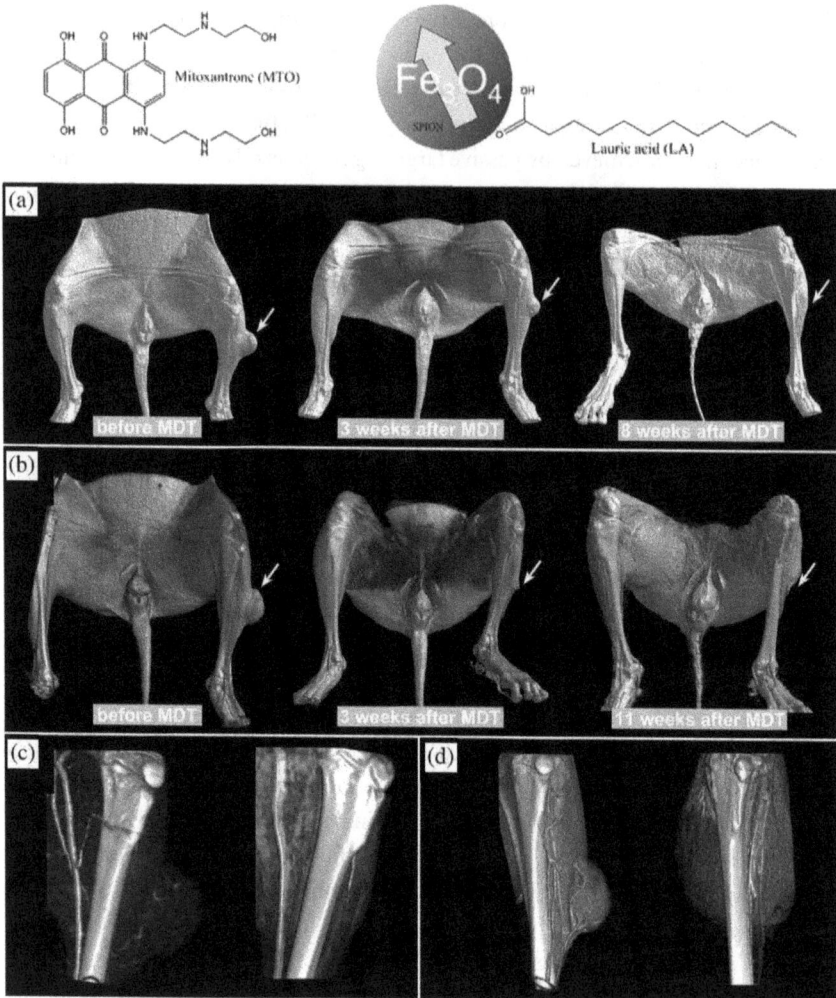

Figure 3.3: Mitoxantrone-coated MNPs for cancer therapy *in vivo*: Outcomes of treatment of VX-2 tumors implanted in the hind limb of rabbits ($n = 38$), with magnetic drug targeting (MDT). Panels (a) and (b) show a 3D-surface view of two different rabbits. The images show the rear view of the rabbit positioned with its back down. Yellow arrows point to the tumor region on the hind limb. After 3 weeks, a clear reduction in tumor size was observed; after 8 and 11 weeks, tumors were no longer visible or palpable. A reduction in muscle volume in the treated region was also be observed. (c) and (d) 3D-angiographic images of the tumor region of two animals in (a) and (b), respectively. Angiographic imaging demonstrates that not only the tumors but also the supporting vessel structures disappeared after treatment, while the main vessels supporting the distal part of the limb remained unaffected. (Reprinted with permission from Tietze *et al.*[62])

Liver cancers are treated by delivering a very high dosage of chemotherapeutics and an embolizing agent. While the embolizing agent creates an anoxic tumor microenvironment that prevents the chemotherapeutic from extravasating, this method is often limited by the chemoembolization of healthy tissues. One solution is MRN whereby the MNPs are steered specifically towards tumors using a modified magnetic resonance imaging system.[63] Magnetic microparticles loaded with DOX and FeCo NPs were steered transversally into the hepatic artery of rabbits with the aid of steering coils of high magnetic field intensity. Precise control over the particle distribution in deep tissues (4 cm below the skin) was made possible with this method.[64]

3.3.1.3   To cross the blood brain barrier

In the case of brain tumors, the BBB formed by tight junctions between endothelial cells acts as an interface between the blood and the extracellular fluid of the brain. The BBB is selectively permeable, allowing only essential nutrients required by neural cells to cross the cell junctions. Most compounds, like lipophilic neurotoxins and neurotherapeutics, are restricted by the BBB. In brain tumor therapy, the main challenge is to transport drugs across the BBB.

MNPs under an applied magnetic field are able to traverse the BBB. Kong *et al.* reported that MNPs guided by an external magnetic field were able to cross the BBB in rat models and were retained for 24 h in the brain parenchyma. No special techniques to disrupt the endothelial junctions were used and the BBB remained intact after MNP extravasation.[65] In another study, magnetically-guided, starch-coated iron-oxide MNPs showed a five-fold increase in bioaccumulation in rat brain tumors.[29] However, in both cases a large amount of MNPs was trapped in the liver and spleen. Also, particles aggregated in the blood vessels; this aggregation might be avoided by modeling the magnetic field topography such that the magnetic force is higher than the drag force of the blood.[66]

3.3.1.4   Magnetic hyperthermia

Mild hyperthermia can help with drug delivery, by conjugating drug molecules to MNPs via thermo-labile linker molecules (e.g. nucleic acid duplex). Upon heat generation due to an external magnetic field, the linkers are cleaved to release the drug at the target site. Hyperthermia-based controlled drug release is termed drug delivery through bond breaking (DBB). For example, Derfus *et al.* conjugated a 30 bp DNA to dextran-coated iron oxide MNPs and added complementary oligonucleotides linked to a model drug (e.g. a fluorophore). Excessive model drugs were removed by trapping the particles on a magnetic column and washing with buffer. The model drugs were pulsatile-released by radiofrequency electromagnetic field (EMF) activation (400 kHz, 1.25 kW).[67]

Another approach would be enhanced permeation, in which the drug molecules along with MNPs are co-encapsulated in liposomes or polymer matrixes.[68,69] In this method, the heat generated is used to create nanosized pores or cracks on the lipid or polymeric carriers, through which the encapsulated drugs diffuse out. This is termed drug delivery through enhanced permeation (DEP). For instance, a self-assembled polymeric hydrogel matrix encapsulating MNPs and DOX (DOX@SMNPs) was synthesized using four different molecular building blocks: adamantane-grafted polyamidoamine dendrimers (Ad-PAMAM), β-CD-grafted branched polyethylenimine (CD-PEI), adamantane-functionalized polyethylene glycol (Ad-PEG), and 6 nm adamantane-grafted $Zn_{0.4}Fe_{2.6}O_4$ MNPs (AdMNP).[70] The embedded MNP (Ad-MNP) served as a built-in transformer that converts radiofrequency external alternating magnetic field (AMF) into heat, allowing for stimuli-responsive drug release from this system [Figure 3.4(a)]. After this system was validated in *in vitro* studies, the complex was tested *in vivo*. The group treated with a double injection (day 0 and day 7) of DOX@SMNPs with AMF showed continued and effective inhibition of tumor growth. The control groups (i.e., DOX@SMNPs w/o AMF, AMF only, and PBS) did not show any statistically significant differences in tumor suppression [Figure 3.4(b)].

3.3.2 *Other Diseases*

3.3.2.1 *Infectious diseases*

Infectious diseases are caused by pathogens such as bacteria, viruses, parasites and fungi. These diseases can be contagious (upon contact with the infected person or their secretions) or transmitted through vectors or sexual transmission. They affect a considerable number of persons regionally (epidemic) or globally (pandemic) and pose a serious global threat. The most common way to treat bacterial infections is by using antimicrobials such as antibiotics, antiseptic ointments/lotions, or disinfectants. The use of antimicrobials, although successful, is threatened by the appearance of antibiotic-resistant pathogens. Although not an antibacterial by itself, MNPs offer a new strategy to overcome this challenge.

Some bacteria adhere to solid surfaces and develop a thick extracellular matrix which is impermeable to antimicrobials. These biofilms are commonly found on biomedical devices such as catheters. Normal sterilization techniques are insufficient to get rid of biofilms. It has been shown that MNPs are capable of penetrating

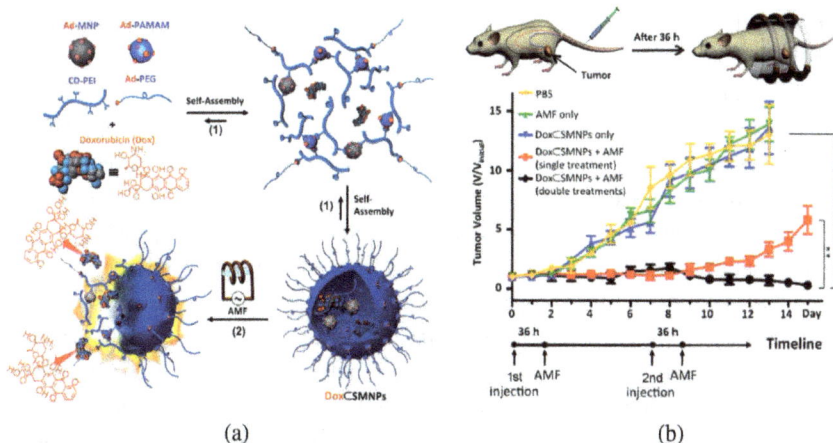

(a)   (b)

Figure 3.4: Molecular design, self-assembly and function of magnetothermally-responsive doxorubicin (DOX)-encapsulated supramolecular magnetic nanoparticles (DOX@SMNPs). 1) The self-assembled synthetic strategy is employed for the preparation of DOX@SMNPs, which is made from a fluorescent anticancer drug (DOX) and four molecular building blocks: Ad-PAMAM, 6 nm Ad-grafted $Zn_{0.4} Fe_{2.6} O_4$ MNPs (Ad-MNP), CD-PEI, and Ad-PEG. 2) The embedded Ad-MNP serves as a built-in heat transformer that triggers the burst release of DOX molecules from the magnetothermally-responsive SMNP vector, achieving on-demand drug release upon the remote application of AMF. (b) Evaluation of *in vivo* therapeutic efficacy: Treatment scheme of DOX@SMNPs and results of the tumor volume change over the course of the treatment (15 days) in DLD-1 xenografted mice ($n = 3$) treated with DOX@SMNPs (w/ and w/o application of AMF) and other controls (AMF only and PBS only). All injections were done on day 0 (and day 7 for the double injection group) when the tumor volume reached $100\,mm^3$; AMF application was performed at 36 h post-injection. The best tumor suppression result was observed in the group treated with a double injection of DOX@SMNPs with AMF application. The group treated with a single injection of DOX@SMNPs with AMF and the other control groups (i.e. treated with DOX@SMNPs only, AMF only and PBS) show either a smaller degree or none of tumor suppression effects (**$p \leq 0.01$; ***$p \leq 0.001$). (Reprinted with permission from Lee *et al.*[70])

the biofilms using an external magnetic field and can be endowed with antimicrobial properties through surface modifications. For example, to prevent biofilm formation in orthopedic implants, Taylor *et al.* exposed biofilm-forming bacteria *Staphylococcus epidermidis* to as-synthesized SPIONs. SPIONs with a concentration of $100\,\mu g/mL$ were found to be toxic to the bacteria and prevented colony formation. Free radical generation and electrostatic interactions between the SPIONs and the host is the mechanism by which SPIONs inhibits biofilm growth.[71] Another work has also demonstrated that MNP-functionalized textile dressing was resistant to fungal biofilm formation compared to non-functionalized dressing.[72] However, *in vivo* studies on whether MNP coatings can eradicate biofilms under physiological conditions have yet to be carried out.

Compared to conventional antimicrobials, MNPs exhibit better penetration into biofilms and this effect is enhanced under a magnetic field. Given this, MNPs themselves can be used as antimicrobials or can serve as drug delivery vehicles for other antimicrobial drugs and metal NPs. In a recent study, SPIONs with different surface chemistries (PEG-coated, anionic and cationic SPIONs) were targeted to treat staphylococcal biofilms produced by gentamicin-susceptible or resistant strains, and the results were compared with that of free gentamicin. SPIONs without magnetic targeting were minimally internalized. Under the influence of an external magnetic field, higher internalization and cytotoxicity was observed. Comparatively, the presence or absence of gentamicin alongside SPIONs had no significant effect on biofilm eradication.[73] This study indicates that MNPs have the potential to induce antimicrobial activity.

3.3.2.2 *Tissue engineering and regenerative medicine*

MNPs have great potential to address concerns in tissue engineering and regenerative medicine. Typical applications include the regeneration of blood vessels, heart, bones, cartilage and neuronal structures. One such application of MNPs is to enhance gene transfection. Human vascular endothelial growth factor (hVEGF) is essential for the repair and revascularization of the heart in patients suffering from ischemic heart disease. Owing to their non-invasiveness and biocompatibility, MNPs were used for targeted delivery of the adenoviral vector (Ad)–encoded hVEGF gene, under the influence of an external magnetic field. *In vivo* studies with MNPs/Ad$_h$ VEGF complexes resulted in neovascularization and improved recovery of the left ventricular function of the heart.[74]

3.3.2.3 *Poisoning*

Hemoperfusion is a technique by which toxic substances are removed from the blood using adsorbents such as activated carbon. MNPs can be employed as an absorbent to remove biological, chemical and radioactive toxic substances from the blood. Once the MNPs coated with antidote substances are injected into the bloodstream, the toxic materials conjugate with the MNPs. The toxins are removed by pumping the blood through a magnetic separator.[75]

Removal of toxic lipophilic substances such as diazepam (a lipophilic drug that may be fatal at high doses) and low density lipoproteins (LDL, cholesterols in plasma that causes diseases like atherosclerosis) has been demonstrated using

β-cyclodextrin (β-CD) and heparin-conjugated MNPs, respectively. β-CD MNPs removed about 50.6% of diazepam from plasma when orally administered in rabbit models. However, these MNPs exhibited non-specific binding to plasma proteins. Hence, further development is required.[76] Heparin-conjugated MNPs, on the other hand, were highly specific and removed up to 67% of LDL *in vitro*. These MNPs were reusable for up to eight cycles and no coagulation or platelet activation was observed.[77]

The presence of bacteria in blood, termed bacteremia, is a potentially fatal condition caused by surgical procedures such as chemotherapy and transplantation. Current methods to remove bacteria from the blood are unsatisfactory; MNPs offer an alternative technique that is both rapid and efficient. Zinc-coordinated bis(dipicolylamine) (bis-Zn-DPA) are highly selective synthetic ligands that bind to anionic phospholipids on bacteria. MNPs conjugated with bis-Zn-DPA removed 95% of *E. coli* bacteria from infected blood in a microfluidic set-up.[78]

3.3.2.4 *Iron deficiency or anemia*

Chronic kidney diseases (CKD) cause severe cases of anemia, which requires intravenous iron replacement therapy. The use of carbohydrate-coated iron particles is often associated with side effects. Recently, MNPs have been proposed as a source of iron to treat iron deficiency. Ferumoxytol (SPIONs coated with polyglucose sorbitol carboxymethyl ether) has been intravenously administered to patients with CKD. Ferumoxytol releases iron components into macrophages in the liver, spleen and bone marrow. The iron is then stored as ferritin or transported into hemoglobin. A significant increase in hemoglobin was observed in patients treated with ferumoxytol compared to those treated with oral iron. Additionally, the drug was better tolerated compared to oral iron.[79,80]

3.4 Perspectives

MNPs are potential candidates for drug carriers as they show promising preclinical results for the treatment of a variety of diseases. However, there is still a huge gap between these preclinical studies and clinical applications, which can only be addressed by understanding and controlling biodistribution, long-term toxicity and MNP-host interactions, as well as standardizing manufacturing procedures.

Size, surface chemistry and morphology are essential elements to consider when designing MNPs. Hollow MNPs and MNPs with large magnetic moments have recently been developed to reduce the concentration of MNPs required for therapeutic efficacy, which in turn minimizes toxic effects and enhances the biocompatibility of MNPs. Knowledge about physiological conditions such as blood flow velocity and drag force, as well as techniques such as magnetic field modeling, can help to improve the efficacy of magnetic targeting. In the future, advanced real-time tracking techniques will provide insights into MNP-host interaction mechanisms. Careful consideration of the various parameters involved in MNP-based drug delivery and addressing issues of long-term toxicity will accelerate the clinical translation of MNPs.

References

1. Farokhzad, O.C., and Langer, R. Impact of nanotechnology on drug delivery. *ACS Nano* 3, 16–20 (2009).
2. Davis, M.E. *et al*. Nanoparticle therapeutics: An emerging treatment modality for cancer. *Nature reviews. Drug Discovery* 7, 771–782 (2008).
3. Park, K. Controlled drug delivery systems: Past forward and future back. *Journal of Controlled Release* 190, 3–8 (2014).
4. Luo, D. *et al*. Nanomedical engineering: Shaping future nanomedicines. *Wiley Interdisciplinary Reviews: Nanomedicine and Nanobiotechnology* (2014).
5. Etheridge, M.L. *et al*. The big picture on nanomedicine: The state of investigational and approved nanomedicine products. *Nanomedicine* 9, 1–14 (2013).
6. Xu, C., and Sun, S. Superparamagnetic nanoparticles as targeted probes for diagnostic and therapeutic applications. *Dalton Transactions*, 5583–5591 (2009).
7. Arruebo, M. *et al*. J. Magnetic nanoparticles for drug delivery. *Nano today* 2, 22–32 (2007).
8. Kumar, C.S., and Mohammad, F. Magnetic nanomaterials for hyperthermia-based therapy and controlled drug delivery. *Advanced Drug Delivery Reviews* 63, 789–808 (2011).
9. Poudyal, N., and Liu, J.P. Advances in nanostructured permanent magnets research. *Journal of Physics D: Applied Physics* 46, 043001 (2013).
10. Colombo, M. *et al*. Biological applications of magnetic nanoparticles. *Chem Soc Rev* 41, 4306–4334 (2012).
11. Xu, J. *et al*. Application of iron magnetic nanoparticles in protein immobilization. *Molecules* 19, 11465–11486 (2014).
12. Gurav, A. *et al*. Aerosol processing of materials. *Aerosol Science and Technology* 19, 411–452 (1993).
13. Hyeon, T. Chemical synthesis of magnetic nanoparticles. *Chemical Communications* 9, 927–934 (2003).
14. Gupta, A.K., and Gupta, M. Synthesis and surface engineering of iron oxide nanoparticles for biomedical applications. *Biomaterials* 26, 3995–4021 (2005).
15. Chomoucka, J. *et al*. Magnetic nanoparticles and targeted drug delivering. *Pharmacological Research* 62, 144–149 (2010).
16. Kim, J., Piao, Y., and Hyeon, T. Multifunctional nanostructured materials for multimodal imaging, and simultaneous imaging and therapy. *Chem Soc Rev* 38, 372–390 (2009).

17. Frey, N.A. *et al.* Magnetic nanoparticles: Synthesis, functionalization, and applications in bioimaging and magnetic energy storage. *Chemical Society reviews* 38, 2532–2542 (2009).
18. Murray, C.B. *et al.* Synthesis and characterization of monodisperse nanocrystals and close-packed nanocrystal assemblies. *Annual Review of Materials Science* 30, 545–610 (2000).
19. Chaubey, G.S. *et al.* Synthesis and stabilization of FeCo nanoparticles. *Journal of the American Chemical Society* 129, 7214–7215 (2007).
20. Sun, S. *et al.* Monodisperse FePt nanoparticles and ferromagnetic FePt nanocrystal superlattices. *Science* 287, 1989–1992 (2000).
21. Chiang, I.C., and Chen, D.H. Synthesis of monodisperse FeAu nanoparticles with tunable magnetic and optical properties. *Advanced Functional Materials* 17, 1311–1316 (2007).
22. Xie, J. *et al.* Production, modification and bio-applications of magnetic nanoparticles gestated by magnetotactic bacteria. *Nano research* 2, 261–278 (2009).
23. Schüler, D. The biomineralization of magnetosomes in *Magnetospirillum gryphiswaldense. Int Microbiol* 5, 209–214 (2002).
24. Bazylinski, D.A. Controlled biomineralization of magnetic minerals by magnetotactic bacteria. *Chemical Geology* 132, 191–198 (1996).
25. Saiyed, Z. *et al.* Application of magnetic techniques in the field of drug discovery and biomedicine. *BioMagnetic Research and Technology* 1(2) (2003).
26. Park, J.Y. *et al.* Highly water-dispersible PEG surface modified ultra small superparamagnetic iron oxide nanoparticles useful for target-specific biomedical applications. *Nanotechnology* 19 (2008).
27. Xie, J. *et al.* Controlled PEGylation of monodisperse Fe_3O_4 nanoparticles for reduced non-specific uptake by macrophage cells. *Advanced Materials* 19, 3163–3166 (2007).
28. Moghimi, S.M. *et al.* Long-circulating and target-specific nanoparticles: Theory to practice. *Pharmacological Reviews* 53, 283–318 (2001).
29. Chertok, B. *et al.* Iron oxide nanoparticles as a drug delivery vehicle for MRI monitored magnetic targeting of brain tumors. *Biomaterials* 29, 487–496 (2008).
30. Sun, Y. *et al.* Magnetic chitosan nanoparticles as a drug delivery system for targeting photodynamic therapy. *Nanotechnology* 20 (2009).
31. Dougherty, T.J. *et al.* Photodynamic therapy. *Journal of the National Cancer Institute* 90, 889–905 (1998).
32. Kumar, A. *et al.* Multifunctional magnetic nanoparticles for targeted delivery. *Nanomedicine*, 1–6 (2009).
33. Purushotham, S., and Ramanujan, R. Thermoresponsive magnetic composite nanomaterials for multimodal cancer therapy. *Acta Biomater* (2009).
34. Veiseh, O. *et al.* Design and fabrication of magnetic nanoparticles for targeted drug delivery and imaging. *Advanced drug delivery reviews* 62, 284–304 (2010).
35. Schellenberger, E.A. *et al.* Optimal modification of annexin V with fluorescent dyes. *ChemBioChem* 5, 271–274 (2004).
36. Lee, J.H. *et al.* Artificially engineered magnetic nanoparticles for ultra-sensitive molecular imaging. *Nature Medicine* 13, 95–99 (2007).
37. Euliss, L.E. *et al.* Cooperative assembly of magnetic nanoparticles and block copolypeptides in aqueous media. *Nano Letters* 3, 1489–1493 (2003).

38. Xu, C. *et al.* Dopamine as a robust anchor to immobilize functional molecules on the iron oxide shell of magnetic nanoparticles. *Journal of the American Chemical Society* 126, 9938–9939 (2004).

39. Yang, X. *et al.* cRGD-functionalized, DOX-conjugated, and 64Cu-labeled superparamagnetic iron oxide nanoparticles for targeted anticancer drug delivery and PET/MR imaging. *Biomaterials* 32, 4151–4160 (2011).

40. Schellenberger, E.A. *et al.* Annexin V-CLIO: A nanoparticle for detecting apoptosis by MRI. *Academic Radiology* 9, S310–S311 (2002).

41. Schellenberger, E.A. *et al.* Magneto/optical annexin V, a multimodal protein. *Bioconjugate Chemistry* 15, 1062–1067 (2004).

42. Funovics, M.A. *et al.* MR imaging of the her2/neu and 9.2.27 tumor antigens using immunospecific contrast agents. *Magnetic Resonance Imaging* 22, 843–850 (2004).

43. Xu, C., and Sun, S. New forms of superparamagnetic nanoparticles for biomedical applications. *Advanced Drug Delivery Reviews* 65, 732–743 (2013).

44. Cheng, K. *et al.* Porous hollow Fe_3O_4 nanoparticles for targeted delivery and controlled release of cisplatin. *Journal of the American Chemical Society* 131, 10637–10644 (2009).

45. Chen, Y. *et al.* Core/shell structured hollow mesoporous nanocapsules: A potential platform for simultaneous cell imaging and anticancer drug delivery. *ACS Nano* 4, 6001–6013 (2010).

46. Pan, Y., *et al.* Colloidosome-based synthesis of a multifunctional nanostructure of silver and hollow iron oxide nanoparticles. *Langmuir* 26, 4184–4187 (2010).

47. Peng, S., and Sun, S. Synthesis and characterization of monodisperse hollow Fe_3O_4 nanoparticles. *Angewandte Chemie — International Edition* 46, 4155–4158 (2007).

48. Shin, J. *et al.* Hollow manganese oxide nanoparticles as multifunctional agents for magnetic resonance imaging and drug delivery. *Angewandte Chemie — International Edition* 48, 321–324 (2009).

49. Wu, H. *et al.* A hollow-core, magnetic, and mesoporous double-shell nanostructure: *In situ* decomposition/reduction synthesis, bioimaging, and drug-delivery properties. *Advanced Functional Materials* 21, 1850–1862 (2011).

50. Szakacs, G. *et al.* Targeting multidrug resistance in cancer. *Nature Reviews. Drug Discovery* 5, 219–234 (2006).

51. Kapse-Mistry, S. *et al.* Nanodrug delivery in reversing multidrug resistance in cancer cells. *Frontiers in Pharmacology* 5, 159 (2014).

52. Wang, X. *et al.* The application of Fe_3O_4 nanoparticles in cancer research: A new strategy to inhibit drug resistance. *Journal of Biomedical Materials Research Part A* 80, 852–860 (2007).

53. Cheng, J. *et al.* Effect of magnetic nanoparticles of Fe_3O_4 and 5-bromotetrandrine on reversal of multidrug resistance in K562/A02 leukemic cells. *International Journal of Nanomedicine* 4, 209–216 (2009).

54. Chen, B. *et al.* Reversal in multidrug resistance by magnetic nanoparticle of Fe_3O_4 loaded with adriamycin and tetrandrine in K562/A02 leukemic cells. *International Journal of Nanomedicine* 3, 277–286 (2008).

55. Brigger, I. *et al.* Nanoparticles in cancer therapy and diagnosis. *Advanced Drug Delivery Reviews* 54, 631–651 (2002).

56. Cheng, Z. *et al.* Multifunctional nanoparticles: Cost versus benefit of adding targeting and imaging capabilities. *Science* 338, 903–910 (2012).

57. Mura, S. *et al.* Stimuli-responsive nanocarriers for drug delivery. *Nat Mater* 12, 991–1003 (2013).
58. Cole, A.J. *et al.* Polyethylene glycol modified, cross-linked starch-coated iron oxide nanoparticles for enhanced magnetic tumor targeting. *Biomaterials* 32, 2183–2193 (2011).
59. Yang, X. *et al.* Folate-encoded and $Fe_3 O_4$ -loaded polymeric micelles for dual targeting of cancer cells. *Polymer* 49, 3477–3485 (2008).
60. Kuznetsov, A.A. *et al.* Application of magnetic liposomes for magnetically guided transport of muscle relaxants and anti-cancer photodynamic drugs. *Journal of Magnetism and Magnetic Materials* 225, 95–100 (2001).
61. Krukemeyer, M.G. *et al.* Mitoxantrone-iron oxide biodistribution in blood, tumor, spleen, and liver — magnetic nanoparticles in cancer treatment. *Journal of Surgical Research* 175, 35–43 (2012).
62. Tietze, R. *et al.* Efficient drug-delivery using magnetic nanoparticles — biodistribution and therapeutic effects in tumour bearing rabbits. *Nanomedicine: Nanotechnology, Biology and Medicine* 9, 961–971 (2013).
63. Pouponneau, P. *et al.* Annealing of magnetic nanoparticles for their encapsulation into microcarriers guided by vascular magnetic resonance navigation. *Journal of Nanoparticle Research* 14, 1–13 (2012).
64. Pouponneau, P. *et al.* Co-encapsulation of magnetic nanoparticles and doxorubicin into biodegradable microcarriers for deep tissue targeting by vascular MRI navigation. *Biomaterials* 32, 3481–3486 (2011).
65. Kong, S.D. *et al.* Magnetic targeting of nanoparticles across the intact blood–brain barrier. *Journal of Controlled Release* 164, 49–57 (2012).
66. Chertok, B. *et al.* Brain tumor targeting of magnetic nanoparticles for potential drug delivery: Effect of administration route and magnetic field topography. *Journal of controlled release: Official journal of the Controlled Release Society* 155, 393–399 (2011).
67. Derfus, A.M. *et al.* Remotely triggered release from magnetic nanoparticles. *Advanced Materials* 19, 3932–3936 (2007).
68. Bonnaud, C. *et al.* Spatial SPION localization in liposome membranes. *Magnetics, IEEE Transactions on* 49, 166–171 (2013).
69. Bannwarth, M.B. *et al.* Tailor-Made Nanocontainers for combined magnetic-field-induced release and MRI. *Macromolecular Bioscience* 14, 1205–1214 (2014).
70. Lee, J.-H. *et al.* On-demand drug release system for *in vivo* cancer treatment through self-assembled magnetic nanoparticles. *Angewandte Chemie* 125, 4480–4484 (2013).
71. Taylor, E.N., and Webster, T.J. The use of superparamagnetic nanoparticles for prosthetic biofilm prevention. *International Journal of Nanomedicine* 4, 145–152 (2009).
72. Anghel, I. *et al.* Magnetite nanoparticles for functionalized textile dressing to prevent fungal biofilms development. *Nanoscale research letters* 7, 1–6 (2012).
73. Subbiahdoss, G. *et al.* Magnetic targeting of surface-modified superparamagnetic iron oxide nanoparticles yields antibacterial efficacy against biofilms of gentamicin-resistant staphylococci. *Acta Biomaterialia* 8, 2047–2055 (2012).
74. Zhang, Y. *et al.* Targeted delivery of human VEGF gene via complexes of magnetic nanoparticle-adenoviral vectors enhanced cardiac regeneration. *PLoS ONE* 7, e39490 (2012).

75. Jain, K.K. *The Handbook of Nanomedicine*. Springer Science & Business Media (2012).
76. Cai, K. *et al.* [small beta]-Cyclodextrin conjugated magnetic nanoparticles for diazepam removal from blood. *Chemical Communications* 47, 7719–7721 (2011).
77. Li, J. *et al.* Recyclable heparin and chitosan conjugated magnetic nanocomposites for selective removal of low-density lipoprotein from plasma. *J Mater Sci: Mater Med* 25, 1055–1064 (2014).
78. Lee, J.-J. *et al.* Synthetic ligand-coated magnetic nanoparticles for microfluidic bacterial separation from blood. *Nano Letters* 14, 1–5 (2014).
79. Spinowitz, B.S. *et al.* Ferumoxytol for treating iron deficiency anemia in CKD. *Journal of the American Society of Nephrology* 19, 1599–1605 (2008).
80. Provenzano, R. *et al.* Ferumoxytol as an intravenous iron replacement therapy in hemodialysis patients. *Clinical Journal of the American Society of Nephrology* 4, 386–393 (2009).

Dendritic Nanocarrier Platforms for Gene and Drug Delivery Applications

4

Hong Y. Cho, Jason Bugno and Seungpyo Hong

Department of Biopharmaceutical Sciences, College of Pharmacy
University of Illinois, Chicago, IL 60612, USA

Abstract

Dendrimers are a unique class of polymers possessing three-dimensional, chemically well-defined, hyperbranched architecture, along with conformational flexibility and high surface area topology. These characteristic features, as well as their modularity—by precisely controlling their nanoscale size and surface functional groups, distinguish them from conventional linear polymers and make them well suited for a variety of biomedical applications. In this chapter, we highlight the synthetic routes and recent biomedical applications of several representative dendrimers and dendritic nanoparticles, with a focus on their use in targeted drug and gene delivery. In addition, we also discuss recently reported hybridization approaches that integrate dendritic polymers with other types of polymers in an effort to overcome current hurdles that have hindered the rapid clinical translation of dendritic polymers.

4.1 Introduction

Over the past few decades, advances in polymer science have led to the development of novel macromolecules exhibiting a wide range of compositional and architectural features. Among these, dendrimers have shown particular promise in biomedical applications due to their advantageous properties, including tunable structures and chemical composition and the ability for surface functionalization. Dendrimers are a unique class of hyperbranched, spherical synthetic polymers with well-defined

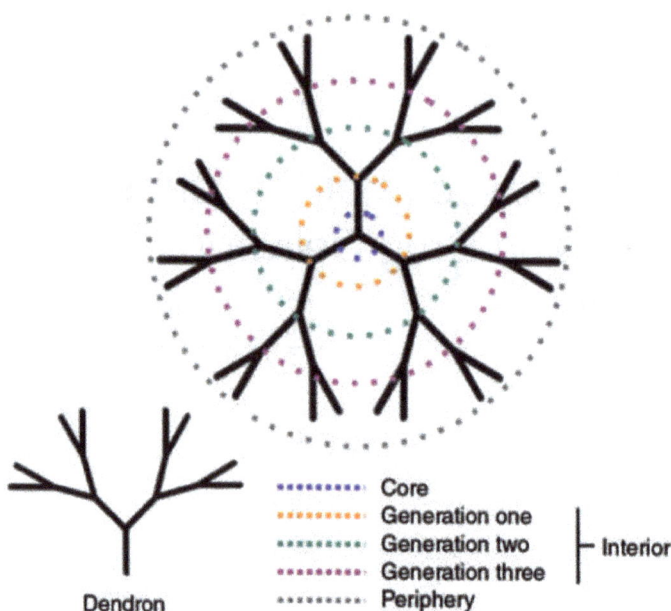

Figure 4.1: Basic dendrimer structure.[10] A dendrimer and a dendron are represented with solid lines. The colored, dotted lines identify the various key regions of the dendrimer. (Copyright ©2005, rights managed by Nature Publishing Group.)

chemical structures (Figure 4.1). These features have led to significant interest in dendrimers, particularly for their use as small (<20 nm) nanocarriers.[1,2] Since they were first introduced in the late 1970s, intensive research into dendrimers has led to several noteworthy advances in their synthesis, characterization and use in biomedical applications.[3]

The structure of dendrimers can be broken down into three basic components: a multi-functional core, repeating "n" branched units of the monomer known as the generation (G_n),[4] and surface-presenting peripheral end groups.[5] These unique characteristics allow researchers to design dendrimers that possess a multitude of additional functionality, including tailoring of their surface chemistry, resulting in controlled cellular interactions, chemical conjugation of ligands and therapeutic agents, and trans-epithelial/endothelial transport.[6–9] Their size, shape and topology, in addition to flexibility and surface functionality, can be precisely tuned at the molecular level, making them a useful platform for various biomedical applications. In particular, all of these features make dendrimers well suited for applications such as targeted drug/gene delivery, diagnostics and biological imaging.

(a)

(d)

(b)

(e)

(c)

(f)

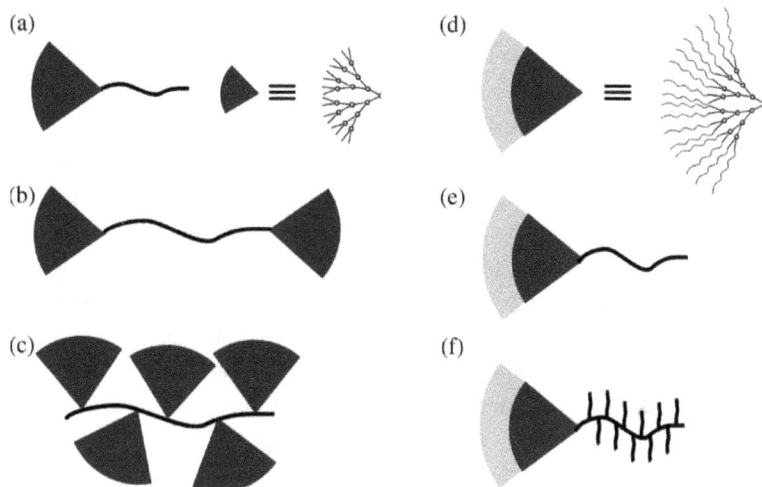

Figure 4.2: A variety of dendron-based architectures for drug delivery. (a) A linear polymer linked to the focal point of a dendron, (b) and (c) a linear polymer bearing two or more dendrons at the end or backbone, (d)–(f) multiple polymer chains linked to a focal point or the periphery of a dendron.[11]

The hyperbranched architecture of dendrimers is another key feature associated with several advantages for biomedical use. One frequently utilized feature is the high density of chemically-modifiable peripheral groups. This property allows for charge manipulation, chemical ligation with drugs and targeting ligands, and conjugation to other polymers, including the commonly used poly(ethylene glycol) (PEG). Additionally, their well-controlled, close-to-monodispersed structure limits batch-by-batch variations compared to other linear polymers. Because of this, several groups have utilized dendrons, i.e., individual branched dendrimer segments (Figure 4.1), and merged them with other polymers, producing dendronized polymers. For instance, linear polymers can be conjugated to dendrons to generate various polymeric topologies (Figure 4.2). These hybrids combine the advantageous structural features and properties of each component, and, as such, have generated a great deal of scientific interest. While the use of dendrons and dendron-based nanomaterials has not been as extensively studied as dendrimers for biomedical applications, they have rapidly emerged as potential delivery platforms due to their ability to hybridize dendritic functionality and multivalent properties with other organic and inorganic systems.[11–13]

Despite these advantages, dendrimers and dendritic nanoparticles have drawbacks, such as complex, multistep synthesis, insufficient drug loading, and rapid systemic clearance, which have hindered their clinical translation. Due to these concerns, dendrimer use has to be tailored to each application, requiring an understanding of their underlying interactions with biological systems. To date, only two dendrimer formulations have made it into clinical trials: one is a topical formulation for the prevention of HIV, and the other is an intravenous formulation to deliver the anticancer drug docetaxel.[3] Nevertheless, dendrimers still present a promising platform as supported by recent advances in both fundamental and *in vivo* findings. This chapter will outline these advances beginning with the synthesis of dendrimers and dendritic polymers, followed by their interactions with biological systems, with a particular emphasis on their applications in targeted drug and gene delivery. Lastly, we will also discuss recently introduced hybridization strategies that augment the benefits of dendrimers and dendritic structure while minimizing their drawbacks.

4.2 Synthesis and Characterization of Dendrimers

Dendrimer synthesis differs significantly from that of other conventional polymer architectures, leading to their symmetric, hyperbranched, spherical 3D shape, and near-monodispersity. This tight structural control stems from the use of ordered sequential reaction steps. Synthesis involves sequential radial growth, at its simplest utilizing alternating activation and condensation reactions, with each round resulting in approximately a doubling of molecular weight and increased diameter. For instance, the diameter of commonly used poly(amidoamine) (PAMAM) dendrimers can range from 1.1 nm for G1 PAMAM dendrimers to 9.0 nm for G8 PAMAM dendrimers.[14] The majority of dendrimer synthesis can be classified by one of two routes: divergent (inside out) synthesis or convergent (outside in) synthesis (Scheme 4.1). In addition to these methods, several other approaches have been utilized to produce a variety of dendrimers with different compositions, all which have been reviewed previously.[15,16]

4.2.1 *Divergent Synthesis*

The divergent approach, first described indirectly by Vögtle[17] and later used by Tomalia *et al.*,[18] involves the radial growth of branched segments from a multifunctional core molecule. At the most basic, every two reactions result in a new generation and the doubling of molecular weight. Firstly, the outermost peripheral groups are activated, followed by reaction coupling and addition of branching monomer units.[19] Thus, the dendrimer is synthesized from a poly-functional

Divergent strategy

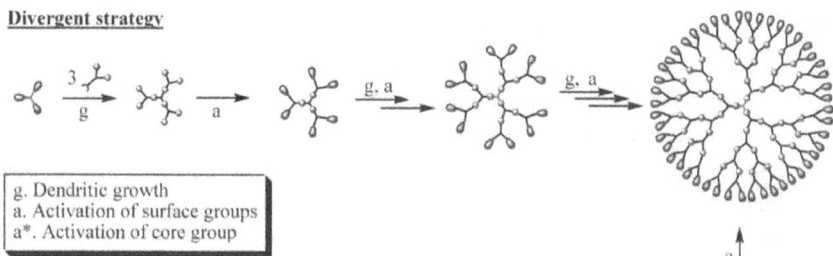

g. Dendritic growth
a. Activation of surface groups
a*. Activation of core group

Convergent strategy

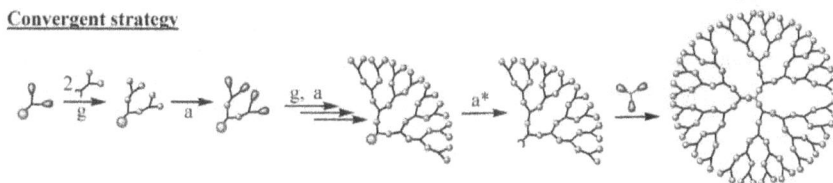

Scheme 4.1: Synthetic routes of a G4 dendrimer: divergent (top) and convergent (bottom) synthesis.[30] (Copyright ©2009, all rights reserved by Royal Society of Chemistry.)

core using a reiterative sequence consisting of activation and condensation reactions. This process is repeated several times until a dendrimer of the desired size is formed. PAMAM starburst dendrimers are typically prepared by this method.[18] Compared to other synthetic methods, the divergent approach has several advantages. It includes the ability to modify the surface of dendrimers by changing the end groups at the peripheral layer, which allows the chemical and physical properties of the dendrimer to be configured to meet specific needs.[20,21] Although the divergent approach is the most widely adopted method, it is prone to branching defects, especially for higher generation dendrimers, due to inefficient reaction coupling, steric hindrances at the periphery, and problems with purification.[22] This method has largely been utilized for commercial dendrimer production, most commonly PAMAM or poly(propyleneimine) (PPI), as the reaction can be stopped at any step allowing for dendrimers to be collected at each generation.[15]

4.2.2 Convergent Synthesis

As the divergent method is associated with branching defects and purification difficulties, many opt to use the convergent approach, first introduced by Hawker and Fréchet in 1990.[23] In this approach, the individual dendrons are first synthesized and then coupled to a multifunctional core molecule. Convergent methods contain two main stages: stepwise sequential growth of a dendron from a protected focal point, and subsequent coupling of multiple dendrons to a core molecule.

This method provides multiple advantages:[15] First, purification is efficient due to greater structural differences among incompletely reacted dendrimers, starting materials, and desired products. Secondly, dendrimers containing functional cores can be generated. Thirdly, whereas the core must be stable throughout the reaction series in divergent synthesis, more sensitive cores can be added in the last step using the convergent approach. In addition, the coupling of multiple dendrons to the core molecule also produces tighter size control, while allowing the dendrimers to have multiple different surface groups.[23] However, compared to divergent methods, convergent approaches also have limitations. While dendrimers can be produced with fewer structural defects at lower generations, it is very difficult to produce dendrimers without defects at high generations, due to steric hindrances.

4.2.3 *Characterization of Dendrimers*

The physicochemical properties of dendrimers can be characterized using several instruments,[24,25] including nuclear magnetic resonance (NMR), electron paramagnetic resonance (EPR), mass spectroscopy, spectral techniques, light scattering, size exclusion chromatography and microscopy.[26] For example, NMR is one of the most routinely used methods for monitoring dendrimer synthesis, purity and chemical modifications. Moreover, using special NMR techniques, the three-dimensional structure, molecular dynamics and drug inclusion can be studied. Size-exclusion chromatography (SEC), also known as gel permeation chromatography (GPC), allows for the separation of macromolecules based on molecular weight, and is a valuable tool for following size changes with radial growth. However, it should be noted that this method has limitations as SEC calibration curves are commonly made using linear polymer standards that are morphologically different from dendrimers, leading to inaccurate molecular weight measurements. Microscopy techniques, including atomic force microscopy (AFM) and transmission electron microscopy (TEM), are valuable for visualizing dendrimers and evaluating size distributions.

Several techniques can be used to assess dendrimer purity. For instance, matrix-assisted laser desorption ionization time-of-flight mass spectroscopy (MALDI-TOF) provides a valuable tool for characterizing the molecular weight distribution and purity of dendritic samples.[27] For instance, mass spectroscopy has been used by Hummelen *et al.* to identify a number of structural defects in G1-G5 PPI dendrimers resulting from incomplete reaction series and undesired cyclization reactions in divergent synthesis.[28] The authors found that G5 PPI dendrimers commonly exhibited purities of approximately 20% despite their polydispersity of

1.002. Similarly, dendrimers can exhibit significant variation in the distribution of their surface ligands. Using HPLC, NMR, SEC and titration techniques, Mullen et al. have characterized the heterogeneity of ligand-conjugated PAMAM dendrimers.[29] Interestingly, while the median and mode of dendrimer-ligand distributions is rarely reported, they can differ significantly from the average, especially in systems containing relatively low ligand conjugation means. Furthermore, the authors found that the samples are inherently heterogeneous, fitting to a skewed Poisson distribution. Notably, these disparities are not usually controlled in a number of reported studies. Thus, it is interesting to consider the impact of different dendrimer-ligand populations on their interactions with biological systems. These findings suggest that investigations into nanoparticle-ligand interactions should also assess for ligand distribution and the impact of different populations on the experiment.

4.3 Representative Dendrimers for Biomedical Applications

Dendrimers encompassing a wide range of chemical compositions and properties have been synthesized.[3] Of these, three of the most commonly used dendrimers for biomedical use include PAMAM, PPI, and poly(L-lysine) (PLL) dendrimers. A summary of their various applications along with corresponding references is listed in Table 4.1. Each of these dendrimers present amine-terminated end groups, offering a wide range of utilizable chemical modifications. For example, the amine groups (positively charged in physiological pH) can be easily converted to neutral acetamide and hydroxyl moieties, or to anionic carboxylic acids, in addition to providing reaction handles for conjugation with bioactive molecules. Each of these dendrimers presents distinct advantages and disadvantages, including biocompatibility, drug loading ability, and ease of synthesis.

4.3.1 Poly(amidoamine) (PAMAM) Dendrimers

PAMAM dendrimers have been extensively studied since 1985, and were the first dendrimers to be synthesized, characterized and commercialized.[18,22] They are comprised of repeating branched internal subunits consisting of alternating amides and tertiary amines and terminal groups of surface-presenting primary amines. The well-controlled, tree-like branched architecture forms a 3D globular conformation with low polydispersity (typically <1.08 for G5-G10 PAMAM dendrimers).[60] Commonly used cores where the repetitive synthesis departs from include amine,

Table 4.1: Examples of biomedical applications using common dendrimers.

Dendrimer	Active Agent	Disease/Target	Generation	Reference
PAMAM	5-FU	Tumor	FA-targeted, PEGylated G4	31
	Beclometasone diprionate	Pulmonary nebulization	G3 and G4	32
	Boron-10	Brain tumor	G4	33
	doxorubicin	Tumor	PEGylated G4	34, 35
	DNA	Vaccine delivery	G5	36
	Follicle stimulating hormone	Ovaries and oviduct	G5	37
	NAC and valproate	Neurological degradation	G4	38, 39
	Risperdone	Toxicity evaluation	G4.5	40
	siRNA	Gene delivery	G3–G5	41–45
			Amphiphilic dendron-based polymer	46
PPI	Dendrimer	Inflammation	G2	47
	Maltose	Cancer	G4	48
	Melphalan	Cancer	G3–G5	49
	siRNA	Tumor	G5	50
PLL	Camptothecin	Tumor	G1	51
	Dendrimer	Lymph	G4	52
	DNA	Gene delivery	G3–G4 G5–G6	53–55 53, 54
	Doxorubicin	Tumor	G5	56, 57
	Methotrexate	Tumor	G3–G5	58
	siRNA	Gene delivery	G6	59

Notes: 5-fluorouracil, 5-FU; poly(amidoamine), PAMAM; folic acid, FA; poly(ethylene glycol), PEG; *N*-acetyl cysteine, NAC; poly(propyleneimine), PPI; poly(L-lysine), PLL.

ethylenediamine and triethanolamine. PAMAM dendrimers containing an amide-based backbone structure typically exhibit higher biocompatibility than other types of dendrimers due to the similarity of their chemical composition (peptide-based) with globular proteins.[18] These advantages have led to PAMAM dendrimers becoming one of the most heavily investigated dendrimers to date for biomedical applications, specifically as nanocarriers for drug and gene delivery.[18,22,61,62] More specific examples of those applications are detailed in subsequent sections of this chapter.

4.3.2 Polypropylenimine (PPI) Dendrimers

PPI dendrimers, first introduced in 1993, are based on the pioneering work of Vögtle and colleagues.[17] PPI dendrimers are amine-terminated hyperbranched macromolecules, synthesized mainly by the divergent method. To synthesize PPI dendrimers, a series of double Michael additions of acrylonitrile to primary amines are performed, followed by a heterogeneously catalyzed hydrogenation of nitriles.[63] With each subsequent repetition, the number of primary amines doubles. PPI dendrimers contain both peripheral primary amines and internal tertiary amines. Similar to PAMAM dendrimer, the primary and tertiary amines of PPI dendrimers have distinct pK_a values of ~10 and 6–9, respectively, allowing for effective complexation with oligonucleotides at physiological pH and efficient endosomal escape, or the "sponge effect".[64,65] This feature makes PPI dendrimers particularly well suited for gene delivery applications.

4.3.3 Poly (L-lysine) (PLL) Dendrimers

The idea of synthesizing PLL dendrimers came from an observation made by Ryser's group. They noted that the uptake of radiolabeled serum albumin by sarcoma-180 cell cultures was significantly enhanced in the presence of proteins rich in both lysine and synthetic peptides derived from either lysine (L, D or LD), L-ornithine, or L-histidine.[66] Since Denkewalter patented the synthesis of PLL dendrimers in the early 1980s,[67] a large number of synthetic PLL dendrimer-based products with molecular weights of 3-60 kDa have become commercially available. Using a divergent strategy, conventional peptide chemistry was applied using a Boc-L-Lys(Boc)-OH monomer, producing lysine-based dendrimers. The use of asymmetric L-lysine residues as branching units clearly distinguish dendritic poly(L-lysine) from classical, highly symmetrical "starburst" dendrimers.[68] Several peptide-based dendrimers based on polylysine have been developed as antiviral and antibacterial compounds, demonstrating the ability to enhance immunological responses to vaccines.[69,70]

4.4 Biological Interactions of Dendrimers

4.4.1 *Cellular Interactions of Dendrimers*

As noted above, dendrimers have received particular attention for their use in biomedical applications, due to the ability to precisely modify both their size and surface charge, leading to controllable cellular interactions. The functional end groups exposed on their surfaces also allow for facile multi-functionalization with targeting, imaging and therapeutic agents for a variety of biomedical purposes. Both size and surface characteristics have been found to govern their binding behaviors with cell membranes and affect cell uptake/internalization.[71,72] For instance, Hong *et al.* demonstrated that cationic PAMAM dendrimers strongly interact with cellular membranes in a generation-dependent manner. Larger (G7) PAMAM dendrimers form nanoscale holes in KB and Rat2 cell membranes as a result of strong, multivalent interactions between positively-charged surface amine groups on dendrimers and anionic phospholipids on cell membranes.[73,74] This effect is greater at higher generations of dendrimers, likely due to the greater number of the charged groups on larger dendrimers compared to their smaller counterparts. Using supported lipid bilayers, observations by AFM also revealed that G7 PAMAM dendrimers rapidly formed holes in lipid bilayers, whereas smaller G5 dendrimers did not form new holes but instead expanded pre-existing defects. Nanoscale hole formation caused by positively-charged PAMAM dendrimers is generally considered to be the cause of toxicities associated with larger PAMAM dendrimers.

One approach to overcoming these associated toxicities is to modify the dendrimer surface with charge neutral (acetamide/hydroxylate) or anionic (carboxylate) constituents.[72] While dendrimer toxicities are highly dependent on the cell line and local environment, positively-charged amine-terminated dendrimers are generally associated with increased toxicities. For instance, Roberts *et al.* have demonstrated that cationic PAMAM dendrimers are able to produce greater than 90% cell death at 1 mM, 10 μM and 100 nM concentrations of G3, G5 and G7, respectively, in V79 fibroblasts.[75] In contrast, others have demonstrated carboxylated PAMAM dendrimers were not toxic at concentrations up to 5 mg/ml following 72 hours of treatment.[76] Moreover, three-arm star PEGylated polyester dendrons have displayed IC$_{50}$ values as high as 20 mg/ml.[71,77] Furthermore, cationic G4 PAMAM dendrimers have been observed to cause non-hemolytic changes in red blood cell morphologies even at concentrations of 10 μg/ml.[71] These findings suggest that in order to use dendrimers for biomedical applications, the positively-charged surface of dendrimers needs to be modified to avoid any undesired toxic effects.[71,77–80]

The cellular uptake of dendrimers is a complex process, influenced by a number of materials-based and cell-specific factors. Findings have differed between reported results, and the observations seem to be dependent on individual experimental conditions. For instance, in 2007, Kitchens *et al.* demonstrated in Caco-2 cells that G4 amine-terminated PAMAM dendrimers were internalized via an endocytic pathway.[81] A year later, Permal and coworkers showed in A549 lung epithelial cells that whereas anionic dendrimers of the same size were taken up through a caveolae-mediated mechanism, both cationic and neutral dendrimers were internalized through non-clathrin, non-caveolae pathways.[82] Interestingly, Albertazzi *et al.* demonstrated in HeLa cells that dendrimers of multiple generations were internalized through a combination of clathrin-mediated endocytosis and micropinocytosis, in agreement with the observation originally seen by Kitchens and coauthors.[83] These findings suggest that several underlying mechanisms are present that govern the cellular uptake of dendrimers. Notably, Hong *et al.* demonstrated that amine-terminated G7 PAMAM dendrimers were internalized at least partially through energy-independent nanoscale hole formation in KB and Rat2 cells, given the observed cell internalization following incubation at 4°C.[84] Inconsistencies in the literature indicate that the cell entry of PAMAM dendrimers is not governed by a single pathway, but that there are multiple mechanisms in operation. Nonetheless, it is clear that the unique properties of dendrimers, such as their controllable size and surface modularity, are critical factors in determining how they interact with biological systems.

4.4.2 *Pharmacokinetics of Dendrimers*

The size and surface characteristics of dendrimers not only dictate their cellular interactions *in vitro*, but also play a critical role in governing their *in vivo* profiles. Following systemic administration, nanoparticles face several barriers before reaching their desired site of action, including protein adsorption, hepatic and renal clearance, metabolism, and uptake by the organs of the reticuloendothelial system (RES). The relatively short circulation times and rapid systemic elimination of dendrimers upon intravenous injection are considered to be a major hurdle that has hindered the clinical translation of dendritic nanoparticles.[11] The relationship between dendrimer size and their biodistribution and circulation profiles have been described in detail by Margerum and coworkers.[85] Typically, larger dendrimers are associated with prolonged circulation times, compared to their smaller counterparts. For instance, the elimination half-lives of gadolinium (III) (Gd)-coupled dendrimers in rats ranged from 11 ± 5 min for G3 to 115 ± 8 min for G5 dendrimers.

Furthermore, whereas G3 and G4 dendrimers were largely excreted in the urine with a relatively low level of accumulation in the liver, G5 dendrimers exhibited significantly greater accumulation in the liver. While smaller dendrimers are typically associated with renal clearance, larger dendrimers tend to display notable hepatic accumulation.[86,87] Kobayashi and coworkers investigated the pharmacokinetics of G3-G9 Gd-labeled dendrimers. Relative to G3-G6 dendrimers, G6 exhibited significantly longer circulation times compared to any of the smaller dendrimers, with G3 and G4 displaying greater renal accumulation.[87]

In a concurrent investigation employing G6-G9 dendrimers, the authors observed significantly increased hepatic uptake of G8 and G9 dendrimers, suggesting their clearance and accumulation via organs of the RES, mainly the liver and spleen. The greatest amount of dendrimers that remained in the bloodstream was found with G7 dendrimers, alluding that these dendrimers have the ability to avoid renal clearance seen with lower generations, while at the same time evading significant RES uptake that plagues larger G8 and G9 dendrimers. Similarly, Kaminskas *et al.* have demonstrated that larger generation PLL dendrimers are opsonized quickly following systemic injection and similarly accumulate in the organs of the RES.[88] Despite this, smaller generation PLL dendrimers are more vulnerable to metabolic degradation and renal elimination. Taken together, these reports suggest that dendrimer structures have to be finely tuned in order to obtain favorable circulation and pharmacokinetic profiles. However, the rapid elimination of dendrimers still remains one of the greatest obstacles to their clinical utility.

The surfaces of dendrimers play a vital role in dictating their systemic circulation and biodistribution. Specifically, dendrimer surface charges have a major impact on their circulation half-life.[89] For instance, the significantly positive charge of amine-terminated, cationic dendrimers at physiological pH can directly impact their circulation time, most likely by causing them to rapidly interact non-specifically with vascular linings and tissues.[90,91] This is consistent with findings from other groups in which neutral and anionic dendrimers are excreted more slowly and to a greater degree in the urine compared to their cationic counterparts.[71] Moreover, Nigavekar *et al.* have demonstrated that compared to cationic G5 PAMAM dendrimers, neutral dendrimers are excreted far more in the urine and feces, suggesting that the volume of distribution of cationic dendrimers is greater, and which likely indicates greater non-specific uptake of amine-terminated dendrimers.[91] These findings suggest that surface modification, either through neutralization or anionic modification, is necessary to increase the plasma half-life of dendrimers, and can be utilized to minimize off-target binding and non-specific interactions.[88]

Another approach to overcoming the limited circulation profiles of dendrimers and nanoparticles is to conjugate them to water-soluble, non-fouling linear polymers such as poly(ethylene glycol) (PEG).[92] PEGylation of dendrimers can increase their sizes above the renal clearance threshold while limiting recognition by the immune system.[92] For example, Kaminskas et al. found that larger (<30 kDa) PEGylated PLL dendrimers exhibited longer circulation times compared to smaller PEGylated dendrimers. Furthermore, PEGylated dendrimers displayed decreased renal clearance and increased uptake by the RES system. PEGylated dendrimers were also more resistant to biodegradation.[93] Similar findings were made by Guillaudeu et al. using PEGylated polyester dendrimers. In their case, the longer circulation times led to increased tumor accumulation with decreased accumulation in healthy, off target organs.[94] PEGylation can also be used to improve the bioavailability of dendrimers. For example, Kaminskas et al. have shown that PEGylated PLL dendrimers are better absorbed systemically than non-PEGylated dendrimers following subcutaneous injection. Moreover, increased PEG molecular weight enhances lymphatic transport compared to PLL dendrimers modified using an anionic benzene sulphonate-cap.[52] Wu and coauthors have introduced an interesting scheme to enhance the delivery of the anticancer drug gemcitabine (GEM) using a series of novel G4 oligo(ethylene glycol) (OEG)-based dendrimers.[95] The authors synthesized a set of OEG dendrimers, each containing a different molecular weight of OEG in the last generation of the dendrimer. By varying the molecular weight of the OEG outer layer, the authors were able to enhance dendrimer circulation time and biodistribution. Dendrimers containing the greatest molecular weight OEG (900 Da) had the longest circulation time, leading to greater accumulation at the tumor mass. Interestingly, the largest dendrimer also displayed the greatest tumor penetration in vivo, likely due to a steeper concentration gradient resulting from enhanced accumulation. The high molecular weight OEG formulated with GEM also displayed the greatest antitumor activity and growth inhibition.

4.4.3 *Tissue Permeation of Dendrimers*

The relatively small size (<20 nm) and high degree of surface functionality of dendrimers set them apart from other nanoparticles.[11] These are useful characteristics for tissue permeation, and several groups have used them as an advantage for efficient tumor permeation. For instance, Dhanikula and coauthors have generated polyether-*co*-polyester (PEPE) dendrimers loaded with methotrexate (MTX) for the treatment of gliomas.[96] Dendrimer surfaces modified with D-glucosamine enhanced the permeation of MTX across the blood brain barrier and resulted in more homogeneous tumor distribution. Similarly, Waite *et al.* conjugated RGD peptides to PAMAM dendrimers. By interfering with the integrin-ECM network, the dendrimers were able to enhance the permeation of short interfering RNA (siRNA) throughout U87 multicellular tumor spheroids, demonstrating their potential as tumor penetration enhancers.[97] Al-Jamal and coworkers have developed cationic G6 PLL dendrimers complexed with doxorubicin (DOX), which exhibited greater penetration into prostate cancer multicellular tumor spheroids compared to free DOX.[98] Dendrimer-DOX complexes also achieved greater efficacy compared to free DOX. Similarly, Hong and coworkers have demonstrated that folate (FA)-targeted dendrimers are able to better permeate KB folate receptor (FR)-expressing multicellular tumor spheroids compared to larger PEGylated nanoparticles.[99] Targeted G4 dendrimers fully permeated the spheroids while larger PEGylated particles were restricted to the surface of the spheroids.

Dendrimers have also been demonstrated to permeate the skin. Yang *et al.* have shown that small, neutral or anionic G2 PAMAM dendrimers are able to effectively permeate the skin better than larger G4 PAMAM dendrimers.[100] Using confocal approaches, G2 PAMAM dendrimers conjugated to the fluorescent dye rhodamine were found to localize to the stratum corneum (SC), the outermost layer of skin, and the viable epidermis following 24 hours of treatment. Larger, G4 PAMAM dendrimers were retained in the topmost layer of the SC. Similarly, Venuganti *et al.* have also demonstrated an inverse correlation between dendrimer size and their ability to permeate the skin.[101] Surface modification also plays a critical role in governing the distribution of PAMAM dendrimers in the skin layers.[100] G2 PAMAM dendrimers modified to contain either neutral or anionic surface charges displayed the greatest skin permeability with limited cellular uptake, indicating that they likely diffused into the skin via an extracellular pathway. Conversely, amine-terminated, cationic G2 PAMAM dendrimers displayed significant cellular uptake and retention in the skin, suggesting that they likely diffused through a more transcellular pathway. Similarly, by increasing the log P of G2 dendrimers through conjugation with oleic acid, dendrimer skin absorption and retention were increased. These findings suggest that dendrimer size and surface characteristics can be tailored to modify their permeability into tissues such as solid tumors and across strong penetration barriers such as the skin.

4.5 Gene Delivery Applications of Dendritic Polymers

The successful delivery of genetic material into the cytoplasm and nucleus of eukaryotic cells using cationic polymers highlights their potential for use in the treatment of a variety of disorders.[102,103] The concept of gene therapy derives from the observation that certain diseases are caused by the inheritance of functionally defective genes.[104] Theoretically, diseases caused by known genetic defects can potentially be treated by the insertion and expression of a normal copy of the mutant or deleted gene in host cells.[105] Gene-replacement therapy may represent the future of how genetic diseases are treated.[106]

However, naked gene transfection *in vitro* and *in vivo* is intrinsically limited due to electrostatic repulsion between the polyanionic phosphate backbone of oligonucleotides and cellular membranes. Moreover, the abundant nucleases in biological systems degrade free oligonucleotides before they can reach their site of action and elicit a response. Various nucleic acid (NA) delivery vehicles have been investigated to increase cell permeability or decrease electrostatic repulsion between the cell membrane and oligonucleotides. Gene delivery vectors are typically classified into one of two groups: (i) viral and (ii) non-viral systems. To date, viral-mediated delivery has played a major role in gene therapeutics.[107,108] The first laboratory tests on viral-based gene therapy were carried out in the mid-1980s.[109] Early clinical trials were initiated in the 1990s, during which gene therapy was regarded as a potential treatment for genetic diseases such as inherited blindness[110–112] and immune deficiency.[113] Yet, the Food and Drug Administration (FDA) issued a warning in September 2000, saying "the hyperbole has exceeded the results" and "little has worked."[114] Despite the high transfection efficiencies of viral gene delivery systems, there are significant safety concerns, specifically with regards to endogenous virus recombination, oncogenic effects and unexpected immune responses.[115] These fears culminated with the death of 18-year-old Jesse Gelsinger in a 1999 gene therapy clinical trial, which initiated a chain of events that disrupted progress in the field.[109]

These limitations have encouraged investigation into synthetic gene delivery vectors (e.g. non-viral vectors). Non-viral delivery systems include peptides,[116] lipids,[117] polymers,[64,118] calcium phosphate[119] and nanoparticles.[120] Of particular interest are cationic polymers, which offer strong binding of oligonucleotides and facilitate efficient delivery into cells *via* adsorptive endocytosis.[121,122] Compared to viral vectors, cationic polymer-based non-viral delivery systems are easy to produce, are safe and stable, and can be manufactured in different sizes and shapes with additional functionalities for targeted delivery.[103] Among the various cationic polymers available, dendrimers are considered to be one of the most promising platforms for non-viral gene delivery due to their unique molecular architecture and distinctive properties, including a well-defined size and structure and low polydispersity index.[9] Although non-viral systems have not shown the high transfection efficiencies achieved by viral counterparts, recent synthetic advances display promise in overcoming these limitations.[9] For instance, one interesting approach is the use of polycationic dendronitic polymers for DNA complexation, which mimics the condensation of DNA by histone proteins, a process that naturally occurs in chromatin.[123]

4.5.1 *General Mechanisms of Cationic Dendrimers for Gene Delivery Systems*

Amine-terminated, cationic dendrimers, including PPI, PAMAM and polyethylen-imine (PEI), are among the most widely used transfection enhancers for the delivery of oligonucleotides, including both DNA and RNA [i.e. siRNA[61,124] and microRNA (miRNA)[125,126]]. Their structural flexibility and surface presentation of positive charges allow them to form condensed complexes with the anionic phosphate backbone of oligonucleotides, known as dendriplexes.[103,127] Dendriplexes are able to prevent rapid nucleic acid degradation by blocking enzymatic degradation and enhancing circulation times.[64,128] Following endocytosis and endosomal escape, the dendriplexes destabilize, releasing the genetic material and allowing for nuclear entry (Figure 4.3).[105]

Cationic amine-terminated PAMAM dendrimers are one of the most heavily investigated systems for their high transfection efficiencies, which is due to their well-defined shape, high flexibility, and the presence of protonatable amines with low pK_a's (3.9 and 6.9). Initial work by Haensler and Szoka showed that the 3D spherical structure of PAMAM dendrimers (G5) could lead to the significantly enhanced transfection efficiencies of naked plasmid DNA.[124] Interestingly, the high flexibility of dendrimers and surface functionality has also been shown to be important to achieving high transfection efficiencies.[129–134] For instance, partially degraded PAMAM dendrimers were reported to have more flexible structures than intact dendrimers and therefore can interact more efficiently with DNA.[124] Another approach for enhancing the complexation ability of dendrimers is to increase their flexibility by modifying their core structure. For example, Zhou *et al.* have built PAMAM dendrimers using a triethanolamine (TEA) core, as opposed to the more commonly used amine core.[129] Based on computational and experimental findings they demonstrated that the flexible structures of these modified dendrimers increase the stability of dendriplexes and enhance transfection.[133,134] The presence of low pK_a amines also plays a role in transfection, as they can participate in initial DNA complexation and promote polymeric swelling, which destabilizes the dendrimer-DNA complex.[128] Moreover, internal tertiary amine groups act as proton sponges in the endosome, enhancing DNA release and endosomal escape.[128]

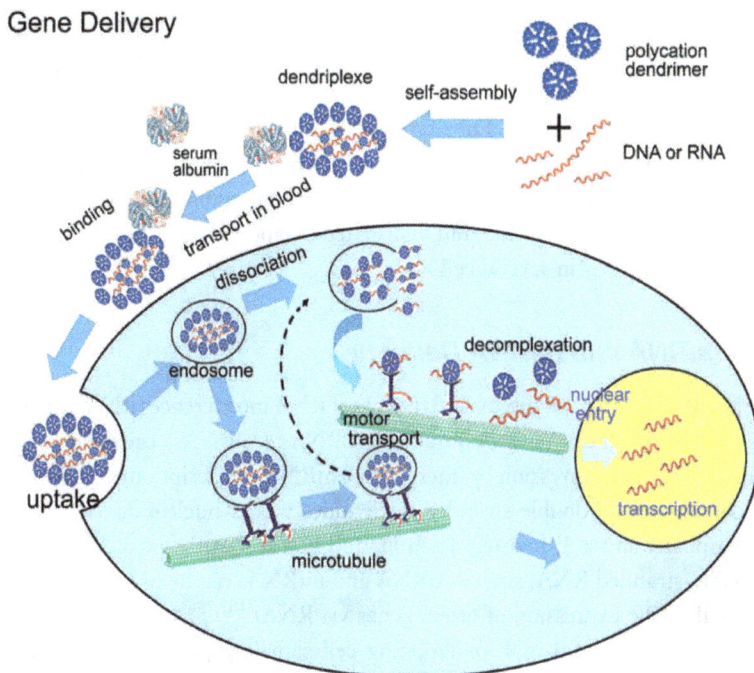

Figure 4.3: Gene transfection mediated by dendrimers.[135] (Copyright ©2013, all rights reserved by Royal Society of Chemistry.)

4.5.2 *DNA Delivery*

Wong *et al.* recently reported the use of PAMAM dendrimers with dual binding motifs (DNA intercalation and receptor recognition) for targeted DNA delivery.[136] A series of dendrimer conjugates derived from G5 PAMAM dendrimers were prepared with multiple folate (FA) or riboflavin (RF) ligands for cell receptor targeting, and with 3,8–diamino–6–phenylphenanthridinium (DAPP)-derived ligands for anchoring a DNA payload. Surface plasmon resonance (SPR) studies investigating the adhesion of the polyplex to a model surface immobilized with folate binding protein (FBP) demonstrated that the DNA payload had a minimal effect on the receptor binding activity of the polyplex: $K_D = 0.22$ nM for G5(FA)(DAPP) versus 0.98 nM for the polyplex. These findings suggest that DNA–ligand intercalation can act as a motif in the design of multivalent dendrimer vectors for targeted gene delivery. Alternatively, Chang *et al.* used an aliphatic hydrocarbon-cored PAMAM dendrimer with dendritic morphologies similar to those shown in Figure 4.2(b) for gene delivery.[137]

Diaminododecane-cored generation 4 (C12G4) PAMAM dendrimers showed dramatically higher luciferase and EGFP gene transfection efficiency compared to diaminoethane-cored generation 4 (C2G4) and diaminohexane-cored generation 4 (C6G4) PAMAM dendrimers. The significantly improved gene transfection efficacy of C12G4 was attributed to its hydrophobic core, which facilitates the cellular uptake of dendriplexes. Further modification of C12G4 with functional ligands such as arginine, 2,4-diamino-1,3,5-triazine, and fluorinated compounds significantly increased transfection efficiency in several cell lines.

4.5.3 *siRNA and miRNA Delivery*

In the late 1990s, Fire *et al.* discovered the role of RNA interference (RNAi), in which small interfering RNAs (siRNAs) and microRNAs (miRNAs) cause site-specific suppression of gene expression by interrupting mRNA transcription.[102] siRNA typically contains 21-23 double-stranded nucleotides with 2-nucleotide 3'-overhangs, a 5'-phosphate, and a 3'-hydroxy termini, and is a key mediator of RNAi. Short and double-stranded RNA, such as siRNA and miRNA, can be used to specifically downregulate the expression of target genes via RNAi.[102,138] Indeed, these RNAs have become a powerful tool for targeting cell signaling pathways and silencing pathological actions in numerous disease states.[138,139] They have several advantages as therapeutic agents over conventional drugs and plasmid DNA, by providing efficient gene silencing at low intracellular concentrations in a highly sequence-specific manner with low toxicity.[140] These advantages have generated many potential applications for siRNA and miRNA, such as in the treatment of human immunodeficiency virus (HIV),[141–144] Alzheimer's disease,[145–148] hepatitis B virus (HBV),[149–153] and several cancers.[154–157] For example, Khan *et al.* recently reported the synthesis and optimization of PAMAM and PPI-based dendrimers for targeted delivery of siRNA to distinct cell subpopulations within the liver.[158] In this study, PAMAM or PPI dendrimers were modified with alkyl epoxides of various carbon chain lengths using Michael addition chemistry, forming nanoparticles.

Researchers can also use hybridization strategies to fabricate versatile, bio-inspired dendritic nanocarriers. For example, Xu *et al.* prepared dual-functionalized low generation peptide dendrons (PDs) which can self-assemble into nanoparticles via coordination interactions, generating multifunctional supramolecular hybrid dendrons (SHDs).[159] These SHDs possess an arginine-rich peptide corona and fluorescent signaling properties. The bio-inspired supramolecular hybrids enhanced gene transfection efficiency by approximately 50,000-fold compared to single PDs at the same arginine to phosphate ratio, while maintaining minimal cytotoxicity and high serum stability. *In vivo* animal experiments demonstrated that the SHDs provide considerable gene transfection efficiency, highlighting the potential of bio-inspired supramolecular hybrid dendritic systems for biomedical applications both *in vitro* and *in vivo*.

4.6 Dendrimers in Drug Delivery

4.6.1 *Dendrimers as Drug Carriers*

Much of the interest in dendrimers for biomedical applications stems from their potential as drug nanocarriers. Two major methods for cargo loading have been used to incorporate drugs into the dendrimer structure: surface conjugation or physical complexation, of which several detailed reviews have been published discussing their advantages and disadvantages.[56,160] Granted that the drug can be first modified, chemical conjugation to the dendrimer surface provides several benefits, including enhanced stability and a better controlled release profile. Despite this, not all drugs can be modified chemically, and while encapsulation or complexation is typically limited by poor drug loading and uncontrolled release kinetics, there have been several attempts to overcome these issues. For instance, one strategy has been to modify the surface of the dendrimer. Fang *et al.* have encapsulated the anionic dyes Congo red and indocyanine green in PAMAM dendrimers.[161] Acetyl-terminated, charge neutral PAMAM dendrimers were able to more stably encapsulate the dyes compared to positively-charged dendrimers. Moreover, Zhang and coauthors prepared dendrimers in complex with DOX for the treatment of cancer.[162] They found that the release of DOX from dendrimers is governed by surface functionality. Anionic, carboxylated PAMAM dendrimers showed a more rapid drug release profile compared to neutral dendrimers, which exhibited a more extended release profile. While the complexation of drugs is not usually ideal for systemic administration, it is occasionally unavoidable due to the inability to chemically ligate certain agents. When this happens, modification of the dendrimer surface may provide an avenue for these therapeutic agents to be delivered.

When chemical conjugation is an option, there exist many tools for investigators to use. Stimuli-responsive linkages allow for more control over the kinetics and localization of drug release. For instance, Satsangi *et al.* have demonstrated the potential for drugs to be conjugated to PAMAM dendrimers by a cathepsin B cleavable tetrapeptide.[163] By conjugating PAMAM dendrimers to the anticancer drug paclitaxel (PTX), the conjugate demonstrated anticancer activity to cells specifically overexpressing cathepsin B. Enhanced efficacy and decreased tumor size was observed in a xenograft mouse model.

4.6.2 *Targeted Drug Delivery*

Targeted drug delivery can potentially enhance therapeutic efficacy by localizing the drug to a site-specific cell type or tissue, while limiting its accumulation in off-target healthy tissues, often the liver, spleen, and kidneys. In cancer treatment, two main strategies are used to target nanocarriers to the tumor site: passive and ligand-mediated targeting. Passive targeting utilizes the enhanced permeability and retention (EPR) phenomenon in tumors,[164] in which leaky microvasculature and lymphatic drainage leads to the selective retention and accumulation of nanoparticles. Following passive accumulation at the tumor site, the drug can be released resulting in localized high therapeutic concentrations. In contrast, ligand-mediated approaches utilize the expression of receptors or antigens specific to the target cell. By coating nanoparticles with the respective ligands or antibodies, they are able to specifically recognize target cells.[165] As multivalent, nanosized particles, dendrimers can be used for both passive and ligand-mediated targeting approaches.[10,61] Dendrimers provide a good platform for ligand-mediated targeting due to their high flexibility, unique modifiable surface functionality, and well-defined structure, and as such have emerged as promising nanocarriers.[5,61,166,167] Moreover, their highly branched structure and molecular flexibility increases the local ligand density for a multivalent binding effect, which is the simultaneous binding of multiple ligands and receptors. Multivalent ligand-functionalized dendrimers have been applied successfully *in vivo* for anticancer targeting in several cases.[10,13,37,168,169]

One of the most heavily explored ligand-mediated targeting strategies has been the use of FA-conjugated dendrimers for the tumor targeting of anti-cancer drugs via the FA receptor (FR).[170,171] FA-based targeting provides a good model as the FR is overexpressed on a variety of cancer cells.[10] Hong *et al.* reported multivalent interactions between FA-functionalized G5 PAMAM dendrimers and FA binding protein (FBP)-functionalized surfaces using surface plasmon resonance (SPR).[172] The binding of polyvalent PAMAM G5 PAMAM dendrimers resulted in binding avidities 170,000-fold greater than that of free FA. Notably, this enhancement was not seen when greater than five FA molecules were conjugated to the dendrimers, suggesting that there is an optimal conjugation ratio for ligand to dendrimer. Elsewhere, multifunctional G5 PAMAM dendrimers conjugated to MTX, FA, and imaging (radiolabel or fluorophore) agents were evaluated in a mouse tumor xenograft model using FR-overexpressing KB cells.[173] The targeted dendrimer conjugates displayed minimal adverse effects along with enhanced antitumor activity. Another example used FA-targeted PEGylated PAMAM dendrimers to deliver 5-flurouracil (5-FU). 5-FU was encapsulated within dendrimers and evaluated for anti-tumor efficacy.[171] The PEGylated dendrimers displayed prolonged retention of

5-FU ($t_{max} = 2.5-5$ h) compared to non-PEGylated dendrimers ($t_{max} = 1-2.5$ h), while minimizing adverse effects and enhancing anti-tumor efficacy compared to free 5-FU.

More recent investigations suggest that the enhanced ligand-mediated binding event is more complicated than originally thought. While it was initially believed by some that enhanced binding avidity was due to statistical rebinding events from localized ligand concentrations, new findings have alluded that this may not be the only underlying mechanism. Van Dongen and coauthors have prepared dendrimers containing precise amounts of FA, a limitation in previous experiments due to inherent polydispersity.[174] Notably, only a small multivalent effect was observed on the scale of the SPR experiments. In contrast, slower, strong binding events were observed, suggesting that polymer-protein interactions play a major role in enhanced binding relative to free FA. This mechanism was originally proposed by Licata and Tkachenko.[175]

Another approach for ligand-mediated targeting is to utilize hormone-receptor or antibody-antigen interactions. Baker and coworkers explored the use of mono-clonal antibodies for site-specific delivery to prostate and non-prostate tumors by targeting the prostate specific membrane antigen (PSMA) J591 using an anti-PSMA G5 PAMAM dendrimer conjugate.[176] Similarly, Modi *et al.* have utilized the inter-action between the follicle stimulating hormone (FSH) and its receptor, which is specifically overexpressed on tumorigenic ovarian cancer cells but not on imma-ture healthy primordial follicles.[37] FSH-targeted G5 PAMAM dendrimers demon-strated enhanced accumulation in the ovary and oviduct following intraperitoneal administration. Interestingly, binding to the FSH receptor *in vitro* downregulated the anti-apoptotic protein survival, likely through receptor mediated action. Den-drimers can also be modified with sugar moieties, e.g., glycosylated dendrimers, which are crucial in biological recognition and a commonly used strategy for tar-geted drug delivery.[177] These sugar moiety-conjugated dendrimers have shown the potential to block metastatic sites of active tumor cells. For example, Lagnoux *et al.* prepared glyco-dendrimers conjugated with colchicine, which exhibited 20–100 times more selective inhibition of HeLa cells than normal cells.[178,179]

In addition, engineered dendrimers offer an opportunity for multifunctional-ization. Jia and coworkers prepared G4 PAMAM dendrimers with PEG, transferrin (Tf) and wheat germagglutinin (WGA) conjugated onto the surface, and doxoru-bicin (DOX) loaded into the core.[180] PAMAM-PEG-WGA-Tf reduced the cyto-toxicity of DOX and inhibited the growth rate of C6 glioma cells. They also prepared pH-sensitive, dual-targeting drug carriers with Tf on the exterior and Tamoxifen (TAM) in the interior using PEGylated G4 PAMAM (G4-DOX-PEG-Tf-TAM) to enhance blood brain barrier (BBB) transportation and improve drug accumulation in glioma cells.[181] The accumulation of DOX in the tumor site was increased due to targeting by both Tf and WGA, whereas the same dendrimers containing either Tf or WGA showed limited efficacy. Jain and coworkers designed PPI dendrimers with two ligands (sialic acid and mannosyl) for dual targeting of the anti-HIV drug Zidovudine.[182] This dual-targeted system was found to have good biocompatibility as well as efficiently target the antiviral drug.

4.7 Hybridized Dendritic Polymers for Drug Delivery

A number of recent approaches have improved the delivery potential of dendrimers by combining dendritic structure with other nanocarrier systems. In this section, we highlight a few examples of hybrid systems that integrate dendrimers (or dendrons) with conventional nanoparticles, overcoming the drawbacks of individual nanocarriers while maximizing their strengths in a synergistic way.

4.7.1 *Dendron-based Polymeric Nanoparticles*

Dendron-based polymers have garnered special interest in drug delivery due to their ability to incorporate advantageous characteristics of dendrons with those of other polymers. For instance, several groups have generated multi-block polymers composed of linear and dendron structures. By integrating dendron architecture with that of linear polymers, unique branching moieties can be introduced into the backbone of the polymers. For instance, amphiphilic dendronized polymers can undergo self-assembly into polymeric micelles and confer several tunable properties.[70,183,184] Compared to linear copolymer micelles, amphiphilic dendronized polymers offer specific advantages. One such advantage is the increased stability of the dendron micelle.[185,186] Micelle assembly occurs at polymer concentrations greater than the critical micelle concentration (CMC), and is largely governed by the architecture and hydrophilic-lipophilic balance (HLB) of the polymer.[187,188] At high HLBs, molecules have a reduced affinity for assembly due to greater solubility in aqueous media.[189] Assembly of poly(caprolactone)-*b*-polyester dendron-*b*-poly(ethylene glycol) has been shown to occur at far lower concentrations than its linear counterpart.[185,186,190] PEGylated dendron copolymers exhibited CMCs that are 1–2 orders of magnitude less than linear micelles at similar HLBs. This enhanced assembly is likely due to the conical architecture of dendronized polymers that are pre-organized to form spherical micelles, while paying minimal entropy cost. Interestingly, the surface of dendrons displays unique characteristics as well.

Whereas amine-terminated, cationic dendrimers exhibit strong non-specific interactions, amine-terminated micelles had negligible cellular interactions. These findings are in part due to the ability of PEG chains to back fold, introducing hydrogen bond interactions between the chain and the peripheral surface group.[186,190] Notably, PEG chain length can also be tailored to control cellular interactions with micelles. These results suggest that the incorporation of dendritic architecture into polymeric nanoparticles can lead to specific benefits for drug delivery.

In comparison to dendrimers, polymeric micelles exhibit significantly longer circulation times, likely due to PEG surface coverage and their large sizes. Thus, while the introduction of dendritic architecture into the micelle structure can enhance stability and confer controlled cellular interactions, micellar formulations themselves have the potential to increase circulation times and drug loading capabilities. Micelles are typically able to achieve higher encapsulation efficiencies compared to dendrimers. For instance, Qiao *et al.* have prepared amphiphilic semi-PAMAM-*b*-poly(lactide)-based dendron micelles for the encapsulation and delivery of the anticancer drug docetaxel (DTX).[191] Dendron-based micelles significantly limited rapid DTX release, increasing circulation times and prolonging clearance. In combination with the benefits of dendritic architecture, these observations demonstrate that incorporating different structures is a viable method to overcoming the limitations of either system.

4.7.2   Hybrid Nanoparticle Formulations

As mentioned above, while dendrimers are able to transit tissues and tumors well, they are limited by their rapid clearance from circulation and inefficient tumor accumulation through passive targeting. To overcome these barriers, Sunoqrot and coauthors prepared a series of hybrid nanoparticles in which smaller, FA-targeted PAMAM dendrimers were encapsulated within larger poly(ethylene glycol)-*b*-poly (lactide) (PEG-PLA) nanoparticles.[99,192,193] In an *in vivo* xenograft model, they demonstrated that the targeted dendrimers encapsulated within the hybrid nanoparticles had the advantageous characteristics of the larger polymeric nanoparticles ($\sim$100 nm).[193] These included a significantly enhanced circulation time and greater tumor accumulation. Additionally, by varying the molecular weight of the encapsulating polymer, kinetic control over the cellular interactions of the FA-targeted dendrimers with FR-overexpressing KB cells was achieved.[192] Furthermore, whereas the larger nanoparticles were restricted to only the very periphery in a multicellular tumor spheroid model, the targeted dendrimers were released in a kinetically-controlled manner and efficiently permeated the entire spheroid.[99]

4.8 Dendritic Nanomaterials for Theranostics

Dendritic nanomaterials have also been used for theranostic applications, highlighting their potential for real time diagnostic monitoring of disease progression and visualization of drug delivery, release and treatment.[194,195] Recent advances in polymer chemistry and imaging have led to the development of novel polymer-based imaging probes for simultaneous diagnosis and treatment of different diseases.[196] Key design requirements for the targeted delivery of imaging agents *in vivo* are: (i) stealth properties preventing clearance before target cells are reached, (ii) the ability to penetrate into cells efficiently, (iii) minimal safety concerns (cytotoxicity, biocompatibility and immunogenicity), and (iv) compatibility with external activation sources, such as a magnetic field, ultrasound, X-ray or optics, to trigger therapeutic release (Figure 4.4).[5]

Dendrimer-based nanoparticles can be precisely tuned to deliver bioactive compounds and imaging agents. For example, the use of dendrimers as nanocarriers for MRI contrast agents has generated a tremendous amount of attention due to

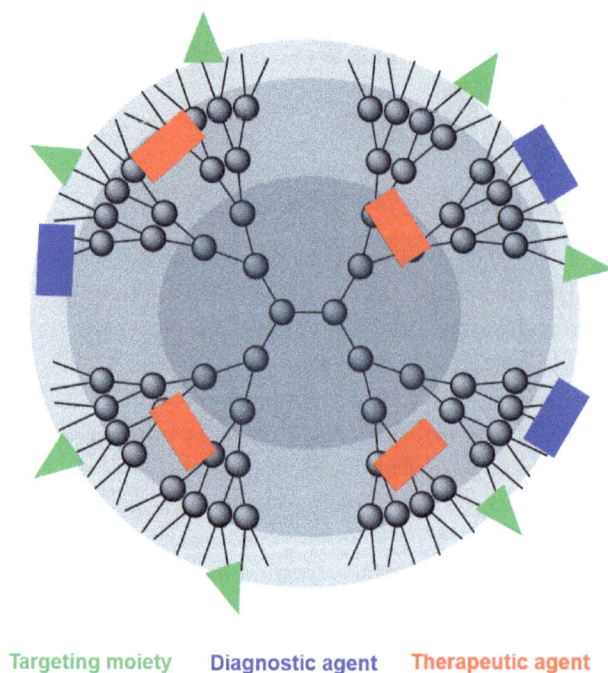

Targeting moiety   Diagnostic agent   Therapeutic agent

Figure 4.4: Schematic representation of multifunctional dendrimers with targeting ligands, drug molecules and imaging probes for theranostic applications.

their unique characteristics and multivalent properties.[197–200] The first dendrimer-based contrast agents were reported in 1994 using G2 and G6 PAMAM dendrimers with Gd^{3+}-1B4M.[201] Gd-DOTA has also been covalently grafted onto the dendrimer surfaces for molecular imaging.[202,203] Besides these covalent approaches, dendrimers have also been widely used to encapsulate inorganic nanoparticles for molecular imaging.[204] It was demonstrated that acetylated dendrimer-entrapped gold nanoparticles were suitable for *in vitro* and *in vivo* computed tomography (CT) imaging of cancer cells.[205] Several other imaging probes, including iron oxide,[206,207] fluorescent quantum dots[208,209] and carbon nanotube,[210,211] have been used successfully with dendrimers for cellular tracking and visualization. Tomalia *et al.* have proposed the use of ~5 nm sodium salt-PAMAM dendrimer (G4.5) nanoparticles possessing a 1,4-diaminobutane core for the co-delivery of drugs and diagnostic MRI contrast agents, such as Magnevist®-functionalized dendrimers, to tumors.[212] All of these studies demonstrate the potential of dendrimers to function as theranostic agents for simultaneous detection, diagnosis and treatment of diseases such as cancer.

4.9 Conclusions and Perspectives

Dendrimers and dendritic nanoparticles are considered to be among the most promising polymeric systems to date for biomedical applications. However, several factors have delayed their translation into the clinic. One such barrier is the significant toxicities observed with amine-terminated dendrimers, commonly used for gene and drug delivery, and the need for surface modification. Their small size and surface-mediated non-specific interactions limit their circulation time in plasma. Another major complication is overcoming their poor drug loading capabilities. Drug loading into the dendrimer interior is inefficient, and encapsulation generally results in rapid, uncontrolled release kinetics. Thus, the preferred method for drug incorporation into dendrimers is typically covalent conjugation. However, given that surface conjugation is dependent upon the chemical structure and availability of chemically-reactive sites on the drug molecules, the types and amounts of drug that can be coupled are limited. One attempt to overcome these limitations has been to focus on the targeting abilities of dendritic systems. Dendrimers have several advantageous characteristics that allow them to exhibit exceptional targeting abilities. Their high degree of flexibility and multivalency enhance ligand-mediated targeting kinetics, while their surface can be modified with non-fouling polymers to enhance circulation *in vivo*. In an attempt to overcome these obstacles, several groups have focused on hybridizing dendritic architecture with other polymers.

These dendron-based or hybrid nanomaterials combine advantageous dendritic characteristics, such as close-to-monodispersity, controlled structure, mulitvalency, and available chemistry functionalities, with those of linear polymer-based nanoconstructs, including enhanced circulation time. These properties have led to a surge in dendrimer applications over the past 20 years, with more recent developments focused on their use in targeted drug/gene delivery and diagnostic medicine. With careful design optimization and engineering, dendritic nanomaterials hold promise as a highly versatile nano-carrier, well suited for a new era of personalized medicine.

References

1. Sahoo, S.K., and Labhasetwar, V. Nanotech approaches to drug delivery and imaging. *Drug Discovery Today* 8(24), 1112–1120 (2003).
2. Agrawal, U., *et al.* Is nanotechnology a boon for oral drug delivery? *Drug Discovery Today* (2014).
3. Bugno, J., *et al.* Recent advances in targeted drug delivery approaches using dendritic polymers. *Biomater. Sci.* (2014).
4. Tomalia, D.A. Birth of a new macromolecular architecture: Dendrimers as quantized building blocks for nanoscale synthetic polymer chemistry. *Progress in Polymer Science* 30(3–4), 294–324 (2005).
5. Oliveira, J.M., Salgado, A.J., Sousa, N., Mano, J.F., and Reis, R.L. Dendrimers and derivatives as a potential therapeutic tool in regenerative medicine strategies — A review. *Progress in Polymer Science* 35(9), 1163–1194 (2010).
6. Wolinsky, J.B., and Grinstaff, M.W. Therapeutic and diagnostic applications of dendrimers for cancer treatment. *Advanced Drug Delivery Reviews* 60(9), 1037–1055 (2008).
7. Wang, Z., Chen, C., Liu, R., Fan, A., Kong, D., and Zhao, Y. Two birds with one stone: dendrimer surface engineering enables tunable periphery hydrophobicity and rapid endosomal escape. *Chemical Communications* 50(90), 14025–14028 (2014).
8. Hubbard, D., Ghandehari, H., and Brayden, D.J. Transepithelial transport of PAMAM dendrimers across isolated rat jejunal mucosae in Ussing chambers. *Biomacromolecules* 15(8), 2889–2895 (2014).
9. Kesharwani, P., Jain, K., and Jain, N.K. Dendrimer as nanocarrier for drug delivery. *Progress in Polymer Science* 39(2), 268–307 (2014).
10. Lee, C.C., MacKay, J.A., Frechet, J.M., and Szoka, F.C. Designing dendrimers for biological applications. *Nat. Biotechnol.* 23(12), 1517–26 (2005).
11. Pearson, R.M., Hsu, H.-J., Bugno, J., and Hong, S. Understanding nano-bio interactions to improve nanocarriers for drug delivery. *MRS Bulletin* 39(03), 227–237 (2014).
12. Pearson, R.M., Bae, J.W., Hong, S., and Gad, S.C. Multifunctional dendritic nanocarriers: The architecture and applications in targeted drug delivery. In *Pharmaceutical Sciences Encyclopedia*, John Wiley & Sons, Inc. (2010).
13. Pearson, R.M., Sunoqrot, S., Hsu, H.-J., Bae, J.W., and Hong, S. Dendritic nanoparticles: The next generation of nanocarriers? *Therapeutic Delivery* 3(8), 941–959 (2012).

14. Shi, X., Bányai, I., Rodriguez, K., Islam, M.T., Lesniak, W., Balogh, P., Balogh, L.P., and Baker, J.R. Electrophoretic mobility and molecular distribution studies of poly(amidoamine) dendrimers of defined charges. *Electrophoresis* 27(9), 1758–1767 (2006).
15. Sowinska, M., and Urbanczyk-Lipkowska, Z. Advances in the chemistry of dendrimers. *New Journal of Chemistry* 38(6), 2168–2203 (2014).
16. Boas, U., Christensen, J.B., and Heegaard, P.M.H. Dendrimers: design, synthesis and chemical properties. *Journal of Materials Chemistry* 16(38), 3785–3798 (2006).
17. Buhleier, E., Wehner, W., and Vögtle, F. "Cascade"- and "nonskid-chain-like" syntheses of molecular cavity topologies. *Synthesis* 1978(02), 155–158 (1978).
18. Tomalia, D.A., Baker, H., Dewald, J., Hall, M., Kallos, G., Martin, S., Roeck, J., Ryder, J., and Smith, P. A new class of polymers: Starburst-dendritic macromolecules. *Polym. J.* 17(1), 117–132 (1985).
19. Tomalia, D.A., Berry, V., Hall, M., and Hedstrand, D.M. Starburst dendrimers. 4. Covalently fixed unimolecular assemblages reiminiscent of spheroidal micelles. *Macromolecules* 20(5), 1164–1167 (1987).
20. Ong, K.K., Jenkins, A.L., Cheng, R., Tomalia, D.A., Durst, H.D., Jensen, J.L., Emanuel, P.A., Swim, C.R., and Yin, R. Dendrimer enhanced immunosensors for biological detection. *Analytica Chimica Acta* 444(1), 143–148 (2001).
21. Islam, M.T., Majoros, I.J., and Baker Jr, J.R. HPLC analysis of PAMAM dendrimer based multifunctional devices. *Journal of Chromatography B* 2005, 822(1–2), 21-26.
22. Esfand, R., and Tomalia, D.A. Poly(amidoamine) (PAMAM) dendrimers: From biomimicry to drug delivery and biomedical applications. *Drug Discovery Today* 6(8), 427–436 (2001).
23. Hawker, C.J., and Frechet, J.M.J. Preparation of polymers with controlled molecular architecture. A new convergent approach to dendritic macromolecules. *Journal of the American Chemical Society* 112(21), 7638–7647 (1990).
24. Crooks, R.M., Zhao, M., Sun, L., Chechik, V., and Yeung, L.K. Dendrimer-encapsulated metal nanoparticles: synthesis, characterization, and applications to catalysis. *Accounts of Chemical Research* 34(3), 181–190 (2000).
25. Fréchet, J.M., and Tomalia, D.A. *Dendrimers and Other Dendritic Polymers.* Wiley (2001).
26. Caminade, A.-M., Laurent, R., and Majoral, J.-P. Characterization of dendrimers. *Advanced Drug Delivery Reviews* 57(15), 2130–2146 (2005).
27. Grayson, S.M., and Fréchet, J.M.J. Convergent dendrons and dendrimers: from synthesis to applications. *Chemical Reviews* 101(12), 3819–3868 (2001).
28. Hummelen, J.C., Van Dongen, J.L.J., and Meijer, E.W. Electrospray mass spectrometry of poly(propylene imine) dendrimers — the issue of dendritic purity or polydispersity. *Chemistry — A European Journal* 3(9), 1489–1493 (1997).
29. Mullen, D.G., Fang, M., Desai, A., Baker, J.R., Orr, B.G., and Banaszak Holl, M.M. A quantitative assessment of nanoparticle–ligand distributions: implications for targeted drug and imaging delivery in dendrimer conjugates. *ACS Nano* 4(2), 657–670 (2010).
30. Carlmark, A., Hawker, C., Hult, A., and Malkoch, M. New methodologies in the construction of dendritic materials. *Chemical Society Reviews* 38(2), 352–362 (2009).

31. Singh, P., Gupta, U., Asthana, A., and Jain, N.K. Folate and Folate–PEG–PAMAM Dendrimers: Synthesis, Characterization, and Targeted Anticancer Drug Delivery Potential in Tumor Bearing Mice. *Bioconjugate Chemistry* 19(11), 2239–2252 (2008).

32. Nasr, M., Najlah, M., D'Emanuele, A., and Elhissi, A. PAMAM dendrimers as aerosol drug nanocarriers for pulmonary delivery via nebulization. *Int. J. Pharm.* 461(1–2), 242–50 (2014).

33. Yang, W., Barth, R.F., Wu, G., Huo, T., Tjarks, W., Ciesielski, M., Fenstermaker, R.A., Ross, B.D., Wikstrand, C.J., Riley, K.J., and Binns, P.J. Convection enhanced delivery of boronated EGF as a molecular targeting agent for neutron capture therapy of brain tumors. *J. Neuro-Oncol.* 95(3), 355–365 (2009).

34. Zhu, S., Hong, M., Tang, G., Qian, L., Lin, J., Jiang, Y., and Pei, Y. Partly PEGylated polyamidoamine dendrimer for tumor-selective targeting of doxorubicin: The effects of PEGylation degree and drug conjugation style. *Biomaterials* 31(6), 1360–1371 (2010).

35. Zhu, S., Hong, M., Zhang, L., Tang, G., Jiang, Y., and Pei, Y. PEGylated PAMAM dendrimer-doxorubicin conjugates: *in vitro* evaluation and *in vivo* tumor accumulation. *Pharm. Res.* 27(1), 161–174 (2010).

36. Daftarian, P., Kaifer, A.E., Li, W., Blomberg, B.B., Frasca, D., Roth, F., Chowdhury, R., Berg, E.A., Fishman, J.B., Al Sayegh, H.A., Blackwelder, P., Inverardi, L., Perez, V.L., Lemmon, V., and Serafini, P. Peptide-conjugated PAMAM dendrimer as a universal DNA vaccine platform to target antigen-presenting cells. *Cancer Research* 2011.

37. Modi, D.A., Sunoqrot, S., Bugno, J., Lantvit, D.D., Hong, S., and Burdette, J.E. Targeting of follicle stimulating hormone peptide-conjugated dendrimers to ovarian cancer cells. *Nanoscale* 6(5), 2812–2820 (2014).

38. Mishra, M.K., Beaty, C.A., Lesniak, W.G., Kambhampati, S.P., Zhang, F., Wilson, M.A., Blue, M.E., Troncoso, J.C., Kannan, S., Johnston, M. V., Baumgartner, W.A., and Kannan, R.M. Dendrimer brain uptake and targeted therapy for brain injury in a large animal model of hypothermic circulatory arrest. *ACS Nano* 8(3), 2134–2147 (2014).

39. Kannan, S., Dai, H., Navath, R.S., Balakrishnan, B., Jyoti, A., Janisse, J., Romero, R., and Kannan, R.M. Dendrimer-based postnatal therapy for neuroinflammation and cerebral palsy in a rabbit model. *Sci. Transl. Med.* 4(130) (2012).

40. Prieto, M.J., del Rio Zabala, N.E., Marotta, C.H., Carreño Gutierrez, H., Arévalo Arévalo, R., Chiaramoni, N.S., and Alonso, S.D.V. Optimization and *in vivo* toxicity evaluation of G4.5 PAMAM dendrimer-risperidone complexes. *PLoS One* 9(2), e90393 (2014).

41. Liu, C., Liu, X., Rocchi, P., Qu, F., Iovanna, J.L., and Peng, L. Arginine-terminated generation 4 PAMAM dendrimer as an effective nanovector for functional siRNA delivery *in vitro* and *in vivo*. *Bioconjugate Chemistry* 25(3), 521–532 (2014).

42. Liu, H., Wang, Y., Wang, M., Xiao, J., and Cheng, Y. Fluorinated poly(propylenimine) dendrimers as gene vectors. *Biomaterials* 35(20), 5407–13 (2014).

43. Arima, H., Yoshimatsu, A., Ikeda, H., Ohyama, A., Motoyama, K., Higashi, T., Tsuchiya, A., Niidome, T., Katayama, Y., Hattori, K., and Takeuchi, T. Folate-PEG-appended dendrimer conjugate with alpha-cyclodextrin as a novel cancer cell-selective siRNA delivery carrier. *Mol. Pharm.* 9(9), 2591–2604 (2012).

44. Patil, M.L., Zhang, M., Betigeri, S., Taratula, O., He, H., and Minko, T. Surface-modified and internally cationic polyamidoamine dendrimers for efficient siRNA delivery. *Bioconjugate Chemistry* 19(7), 1396–1403 (2008).

45. Kang, H.M., DeLong, R., Fisher, M.H., and Juliano, R.L. Tat-conjugated PAMAM dendrimers as delivery agents for antisense and siRNA oligonucleotides. *Pharmaceutical Research* 22(12), 2099–2106 (2005).

46. Yu, T.Z., Liu, X.X., Bolcato-Bellemin, A.L., Wang, Y., Liu, C., Erbacher, P., Qu, F.Q., Rocchi, P., Behr, J.P., and Peng, L. An amphiphilic dendrimer for effective delivery of small interfering RNA and gene silencing *in vitro* and *in vivo*. *Angew Chem. Int. Ed.* 51(34), 8478–8484 (2012).

47. Hayder, M., Poupot, M., Baron, M., Nigon, D., Turrin, C.-O., Caminade, A.-M., Majoral, J.-P., Eisenberg, R.A., Fournic, J.-J., Cantagrel, A., Poupot, R., and Davignon, J.-L. A phosphorus-based dendrimer targets inflammation and osteoclastogenesis in experimental arthritis. *Sci. Transl. Med.* 3(81), 81ra35 (2011).

48. Ziemba, B., Franiak-Pietryga, I., Pion, M., Appelhans, D., Munoz-Fernandez, M.A., Voit, B., Bryszewska, M., and Klajnert-Maculewicz, B. Toxicity and proapoptotic activity of poly(propylene imine) glycodendrimers in vitro: considering their contrary potential as biocompatible entity and drug molecule in cancer. *Int. J. Pharm.* 461(1–2), 391–402 (2014).

49. Kesharwani, P., Tekade, R.K., and Jain, N.K. Generation dependent cancer targeting potential of poly(propyleneimine) dendrimer. *Biomaterials* 35(21), 5539–5548 (2014).

50. Taratula, O., Garbuzenko, O.B., Kirkpatrick, P., Pandya, I., Savla, R., Pozharov, V.P., He, H., and Minko, T. Surface-engineered targeted PPI dendrimer for efficient intracellular and intratumoral siRNA delivery. *Journal of Controlled Release: Official Journal of the Controlled Release Society* 140(3), 284–293 (2009).

51. Fox, M.E., Guillaudeu, S., Frechet, J.M. J., Jerger, K., Macaraeg, N., and Szoka, F.C. Synthesis and *in vivo* antitumor efficacy of pegylated poly(L-lysine) dendrimer-camptothecin conjugates. *Molecular Pharmaceutics* 6(5), 1562–1572 (2009).

52. Kaminskas, L.M., Kota, J., McLeod, V.M., Kelly, B.D., Karellas, P., and Porter, C.J.H. PEGylation of polylysine dendrimers improves absorption and lymphatic targeting following SC administration in rats. *J. Control Release* 140(2), 108–116 (2009).

53. Ohsaki, M., Okuda, T., Wada, A., Hirayama, T., Niidome, T., and Aoyagi, H. *In vitro* gene transfection using dendritic poly(L-lysine). *Bioconjugate Chemistry* 13(3), 510–517 (2002).

54. Yamagata, M., Kawano, T., Shiba, K., Mori, T., Katayama, Y., and Niidome, T. Structural advantage of dendritic poly(L-lysine) for gene delivery into cells. *Bioorganic & Medicinal Chemistry* 15(1), 526–532 (2007).

55. Choi, J.S., Joo, D.K., Kim, C.H., Kim, K., and Park, J.S. Synthesis of a barbell-like triblock copolymer, poly(L-lysine) dendrimer-block-poly(ethylene glycol)-block-poly(L-lysine) dendrimer, and its self-assembly with plasmid DNA. *Journal of the American Chemical Society* 122(3), 474–480 (2000).

56. Kaminskas, L.M., McLeod, V.M., Kelly, B.D., Cullinane, C., Sberna, G., Williamson, M., Boyd, B.J., Owen, D.J., and Porter, C.J.H. Doxorubicin-conjugated PEGylated

dendrimers show similar tumoricidal activity but lower systemic toxicity when compared to PEGylated liposome and solution formulations in mouse and rat tumor models. *Mol. Pharm.* 9(3), 422–432 (2012).

57. Kaminskas, L.M., Kelly, B.D., McLeod, V.M., Sberna, G., Owen, D.J., Boyd, B.J., and Porter, C.J.H. Characterisation and tumour targeting of PEGylated polylysine dendrimers bearing doxorubicin via a pH labile linker. *Journal of Controlled Release: Official Journal of the Controlled Release Society* 152(2), 241–248 (2011).

58. Kaminskas, L.M., Kelly, B.D., McLeod, V.M., Boyd, B.J., Krippner, G.Y., Williams, E.D., and Porter, C.J.H. Pharmacokinetics and tumor disposition of PEGylated, methotrexate conjugated poly-L-lysine dendrimers. *Molecular Pharmaceutics* 6(4), 1190–1204 (2009).

59. Inoue, Y., Kurihara, R., Tsuchida, A., Hasegawa, M., Nagashima, T., Mori, T., Niidome, T., Katayama, Y., and Okitsu, O. Efficient delivery of siRNA using dendritic poly(L-lysine) for loss-of-function analysis. *Journal of Controlled Release* 126(1), 59–66 (2008).

60. Cowie, J.M. *Polymers: Chemistry and Physics of Modern Materials*. CRC Press: 1991.

61. Svenson, S., Tomalia, D.A. Dendrimers in biomedical applications — reflections on the field. *Advanced Drug Delivery Reviews* 57(15), 2106–2129 (2005).

62. Astruc, D., Boisselier, E., and Ornelas, C. Dendrimers designed for functions: from physical, photophysical, and supramolecular properties to applications in sensing, catalysis, molecular electronics, photonics, and nanomedicine. *Chemical Reviews* 110(4), 1857–1959 (2010).

63. Koper, G.J.M., van Genderen, M.H.P., Elissen-Román, C., Baars, M.W.P.L., Meijer, E.W., and Borkovec, M. Protonation mechanism of poly(propylene imine) dendrimers and some associated oligo amines. *Journal of the American Chemical Society* 119(28), 6512–6521 (1997).

64. Dufes, C., Uchegbu, I.F., and Schatzlein, A.G. Dendrimers in gene delivery. *Advanced Drug Delivery Reviews* 57(15), 2177–2202 (2005).

65. Sonawane, N.D., Szoka, F.C., and Verkman, A.S. Chloride accumulation and swelling in endosomes enhances DNA transfer by polyamine-DNA polyplexes. *Journal of Biochemistry* 278(45), 44826–44831 (2003).

66. Shen, W.C., and Ryser, H.J. Conjugation of poly-L-lysine to albumin and horseradish peroxidase: a novel method of enhancing the cellular uptake of proteins. *Proceedings of the National Academy of Sciences of the United States of America* 75(4), 1872–1876 (1978).

67. Denkewalter, R.G., Kolc, J., and Lukasavage, W.J. Preparation of lysine based macromolecular highly branched homogeneous compound. In US4360646: 1982.

68. Ravina, M., Paolicelli, P., Seijo, B., and Sanchez, A. Knocking down gene expression with dendritic vectors. *Mini-Reviews in Medicinal Chemistry* 10(1), 73–86 (2010).

69. Sadler, K., and Tam, J.P. Peptide dendrimers: applications and synthesis. *Reviews in Molecular Biotechnology* 90(3–4), 195–229 (2002).

70. Gillies, E.R., and Frechet, J.M. Dendrimers and dendritic polymers in drug delivery. *Drug Discovery Today* 10(1), 35–43 (2005).

71. Malik, N., Wiwattanapatapee, R., Klopsch, R., Lorenz, K., Frey, H., Weener, J.W., Meijer, E.W., Paulus, W., and Duncan, R. Dendrimers: Relationship between structure

and biocompatibility in vitro, and preliminary studies on the biodistribution of 125I-labelled polyamidoamine dendrimers in vivo. *Journal of Controlled Release* 65(1–2), 133–148 (2000).

72. Duncan, R., and Izzo, L. Dendrimer biocompatibility and toxicity. *Advanced Drug Delivery Reviews* 57(15), 2215–2237 (2005).

73. Hong, S., Bielinska, A.U., Mecke, A., Keszler, B., Beals, J.L., Shi, X., Balogh, L., Orr, B.G., Baker, J.R., and Banaszak Holl, M.M. Interaction of poly(amidoamine) dendrimers with supported lipid bilayers and cells: hole formation and the relation to transport. *Bioconjugate Chemistry* 15(4), 774–782 (2004).

74. Hong, S., Leroueil, P.R., Janus, E.K., Peters, J.L., Kober, M.-M., Islam, M.T., Orr, B.G., Baker, J.R., and Banaszak Holl, M.M. Interaction of polycationic polymers with supported lipid bilayers and cells: nanoscale hole formation and enhanced membrane permeability. *Bioconjugate Chemistry* 17(3), 728–734 (2006).

75. Roberts, J.C., Bhalgat, M.K., and Zera, R.T. Preliminary biological evaluation of polyamidoamine (PAMAM) StarburstTM dendrimers. *Journal of Biomedical Materials Research* 30(1), 53–65 (1996).

76. Choi, J.S., Nam, K., Park, J., Kim, J.B., and Lee, J.K. Enhanced transfection efficiency of PAMAM dendrimer by surface modification with L-arginine. *Journal of Controlled Release* 99(3), 445–456 (2004).

77. Ihre, H.R. Padilla De Jesús, O.L., Szoka, F.C., and Fréchet, J.M.J., Polyester dendritic systems for drug delivery applications: design, synthesis, and characterization. *Bioconjugate Chemistry* 13(3), 443–452 (2002).

78. Ihre, H., Padilla De Jesús, O.L., and Fréchet, J.M.J. Fast and convenient divergent synthesis of aliphatic ester dendrimers by anhydride coupling. *Journal of the American Chemical Society* 123(25), 5908–5917 (2001).

79. Winnicka, K., Bielawski, K., Rusak, M., and Bielawska, A. The effect of generation 2 and 3 poly(amidoamine) dendrimers on viability of human breast cancer cells. *Journal of Health Science* 55(2), 169–177 (2009).

80. Jain, K., Kesharwani, P., Gupta, U., and Jain, N.K. Dendrimer toxicity: Let's meet the challenge. *International Journal of Pharmaceutics* 394(1–2), 122–142 (2010).

81. Kitchens, K., Foraker, A., Kolhatkar, R., Swaan, P., and Ghandehari, H. Endocytosis and interaction of poly (amidoamine) dendrimers with caco-2 cells. *Pharmaceutical Research* 24(11), 2138–2145 (2007).

82. Perumal, O.P., Inapagolla, R., Kannan, S., and Kannan, R.M. The effect of surface functionality on cellular trafficking of dendrimers. *Biomaterials* 29(24–25), 3469–3476 (2008).

83. Albertazzi, L., Serresi, M., Albanese, A., and Beltram, F. Dendrimer internalization and intracellular trafficking in living cells. *Molecular Pharmaceutics* 7(3), 680–688 (2010).

84. Hong, S., Rattan, R., Majoros, I.J., Mullen, D.G., Peters, J.L., Shi, X., Bielinska, A.U., Blanco, L., Orr, B.G., and Baker Jr, J.R. The role of ganglioside GM1 in cellular internalization mechanisms of poly (amidoamine) dendrimers. *Bioconjugate Chemistry* 20(8), 1503–1513 (2009).

85. Margerum, L.D., Campion, B.K., Koo, M., Shargill, N., Lai, J.-J., and Marumoto, A., Christian Sontum, P. Gadolinium(III) DO3A macrocycles and polyethylene glycol coupled to dendrimers effect of molecular weight on physical and biological properties

of macromolecular magnetic resonance imaging contrast agents. *Journal of Alloys and Compounds* 249(1–2), 185–190 (1997).

86. Kobayashi, H., Kawamoto, S., Saga, T., Sato, N., Hiraga, A., Konishi, J., Togashi, K., and Brechbiel, M.W. Micro-MR angiography of normal and intratumoral vessels in mice using dedicated intravascular MR contrast agents with high generation of polyamidoamine dendrimer core: Reference to pharmacokinetic properties of dendrimer-based MR contrast agents. *Journal of Magnetic Resonance Imaging* 14(6), 705–713 (2001).

87. Kobayashi, H., Sato, N., Hiraga, A., Saga, T., Nakamoto, Y., Ueda, H., Konishi, J., Togashi, K., and Brechbiel, M.W. 3D-micro-MR angiography of mice using macro-molecular MR contrast agents with polyamidoamine dendrimer core with reference to their pharmacokinetic properties. *Magnetic Resonance in Medicine* 45(3), 454–460 (2001).

88. Kaminskas, L.M., Boyd, B.J., Karellas, P., Henderson, S.A., Giannis, M.P., Krippner, G.Y., and Porter, C.J.H. Impact of surface derivatization of poly-l-lysine dendrimers with anionic arylsulfonate or succinate groups on intravenous pharmacokinetics and disposition. *Molecular Pharmaceutics* 4(6), 949–961 (2007).

89. Kaminskas, L.M., Boyd, B.J., and Porter, C.J.H. Dendrimer pharmacokinetics: the effect of size, structure and surface characteristics on ADME properties. *Nanomedicine* 6(6), 1063–1084 (2011).

90. Boyd, B.J., Kaminskas, L.M., Karellas, P., Krippner, G., Lessene, R., and Porter, C.J.H. Cationic poly-l-lysine dendrimers: pharmacokinetics, biodistribution, and evidence for metabolism and bioresorption after intravenous administration to rats. *Molecular Pharmaceutics* 3(5), 614–627 (2006).

91. Nigavekar, S., Sung, L., Llanes, M., El-Jawahri, A., Lawrence, T., Becker, C., Balogh, L., and Khan, M. 3H dendrimer nanoparticle organ/tumor distribution. *Pharmaceutical Research* 21(3), 476–483 (2004).

92. Owens Iii, D.E., and Peppas, N.A. Opsonization, biodistribution, and pharmacokinetics of polymeric nanoparticles. *International Journal of Pharmaceutics* 307(1), 93–102 (2006).

93. Kaminskas, L.M., Boyd, B.J., Karellas, P., Krippner, G.Y., Lessene, R., Kelly, B., and Porter, C.J.H. The impact of molecular weight and PEG chain length on the systemic pharmacokinetics of PEGylated poly L-lysine dendrimers. *Molecular Pharmaceutics* 5(3), 449–463 (2008).

94. Guillaudeu, S.J., Fox, M.E., Haidar, Y.M., Dy, E.E., Szoka, F.C., and Fréchet, J.M.J. PEGylated dendrimers with core functionality for biological applications. *Bioconjugate Chemistry* 19(2), 461–469 (2008).

95. Wu, W., Driessen, W., and Jiang, X. Oligo(ethylene glycol)-based thermosensitive dendrimers and their tumor accumulation and penetration. *Journal of the American Chemical Society* 136(8), 3145–3155 (2014).

96. Dhanikula, R.S., Argaw, A., Bouchard, J.-F., and Hildgen, P. Methotrexate loaded polyether-copolyester dendrimers for the treatment of gliomas: enhanced efficacy and intratumoral transport capability. *Molecular Pharmaceutics* 5(1), 105–116 (2008).

97. Waite, C.L., and Roth, C.M. PAMAM-RGD conjugates enhance siRNA delivery through a multicellular spheroid model of malignant glioma. *Bioconjugate Chemistry* 20(10), 1908–1916 (2009).

98. Al-Jamal, K.T., Al-Jamal, W.T., Wang, J.T.W., Rubio, N., Buddle, J., Gathercole, D., Zloh, M., and Kostarelos, K. Cationic poly-l-lysine dendrimer complexes doxorubicin and delays tumor growth *in vitro* and *in vivo*. *Acs Nano* 7(3), 1905–1917 (2013).

99. Sunoqrot, S., Liu, Y., Kim, D.-H., and Hong, S. *In vitro* evaluation of dendrimer–polymer hybrid nanoparticles on their controlled cellular targeting kinetics. *Molecular Pharmaceutics* 10(6), 2157–2166 (2013).

100. Yang, Y., Sunoqrot, S., Stowell, C., Ji, J., Lee, C.-W., Kim, J.W., Khan, S.A., and Hong, S. Effect of size, surface charge, and hydrophobicity of poly(amidoamine) dendrimers on their skin penetration. *Biomacromolecules* 13(7), 2154–2162 (2012).

101. Venuganti, V., Sahdev, P., Hildreth, M., Guan, X., and Perumal, O. Structure-skin permeability relationship of dendrimers. *Pharm. Res.* 28(9), 2246–2260 (2011).

102. Fire, A., Xu, S.Q., Montgomery, M.K., Kostas, S.A., Driver, S.E., and Mello, C.C. Potent and specific genetic interference by double-stranded RNA in Caenorhabditis elegans. *Nature* 391(6669), 806–811 (1998).

103. Pack, D.W., Hoffman, A.S., Pun, S., and Stayton, P.S. Design and development of polymers for gene delivery. *Nature Reviews Drug Discovery* 4(7), 581–593 (2005).

104. Roth, J.A., and Cristiano, R.J. Gene therapy for cancer: What have we done and where are we going? *Journal of the National Cancer Institute* 89(1), 21–39 (1997).

105. Luo, D., and Saltzman, W.M. Synthetic DNA delivery systems. *Nat. Biotech.* 18(1), 33–37 (2000).

106. Mulligan, R.C. The basic science of gene-therapy. *Science* 260(5110), 926–932 (1993).

107. Holkers, M., Maggio, I., Liu, J., Janssen, J.M., Miselli, F., Mussolino, C., Recchia, A., Cathomen, T., and Goncalves, M.A.F.V. Differential integrity of TALE nuclease genes following adenoviral and lentiviral vector gene transfer into human cells. *Nucleic Acids Research* 41(5), (2013).

108. Schoggins, J.W., MacDuff, D.A., Imanaka, N., Gainey, M.D., Shrestha, B., Eitson, J.L., Mar, K.B., Richardson, R.B., Ratushny, A.V., Litvak, V., Dabelic, R., Manicassamy, B., Aitchison, J.D., Aderem, A., Elliott, R. M., Garcia-Sastre, A., Racaniello, V., Snijder, E.J., Yokoyama, W.M., Diamond, M.S., Virgin, H.W., and Rice, C.M. Pan-viral specificity of IFN-induced genes reveals new roles for cGAS in innate immunity. *Nature* 505(7485), 691−+ (2014).

109. Wilson, J.M. A history lesson for stem cells. *Science* 324(5928), 727–728 (2009).

110. Bainbridge, J.W. B., Smith, A.J., Barker, S.S., Robbie, S., Henderson, R., Balaggan, K., Viswanathan, A., Holder, G.E., Stockman, A., Tyler, N., Petersen-Jones, S., Bhattacharya, S.S., Thrasher, A.J., Fitzke, F.W., Carter, B.J., Rubin, G.S., Moore, A.T., and Ali, R.R. Effect of gene therapy on visual function in Leber's congenital amaurosis. *New England Journal of Medicine* 358(21), 2231–2239 (2008).

111. Hauswirth, W.W., Aleman, T.S., Kaushal, S., Cideciyan, A.V., Schwartz, S.B., Wang, L.L., Conlon, T.J., Boye, S.L., Flotte, T.R., Byrne, B.J., and Jacobson, S.G. Treatment of leber congenital amaurosis due to RPE65 mutations by ocular subretinal injection of adeno-associated virus gene vector: short-term results of a phase I trial. *Human Gene Therapy* 19(10), 979–990 (2008).

112. Maguire, A.M., Simonelli, F., Pierce, E.A., Pugh, E.N., Mingozzi, F., Bennicelli, J., Banfi, S., Marshall, K.A., Testa, F., Surace, E.M., Rossi, S., Lyubarsky, A., Arruda, V.R., Konkle, B., Stone, E., Sun, J., Jacobs, J., Dell'Osso, L., Hertle, R., Ma, J.-x.,

Redmond, T.M., Zhu, X., Hauck, B., Zelenaia, O., Shindler, K.S., Maguire, M.G., Wright, J.F., Volpe, N.J., McDonnell, J.W., Auricchio, A., High, K.A., and Bennett, J. Safety and efficacy of gene transfer for Leber's congenital amaurosis. *New England Journal of Medicine* 358(21), 2240–2248 (2008).

113. Aiuti, A., Cattaneo, F., Galimberti, S., Benninghoff, U., Cassani, B., Callegaro, L., Scaramuzza, S., Andolfi, G., Mirolo, M., Brigida, I., Tabucchi, A., Carlucci, F., Eibl, M., Aker, M., Slavin, S., Al-Mousa, H., Al Ghonaium, A., Ferster, A., Duppenthaler, A., Notarangelo, L., Wintergerst, U., Buckley, R.H., Bregni, M., Marktel, S., Valsecchi, M.G., Rossi, P., Ciceri, F., Miniero, R., Bordignon, C., and Roncarolo, M. Gene therapy for immunodeficiency due to adenosine deaminase deficiency. *New England Journal of Medicine* 360(5), 447–458 (2009).

114. Thompson, L. Human gene therapy. Harsh lessons, high hopes. *FDA Consumer* 34(5), 19 (2000).

115. Kay, M.A., Glorioso, J.C., and Naldini, L. Viral vectors for gene therapy: the art of turning infectious agents into vehicles of therapeutics. *Nature Medicine* 7(1), 33–40 (2001).

116. Fichter, K.M., Zhang, L., Kiick, K.L., and Reineke, T.M. Peptide-functionalized poly(ethylene glycol) star polymers: DNA delivery vehicles with multivalent molecular architecture. *Bioconjugate Chemistry* 19(1), 76–88 (2008).

117. Felgner, P.L., Gadek, T.R., Holm, M., Roman, R., Chan, H.W., Wenz, M., Northrop, J.P., Ringold, G.M., and Danielsen, M. Lipofection: a highly efficient, lipid-mediated DNA-transfection procedure. *Proc. Natl. Acad. Sci. USA* 84(21), 7413–7 (1987).

118. Kataoka, K., Harada, A., and Nagasaki, Y. Block copolymer micelles for drug delivery: design, characterization and biological significance. *Advanced Drug Delivery Reviews* 47(1), 113–131 (2001).

119. Kakizawa, Y., Furukawa, S., and Kataoka, K. Block copolymer-coated calcium phosphate nanoparticles sensing intracellular environment for oligodeoxynucleotide and siRNA delivery. *J. Control. Release* 97(2), 345–56 (2004).

120. Pankhurst, Q.A., Thanh, N.T.K., Jones, S.K., and Dobson, J. Progress in applications of magnetic nanoparticles in biomedicine. *Journal of Physics D-Applied Physics* 42(22), (2009).

121. Duncan, R. The dawning era of polymer therapeutics. *Nature Reviews Drug Discovery* 2(5), 347–360 (2003).

122. Langer, R., and Tirrell, D.A. Designing materials for biology and medicine. *Nature* 428(6982), 487–492 (2004).

123. Shogren-Knaak, M., Ishii, H., Sun, J.-M., Pazin, M.J., Davie, J.R., and Peterson, C.L. Histone H4-K16 acetylation controls chromatin structure and protein interactions. *Science* 311(5762), 844–847 (2006).

124. Haensler, J., and Szoka Jr, F.C. Polyamidoamine cascade polymers mediate efficient transfection of cells in culture. *Bioconjugate Chemistry* 4(5), 372–379 (1993).

125. Zhang, Y., Wang, Z., and Gemeinhart, R.A. Progress in microRNA delivery. *J Controlled Release* 172 962–974 (2013).

126. Liu, X., Liu, C., Catapano, C.V., Peng, L., Zhou, J., and Rocchi, P. Structurally flexible triethanolamine-core poly(amidoamine) dendrimers as effective nanovectors to delivery RNAi-based thereapeutics. *Biotechnology Advances* 32, 844–852 (2014).

127. Eichman, J.D., Bielinska, A.U., Kukowska-Latallo, J.F., and Baker Jr, J.R. The use of PAMAM dendrimers in the efficient transfer of genetic material into cells. *Pharmaceutical Science & Technology Today* 3(7), 232–245 (2000).

128. Juliano, R., Alam, M.R., Dixit, V., and Kang, H. Mechanisms and strategies for effective delivery of antisense and siRNA oligonucleotides. *Nucleic Acids Research* 36(12), 4158–4171 (2008).

129. Zhou, J., Wu, J., Hafdi, N., Behr, J.-P., Erbacher, P., and Peng, L. PAMAM dendrimers for efficient siRNA delivery and potent gene silencing. *Chemical Communications* (22), 2362–2364 (2006).

130. Wang, R., Zhou, L., Zhou, Y., Li, G., Zhu, X., Gu, H., Jiang, X., Li, H., Wu, J., He, L., Guo, X., Zhu, B., and Yan, D. Synthesis and gene delivery of poly(amido amine)s with different branched architecture. *Biomacromolecules* 11(2), 489–495 (2010).

131. Shcharbin, D., Janaszewska, A., Klajnert-Maculewicz, B., Ziemba, B., Dzmitruk, V., Halets, I., Loznikova, S., Shcharbina, N., Milowska, K., Ionov, M., Shakhbazau, A., and Bryszewska, M. How to study dendrimers and dendriplexes III. Biodistribution, pharmacokinetics and toxicity in vivo. *J Controlled Release* 181, 40–52 (2014).

132. Kala, S., Mak, A.S.C., Liu, X., Posocco, P., Pricl, S., Peng, L., and Wong, A.S.T. Combination of dendrimer-nanovector-mediated small interfering RNA delivery to target akt with the clinical anticancer drug paclitaxel for effective and potent anticancer activity in treating ovarian cancer. *Journal of Medicinal Chemistry* 57(6), 2634–2642 (2014).

133. Karatasos, K., Posocco, P., Laurini, E., and Pricl, S. Poly(amidoamine)-based dendrimer/siRNA complexation studied by computer simulations: effects of pH and generation on dendrimer structure and siRNA binding. *Macromolecular Bioscience* 12(2), 225–240 (2012).

134. Posocco, P., Laurini, E., Dal Col, V., Marson, D., Karatasos, K., Fermeglia, M., and Pricl, S. Tell Me Something I Do Not Know. Multiscale molecular modeling of dendrimer/dendron organization and self-assembly in gene therapy. *Current Medicinal Chemistry* 19(29), 5062–5087 (2012).

135. Tian, W.-D., and Ma, Y.-Q. Theoretical and computational studies of dendrimers as delivery vectors. *Chemical Society Reviews* 42(2), 705–727 (2013).

136. Wong, P.T., Tang, K., Coulter, A., Tang, S., Baker, J.R., and Choi, S. K. Multivalent dendrimer vectors with DNA intercalation motifs for gene delivery. *Biomacromolecules* 15(11), 4134–4145 (2014).

137. Chang, H., Wang, H., Shao, N., Wang, M., Wang, X., and Cheng, Y. Surface-engineered dendrimers with a diaminododecane core achieve efficient gene transfection and low cytotoxicity. *Bioconjugate Chemistry* 25(2), 342–350 (2014).

138. Elbashir, S.M., Harborth, J., Lendeckel, W., Yalcin, A., Weber, K., and Tuschl, T. Duplexes of 21-nucleotide RNAs mediate RNA interference in cultured mammalian cells. *Nature* 411(6836), 494–498 (2001).

139. Soutschek, J., Akinc, A., Bramlage, B., Charisse, K., Constien, R., Donoghue, M., Elbashir, S., Geick, A., Hadwiger, P., Harborth, J., John, M., Kesavan, V., Lavine, G., Pandey, R.K., Racie, T., Rajeev, K.G., Rohl, I., Toudjarska, I., Wang, G., Wuschko, S., Bumcrot, D., Koteliansky, V., Limmer, S., Manoharan, M., and Vornlocher, H.P. Therapeutic silencing of an endogenous gene by systemic administration of modified siRNAs. *Nature* 432(7014), 173–178 (2004).

140. Reynolds, A., Leake, D., Boese, Q., Scaringe, S., Marshall, W.S., and Khvorova, A. Rational siRNA design for RNA interference. *Nat. Biotechnol.* 22(3), 326–330 (2004).

141. Jacque, J.M., Triques, K., and Stevenson, M. Modulation of HIV-1 replication by RNA interference. *Nature* 418(6896), 435–438 (2002).

142. Lee, N.S., Dohjima, T., Bauer, G., Li, H.T., Li, M.J., Ehsani, A., Salvaterra, P., and Rossi, J. Expression of small interfering RNAs targeted against HIV-1 rev transcripts in human cells. *Nat. Biotechnol.* 20(5), 500–505 (2002).

143. Novina, C.D., Murray, M.F., Dykxhoorn, D.M., Beresford, P.J., Riess, J., Lee, S.K., Collman, R.G., Lieberman, J., Shankar, P., and Sharp, P.A. siRNA-directed inhibition of HIV-1 infection. *Nature Medicine* 8(7), 681–686 (2002).

144. White, T.C. Increased mRNA levels of ERG16, CDR, and MDR1 correlate with increases in azole resistance in Candida albicans isolates from a patient infected with human immunodeficiency virus. *Antimicrobial Agents and Chemotherapy* 41(7), 1482–1487 (1997).

145. Miller, V.M., Gouvion, C.M., Davidson, B.L., and Paulson, H.L. Targeting Alzheimer's disease genes with RNA interference: an efficient strategy for silencing mutant alleles. *Nucleic Acids Research* 32(2), 661–668 (2004).

146. Singer, O., Marr, R.A., Rockenstein, E., Crews, L., Coufal, N.G., Gage, F.H., Verma, I.M., and Masliah, E. Targeting BACE1 with siRNAs ameliorates Alzheimer disease neuropathology in a transgenic model. *Nature Neuroscience* 8(10), 1343–1349 (2005).

147. Alvarez-Erviti, L., Seow, Y., Yin, H., Betts, C., Lakhal, S., and Wood, M.J.A. Delivery of siRNA to the mouse brain by systemic injection of targeted exosomes. *Nat. Biotechnol.* 29(4), 341-U179 (2011).

148. Nordberg, A. Nicotinic receptor abnormalities of Alzheimer's disease: Therapeutic implications. *Biological Psychiatry* 49(3), 200–210 (2001).

149. Dorsett, Y., and Tuschl, T. siRNAs: Applications in functional genomics and potential as therapeutics. *Nature Reviews Drug Discovery* 3(4), 318–329 (2004).

150. McCaffrey, A.P., Nakai, H., Pandey, K., Huang, Z., Salazar, F.H., Xu, H., Wieland, S.F., Marion, P.L., and Kay, M.A. Inhibition of hepatitis B virus in mice by RNA interference. *Nat. Biotechnol.* 21(6), 639–644 (2003).

151. Morrissey, D.V., Lockridge, J.A., Shaw, L., Blanchard, K., Jensen, K., Breen, W., Hartsough, K., Machemer, L., Radka, S., Jadhav, V., Vaish, N., Zinnen, S., Vargeese, C., Bowman, K., Shaffer, C.S., Jeffs, L.B., Judge, A., MacLachlan, I., and Polisky, B. Potent and persistent in vivo anti-HBV activity of chemically modified siRNAs. *Nat. Biotechnol.* 23(8), 1002–1007 (2005).

152. Cavanaugh, V.J., Guidotti, L.G., and Chisari, F.V. Interleukin-12 inhibits hepatitis B virus replication in transgenic mice. *Journal of Virology* 71(4), 3236–3243 (1997).

153. Hamasaki, K., Nakao, K., Matsumoto, K., Ichikawa, T., Ishikawa, H., and Eguchi, K. Short interfering RNA-directed inhibition of hepatitis B virus replication. *Febs Letters* 543(1–3), 51–54 (2003).

154. Behlke, M.A. Progress towards in vivo use of siRNAs. *Molecular Therapy* 13(4), 644-670 (2006).

155. Davis, M.E. The first targeted delivery of siRNA in humans via a self-assembling, cyclodextrin polymer-based nanoparticle: from concept to clinic. *Molecular Pharmaceutics* 6(3), 659–668 (2009).

156. Chen, C.D., Welsbie, D.S., Tran, C., Baek, S.H., Chen, R., Vessella, R., Rosenfeld, M.G., and Sawyers, C.L. Molecular determinants of resistance to antiandrogen therapy. *Nature Medicine* 10(1), 33–39 (2004).
157. Davis, M.E., Zuckerman, J.E., Choi, C.H. J., Seligson, D., Tolcher, A., Alabi, C.A., Yen, Y., Heidel, J.D., and Ribas, A. Evidence of RNAi in humans from systemically administered siRNA via targeted nanoparticles. *Nature* 464(7291), 1067–U140 (2010).
158. Khan, O.F., Zaia, E.W., Yin, H., Bogorad, R.L., Pelet, J.M., Webber, M.J., Zhuang, I., Dahlman, J.E., Langer, R., and Anderson, D.G. Ionizable amphiphilic dendrimer-based nanomaterials with alkyl-chain-substituted amines for tunable siRNA delivery to the liver endothelium *in vivo*. *Angewandte Chemie* 2014, n/a-n/a.
159. Xu, X., Jian, Y., Li, Y., Zhang, X., Tu, Z., and Gu, Z. Bio-inspired supramolecular hybrid dendrimers self-assembled from low-generation peptide dendrons for highly efficient gene delivery and biological tracking. *Acs Nano* 8(9), 9255–9264 (2014).
160. Caminade, A.-M., and Turrin, C.-O. Dendrimers for drug delivery. *Journal of Materials Chemistry B* 2(26), 4055–4066 (2014).
161. Fang, M., Zhang, J., Wu, Q., Xu, T., and Cheng, Y. Host–Guest Chemistry of Dendrimer–Drug Complexes: 7. Formation of stable inclusions between acetylated dendrimers and drugs bearing multiple charges. *The Journal of Physical Chemistry B* 116(10), 3075–3082 (2012).
162. Zhang, M., Guo, R., Kéri, M., Bányai, I., Zheng, Y., Cao, M., Cao, X., and Shi, X. Impact of dendrimer surface functional groups on the release of doxorubicin from dendrimer carriers. *The Journal of Physical Chemistry B* 118(6), 1696–1706 (2014).
163. Satsangi, A., Roy, S.S., Satsangi, R.K., Vadlamudi, R.K., and Ong, J.L. Design of a Paclitaxel prodrug conjugate for active targeting of an enzyme upregulated in breast cancer cells. *Molecular Pharmaceutics* 11(6), 1906–18 (2014).
164. Maeda, H., Wu, J., Sawa, T., Matsumura, Y., and Hori, K. Tumor vascular permeability and the EPR effect in macromolecular therapeutics: a review. *J. Controlled Release* 65(1–2), 271–284 (2000).
165. Singh, S.K., Hawkins, C., Clarke, I.D., Squire, J.A., Bayani, J., Hide, T., Henkelman, R.M., Cusimano, M.D., and Dirks, P.B. Identification of human brain tumour initiating cells. *Nature* 432(7015), 396–401 (2004).
166. Mecke, A., Lee, I., Jr, J.R.B., Holl, M.M.B., and Orr, B.G. Deformability of poly(amidoamine) dendrimers. *Eur. Phys. J. E* 14(1), 7–16 (2004).
167. Agarwal, A., Saraf, S., Asthana, A., Gupta, U., Gajbhiye, V., and Jain, N. K. Ligand based dendritic systems for tumor targeting. *International Journal of Pharmaceutics* 350(1–2), 3–13 (2008).
168. Myung, J.H., Gajjar, K.A., Chen, J., Molokie, R.E., and Hong, S. Differential detection of tumor cells using a combination of cell rolling, multivalent binding, and multiple antibodies. *Analytical Chemistry* (2014).
169. Myung, J.H., Gajjar, K.A., Saric, J., Eddington, D.T., and Hong, S. Dendrimer-mediated multivalent binding for the enhanced capture of tumor cells. *Angewandte Chemie International Edition* 50(49), 11769–11772 (2011).
170. Quintana, A., Raczka, E., Piehler, L., Lee, I., Myc, A., Majoros, I., Patri, A., Thomas, T., Mulé, J., and Baker, J., Jr. Design and function of a dendrimer-based therapeutic nanodevice targeted to tumor cells through the folate receptor. *Pharm. Res.* 19(9), 1310–1316 (2002).

171. Singh, P., Gupta, U., Asthana, A., and Jain, N.K. Folate and Folate-PEG-PAMAM Dendrimers: synthesis, characterization, and targeted anticancer drug delivery potential in tumor bearing mice. *Bioconjugate Chemistry* 19(11), 2239–2252 (2008).

172. Hong, S., Leroueil, P.R., Majoros, I.J., Orr, B.G., Baker Jr, J.R., and Banaszak Holl, M.M. The binding avidity of a nanoparticle-based multivalent targeted drug delivery platform. *Chemistry & Biology* 14(1), 107–115 (2007).

173. Kukowska-Latallo, J.F., Candido, K.A., Cao, Z.Y., Nigavekar, S.S., Majoros, I.J., Thomas, T.P., Balogh, L.P., Khan, M.K., and Baker, J.R. Nanoparticle targeting of anticancer drug improves therapeutic response in animal model of human epithelial cancer. *Cancer Res.* 65(12), 5317–5324 (2005).

174. van Dongen, M.A., Silpe, J.E., Dougherty, C.A., Kanduluru, A.K., Choi, S.K., Orr, B.G., Low, P.S., and Banaszak Holl, M.M. Avidity mechanism of dendrimer–folic acid conjugates. *Molecular Pharmaceutics* 11(5), 1696–1706 (2014).

175. Licata, N.A., and Tkachenko, A.V. Kinetic limitations of cooperativity-based drug delivery systems. *Physical Review Letters* 100(15), 158102 (2008).

176. Patri, A.K., Myc, A., Beals, J., Thomas, T.P., Bander, N.H., and Baker, J.R. Synthesis and in vitro testing of J591 antibody-dendrimer conjugates for targeted prostate cancer therapy. *Bioconjugate Chemistry* 15(6), 1174–1181 (2004).

177. Benito, J.M., Gomez-Garcia, M., Mellet, C.O., Baussanne, I., Defaye, J., and Fernandez, J.M.G. Optimizing saccharide-directed molecular delivery to biological receptors: Design, synthesis, and biological evaluation of glycodendrimer — Cyclodextrin conjugates. *Journal of the American Chemical Society* 126(33), 10355–10363 (2004).

178. Lagnoux, D., Darbre, T., Schmitz, M.L., and Reymond, J.L. Inhibition of mitosis by glycopeptide dendrimer conjugates of colchicine. *Chemistry-a European Journal* 11(13), 3941–3950 (2005).

179. Ciolkowski, M., Pałecz, B., Appelhans, D., Voit, B., Klajnert, B., and Bryszewska, M. The influence of maltose modified poly(propylene imine) dendrimers on hen egg white lysozyme structure and thermal stability. *Colloids and Surfaces B: Biointerfaces* 95(0), 103–108 (2012).

180. He, H., Li, Y., Jia, X.R., Du, J., Ying, X., Lu, W.L., Lou, J.N., and Wei, Y. PEGylated poly(amidoamine) dendrimer-based dual-targeting carrier for treating brain tumors. *Biomaterials* 32(2), 478–487 (2011).

181. Li, Y., He, H., Jia, X.R., Lu, W.L., Lou, J.N., and Wei, Y. A dual-targeting nanocarrier based on poly(amidoamine) dendrimers conjugated with transferrin and tamoxifen for treating brain gliomas. *Biomaterials* 33(15), 3899–3908 (2012).

182. Gajbhiye, V., Ganesh, N., Barve, J., and Jain, N.K. Synthesis, characterization and targeting potential of zidovudine loaded sialic acid conjugated-mannosylated poly(propyleneimine) dendrimers. *European Journal of Pharmaceutical Sciences* 48(4–5), 668–679 (2013).

183. Rosen, B.M., Wilson, C.J., Wilson, D.A., Peterca, M., Imam, M.R., and Percec, V. Dendron-mediated self-assembly, disassembly, and self-organization of complex systems. *Chemical Reviews* 109(11), 6275–6540 (2009).

184. Xiao, K., Luo, J., Fowler, W.L., Li, Y., Lee, J.S., Xing, L., Cheng, R.H., Wang, L., and Lam, K.S. A self-assembling nanoparticle for paclitaxel delivery in ovarian cancer. *Biomaterials* 30(30), 6006–6016 (2009).

185. Bae, J.W., Pearson, R.M., Patra, N., Sunoqrot, S., Vukovic, L., Kral, P., and Hong, S. Dendron-mediated self-assembly of highly PEGylated block copolymers: a modular nanocarrier platform. *Chemical Communications* 47(37), 10302–10304 (2011).

186. Pearson, R.M., Patra, N., Hsu, H.-j., Uddin, S., Král, P., and Hong, S. Positively charged dendron micelles display negligible cellular interactions. *ACS Macro Letters* 2(1), 77–81 (2012).

187. Israelachvili, J.N., Mitchell, D.J., and Ninham, B.W. Theory of self-assembly of hydrocarbon amphiphiles into micelles and bilayers. *Journal of the Chemical Society, Faraday Transactions 2: Molecular and Chemical Physics* 72, 1525–1568 (1976).

188. Torchilin, V.P. Structure and design of polymeric surfactant-based drug delivery systems. *Journal of Controlled Release: Official Journal of the Controlled Release Society* 73(2–3), 137–72 (2001).

189. Tanford, C. The hydrophobic effect and the organization of living matter. *Science* 200(4345), 1012–1018 (1978).

190. Hsu, H.-J., Sen, S., Pearson, R.M., Uddin, S., Král, P., and Hong, S. Poly(ethylene glycol) corona chain length controls end-group-dependent cell interactions of dendron micelles. *Macromolecules* (2014).

191. Qiao, H., Li, J., Wang, Y., Ping, Q., Wang, G., and Gu, X. Synthesis and characterization of multi-functional linear-dendritic block copolymer for intracellular delivery of antitumor drugs. *International Journal of Pharmaceutics* 452(1–2), 363–373 (2013).

192. Sunoqrot, S., Bae, J.W., Pearson, R.M., Shyu, K., Liu, Y., Kim, D.-H., and Hong, S. Temporal control over cellular targeting through hybridization of folate-targeted dendrimers and PEG-PLA nanoparticles. *Biomacromolecules* 13(4), 1223–1230 (2012).

193. Sunoqrot, S., Bugno, J., Lantvit, D., Burdette, J.E., and Hong, S. Prolonged blood circulation and enhanced tumor accumulation of folate-targeted dendrimer-polymer hybrid nanoparticles. *Journal of Controlled Release* (0) (2014).

194. Haag, R., and Kratz, F., Polymer therapeutics: Concepts and applications. *Angewandte Chemie International Edition* 45(8), 1198–1215 (2006).

195. Janib, S.M., Moses, A.S., and MacKay, J.A. Imaging and drug delivery using theranostic nanoparticles. *Advanced Drug Delivery Reviews* 62(11), 1052–1063 (2010).

196. van Dongen, M.A., Dougherty, C.A., and Holl, M.M.B. Multivalent polymers for drug delivery and imaging: the challenges of conjugation. *Biomacromolecules* 15(9), 3215–3234 (2014).

197. L. Villaraza, A.J., Bumb, A., and Brechbiel, M.W. Macromolecules, dendrimers, and nanomaterials in magnetic resonance imaging: The interplay between size, function, and pharmacokinetics. *Chemical Reviews* 110(5), 2921–2959 (2010).

198. Fischer, M., and Vögtle, F. Dendrimers: from design to application — a progress report. *Angewandte Chemie International Edition* 38(7), 884–905 (1999).

199. Venditto, V.J., Regino, C.A.S., and Brechbiel, M.W. PAMAM dendrimer based macro-molecules as improved contrast agents. *Molecular Pharmaceutics* 2(4), 302–311 (2005).

200. Langereis, S., Dirksen, A., Hackeng, T.M., van Genderen, M.H., and Meijer, E. Dendrimers and magnetic resonance imaging. *New Journal of Chemistry* 31(7), 1152–1160 (2007).

201. Wiener, E., Brechbiel, M., Brothers, H., Magin, R., Gansow, O., Tomalia, D., and Lauterbur, P. Dendrimer-based metal chelates: A new class of magnetic resonance imaging contrast agents. *Magnetic Resonance in Medicine* 31(1), 1–8 (1994).

202. Kaneshiro, T.L., Jeong, E.-K., Morrell, G., Parker, D.L., and Lu, Z.-R. Synthesis and evaluation of globular Gd-DOTA-monoamide conjugates with precisely controlled nanosizes for magnetic resonance angiography. *Biomacromolecules* 9(10), 2742–2748 (2008).

203. Cyran, C.C., Fu, Y., Raatschen, H.-J., Rogut, V., Chaopathomkul, B., Shames, D.M., Wendland, M.F., Yeh, B.M., and Brasch, R.C. New macromolecular polymeric MRI contrast agents for application in the differentiation of cancer from benign soft tissues. *Journal of Magnetic Resonance Imaging* 27(3), 581–589 (2008).

204. Kim, Y., and Kim, J. Modification of indium tin oxide with dendrimer-encapsulated nanoparticles to provide enhanced stable electrochemiluminescence of Ru (bpy) 32+/tripropylamine while preserving optical transparency of indium tin oxide for sensitive electrochemiluminescence-based analyses. *Analytical Chemistry* 86(3), 1654–1660 (2014).

205. Wang, H., Zheng, L., Peng, C., Guo, R., Shen, M., Shi, X., and Zhang, G. Computed tomography imaging of cancer cells using acetylated dendrimer-entrapped gold nanoparticles. *Biomaterials* 32(11), 2979–2988 (2011).

206. Artemov, D., Mori, N., Okollie, B., and Bhujwalla, Z.M. MR molecular imaging of the Her-2/neu receptor in breast cancer cells using targeted iron oxide nanoparticles. *Magnetic Resonance in Medicine* 49(3), 403–408 (2003).

207. Kobayashi, H., and Brechbiel, M.W. Nano-sized MRI contrast agents with dendrimer cores. *Advanced Drug Delivery Reviews* 57(15), 2271–2286 (2005).

208. Triulzi, R.C., Micic, M., Giordani, S., Serry, M., Chiou, W.-A., and Leblanc, R.M. Immunoasssay based on the antibody-conjugated PAMAM-dendrimer-gold quantum dot complex. *Chemical Communications* (48), 5068–5070 (2006).

209. Lin, C.-A.J., Lee, C.-H., Hsieh, J.-T., Wang, H.-H., Li, J.K., Shen, J.-L., Chan, W.-H., Yeh, H.-I., and Chang, W.H. Synthesis of fluorescent metallic nanoclusters toward biomedical application: recent progress and present challenges. *Journal of Medical and Biological Engineering* 29(6), 276–283 (2009).

210. Pan, B., Cui, D., Sheng, Y., Ozkan, C., Gao, F., He, R., Li, Q., Xu, P., and Huang, T. Dendrimer-modified magnetic nanoparticles enhance efficiency of gene delivery system. *Cancer Res.* 67(17), 8156–8163 (2007).

211. Sahoo, N.G., Rana, S., Cho, J.W., Li, L., and Chan, S.H. Polymer nanocomposites based on functionalized carbon nanotubes. *Progress in Polymer Science* 35(7), 837–867 (2010).

212. Tomalia, D.A., Reyna, L.A., and Svenson, S., Dendrimers as multi-purpose nanodevices for oncology drug delivery and diagnostic imaging. *Biochemical Society Transactions* 35, 61–67 (2007).

Combinatorial Cationic Lipid Library Synthesis for Drug Delivery

5

Sarah Altinoglu* and Qiaobing Xu*,†,‡,§

*Department of Biomedical Engineering, Tufts University, Medford, MA 02155, USA
†Department of Chemical and Biological Engineering, Tufts University, Medford, MA 02155, USA
‡School of Medicine, Tufts University, Boston, MA 02111, USA
§Program in Cell, Molecular and Developmental Biology, Sackler School of Graduate Biomedical Science, Tufts University, Boston, MA 02111, USA

Abstract

Research in the past two decades has shown the high efficiency and efficacy of cationic lipids and liposomal formations for drug delivery. The tediousness of synthesizing conventional lipids and the inefficiency in studying structure-activity relationships, however, have hindered the clinical translation of lipid nanoparticle delivery systems. Combinatorial synthesis of lipid-like nanoparticles ("lipidoids") has recently emerged as an approach to accelerate the development of these delivery platforms. Utilizing a high-throughput screening strategy, libraries of lipidoids are sorted and prime candidates for the intended application can be identified and optimized for the next generation library. In this chapter, we outline methods used for combinatorial lipidoid synthesis, the application of high throughput screening, and current medical applications for candidate lipidoids.

5.1 Introduction

Nanotechnology can be applied to medicine in many ways. One key application is the use of lipid nanoparticles, or liposomes, to facilitate drug delivery. Many drugs require the assistance of a delivery carrier. For example, proteins are often too big to cross the cell membrane, have low stability, and suffer from high clearance rates.[1] Challenges with the delivery of nucleic acids, such as DNA and RNA, include crossing the cell membrane as well as overcoming their rapid degradation and clearance profiles.[2] The use of lipids as nanoparticles for drug delivery addresses these shortcomings by facilitating intracellular delivery, protecting the drug from degradation, and decreasing clearance rate, all resulting in increased drug activity.

Cationic lipids are promising drug delivery materials consisting of positively-charged amphiphilic structures that naturally form bilayers in aqueous solutions. These cationic nanoparticles have been widely used to formulate liposomes with precisely tailored physical and chemical properties, which can entrap a variety of hydrophobic and hydrophilic drugs for intracellular delivery.[3–10] In 1987, Felgner *et al.* designed the first cationic lipid, DOTMA, for therapeutic delivery of RNA and DNA into mammalian cells.[11,12] Since then, many other types of cationic lipids and lipid-like materials have been used to deliver therapeutic molecules, including phospholipids,[13] Gemini lipids,[14] and, more recently, lipidoids.[15]

Despite extensive research into cationic lipid delivery, a carrier capable of high transfection efficacy, mimicking that of viral vectors, is still needed.[16] One explanation for the current limitation in transfection efficacy is the relatively slow expansion and improvement of delivery materials. In response, researchers have developed a new approach to making novel cationic lipids, termed combinatorial synthesis, to more effectively build upon and improve this delivery platform.

Combinatorial synthesis is a systematic synthesis method that utilizes quick chemistries to combine structurally diverse polar head and non-polar tail structures for generating large quantities of unique lipids. Using combinatorial synthesis with high-throughput screening allows for the rapid identification and optimization of novel, superior drug delivery systems.

This chapter will give an overview of the benefits of using cationic lipids in drug delivery. We will focus on combinatorial lipid library synthesis strategies, exploring the uses and advantages of this synthesis method along with a detailed description of high-throughput library screening techniques. The applications of combinatorial lipid libraries in medicine and the future of lipid-based nanotechnology will also be discussed.

5.2 Cationic Lipids for Delivery

5.2.1 *Features of Cationic Lipids for Drug Delivery*

Nanoparticles formulated with cationic lipids offer many characteristics that are beneficial for drug delivery, such as small size, drug protection and intracellular delivery. The size of the drug delivery system plays an important role in both distribution in the body and cellular uptake.[1,17,18] Specifically, particles less than 150 nm in diameter are paramount for therapeutic delivery because the vasculature in diseased tissues (such as inflammation or tumor sites) is leaky and therefore allows for particle extravasation.[8] In addition, smaller particles are able to diffuse through the extracellular matrix to reach some targets.[8] Cationic lipid nanoparticles generally have sizes of <200 nm, but can vary depending on the formulation method.[10,16,17] Most commonly, liposomes are prepared by directly mixing the cationic lipid solution with the cargo.[8] Other popular methods include thin film hydration,[19] detergent dialysis[20] and ethanol injection.[19,21] To more precisely control the size of the nanoparticles, methods such as sonication (the use of energy-induced pressure to break vesicles into more homogenous sizes controlled by sonication time),[22–24] extrusion (forcing the liposome solution through a membrane with a well-defined pore size, resulting in nanoparticles with a similar size),[24] and microfluidics (a less utilized yet successful technology that uses small chambers with controlled flow rates and mixing of reagents to fine tune liposome size and structure)[23,25] have been used.

Cargo protection is a highly important component to consider in drug delivery due to the quick degradation and low stability of many types of cargo. The liposome lipid bilayer provides a high level of protection to the cargo by limiting its interaction with degradation enzymes[16] and harsh physiological conditions.[26] Thus, encapsulation of the drug helps to both protect and deliver the cargo. The positive charge of cationic lipids is an advantage over many other delivery systems because it facilitates the encapsulation and condensation of negatively-charged cargos, such as DNA and RNA, through electrostatic interactions, resulting in a higher encapsulation yields.[4,16] Lipid-based nanoparticles can deliver different types of cargo: they can encapsulate hydrophobic drugs in the lipid bilayer or hydrophilic molecules inside the liposome.[27]

Delivery of the encapsulated cargo into the cellular cytoplasm is a multi-stage process, first involving an interaction with the cell membrane and followed by cellular entry and release of cargo, as shown in Figure 5.1.[17] Many cationic lipid nanoparticles exhibit good intracellular delivery as a consequence of their net positive charge, allowing for direct interaction with the negatively-charged cell membrane.[28] More specifically, cell surface proteins such as heparin sulfate proteoglycans provide a binding site for many cationic delivery systems.[17,29] Once bound to the membrane, cellular entry can occur via different mechanisms but most commonly through endocytosis, resulting in the nanoparticle being endosome-bound.[16,17,30] Finally, release of the cargo must be triggered by endosomal encapsulation to avoid degradation through lysosomal degradation. Disruption of the endosomal membrane can be

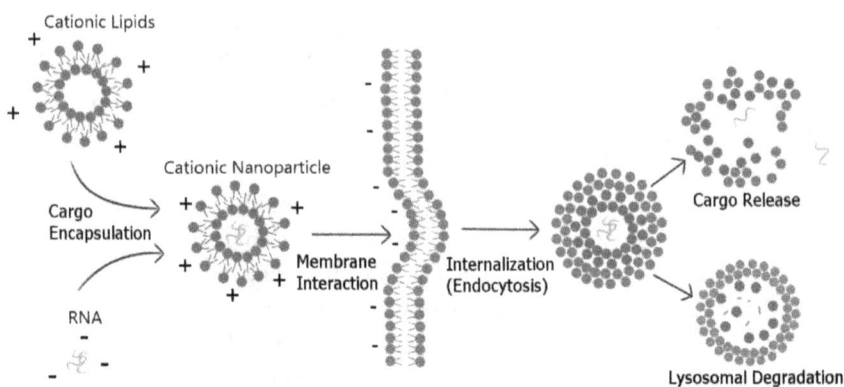

Figure 5.1: The delivery process of cationic lipid nanoparticles involves efficient cargo encapsulation, cell membrane interaction, and internalization and cargo release.

induced in a variety of ways, such as liposome fusion with the membrane, which can be prompted by the addition of dioleoylphosphatidylethanolamine (DOPE) to the nanoparticle formulation, or by lipid translocation via non-lamellar intermediates across the membrane.[17,30]

5.2.2 Advantages of Combinatorial Synthesis of Cationic Lipid Structures for Drug Delivery

Combinatorial synthesis and high-throughput screening has long been utilized by the pharmaceutical industry for drug discovery. The concept is based on assembling building blocks to create new molecules and testing for biological activity against a target.[31,32] This library production and screening method allows for libraries larger than 100,000 molecules to be assembled, tested and ultimately evaluated as drug candidates against certain targets.[31] A similar approach can be applied to finding appropriate delivery vehicles in medicine.

The adaptation of combinatorial synthesis and high-throughput screening to cationic lipid development has many advantages. An important benefit is the vast number of structures that can be synthesized. As detailed in this chapter, libraries of more than 1,200 lipid structures can be produced and screened in a short period of time, which greatly increases the chances of finding an optimal delivery vehicle structure. Another key advantage is that it alleviates the time-consuming approach of designing cationic lipids based purely on structure-activity relationships (SAR), which are not yet well understood. The use of combinatorial synthesis allows for large diversity in the chemical structures of the lipids, while high-throughput screening can rapidly detect their activity levels. Therefore, combinatorial library synthesis paired with high-throughput screening can lead to new and more rapid SAR, which in turn results in improved delivery systems.[33]

Selecting the appropriate lipid structure for the medical application is crucial to ensuring successful delivery. Langer *et al.*[15,34-36] and, more recently, Xu *et al.*[37-39] have used combinatorial lipid library synthesis to create new and diverse lipid structures capable of effective cargo delivery. High-throughput screening of these novel lipid libraries *in vitro* first identifies top performing lipid structures for the delivery of a particular cargo. Next, *in vivo* formulation of these lipids (usually with helper lipids such as DOPE, cholesterol or PEGylated lipids) along with further screening helps to identify and fine-tune the physical properties of the nanoparticle for a specific medical application, such as delivery to a specific organ.[40]

5.3 Strategic Design of a Cationic Lipid Combinatorial Library

The first cationic lipid library built for high-throughput screening was created in 2002, when Lenssen *et al.*[41] synthesized a library of 17 structurally diverse cationic lipids using solid phase synthesis. Since then, advancements in synthesis methods have eliminated some of the rate-limiting aspects of standard cationic lipid production, which usually comprises of multiple, slow and tedious steps such as protection and de-protection of functional groups, solvent exchanges, and serial purification steps.[42-44] Recent synthesis methods include the Michael addition of amine to alkyl acrylamides or acrylates, the ring-opening reaction between epoxide and amine, and click chemistry (Figure 5.2).[45] The chemical structures of lipids that result from using these methods are somewhat different from typical cationic lipids in that there can be more than two tail groups attached to the amine core, and the cationic head and hydrophobic tail elements are less distinct. These new materials are termed "lipidoid" to mean lipid-like.[15,42]

This section will review the chemistries that have been utilized to create diverse libraries of delivery materials.

5.3.1 *Solid-Phase Synthesis*

Solid-phase synthesis is a method introduced by Bruce Merrifield in 1959, in which a starting reagent is bound to a bead and synthesized step by step in a controlled manner by the repeated addition of protected groups followed by a de-protection

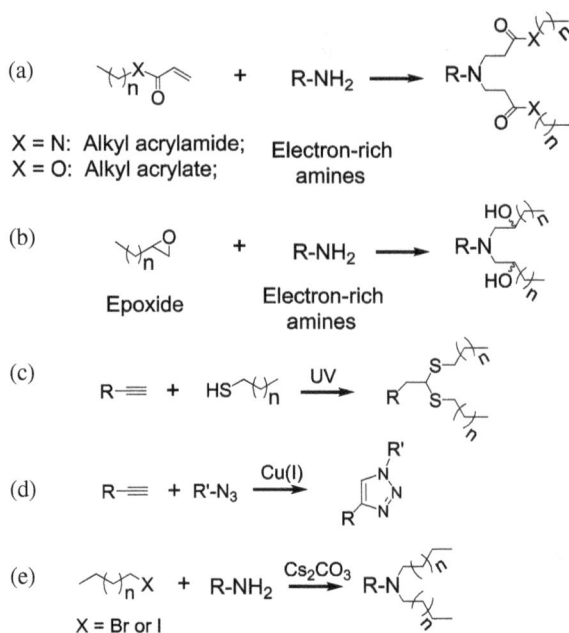

Figure 5.2: Chemical approaches that have been used for the combinatorial synthesis of cationic lipids: (a) Michael addition reaction, (b) ring-opening reaction of epoxide, (c) thiolyne click chemistry, (d) copper-mediated click chemistry, and (e) alkylation addition of amines.[45]

step.[46] Originally used for peptide synthesis, this method was eventually adapted for cationic lipids.

Lessen et al.[41] produced a small cationic lipid library based on one cationic head group, (R)-2,3-epoxy-1-propanol, and six different tail structures with two possible configurations. Synthesis was performed by first immobilizing the head group to a 4-methy-oxytrityl chloride resin and stirring the suspension. Next, an epoxide-amine addition reaction was used to bind the first hydrocarbon group to the head and reductive amination chemistry was used to add the second tail group. To finish the reaction, the tertiary amine was quaternized with methyl iodide and the newly formed cationic lipid was cleaved from the resin. Filtration, washing and solvent exchange steps were carried out between each of these reactions. Using this method, the group was able to make 17 structurally-diverse lipids.[41]

Following Lessen's publication, solid-phase synthesis has been utilized to create many other lipid libraries.[47,48] Liberska et al.[49] synthesized an even larger combinatorial library of 63 two tailed lipid-peptide conjugates. This library contained diversity in all three chemical structures of the head, linker and tail. The authors also investigated the impact of changing separation distance between the head and tail groups by using DdeOH to block specific amine groups during elongation. Through evaluation of this library, the authors found that di-arginine compounds with a diethylenetriamine-based spacer (SP2) showed the highest DNA transfection rates.[49]

Despite continued success in using solid-phase synthesis to create combinatorial lipid libraries, this method is viewed by some to be unattractive due to the labor-intensive and time-consuming steps of protection/de-protection, filtration, washing and purification.

5.3.2 *Michael Addition of Amines and Acrylates/Acrylamides*

Although Michael addition chemistry has been utilized in organic synthesis for more than 125 years,[50] it was only recently in 2008 that Langer and Anderson applied this chemistry to simplify the synthesis of cationic lipid-like materials.[15] The Michael addition chemistry used here is the conjugation addition of an α, β unsaturated carbonyl on a hydrocarbon tail to an electron-rich amine group on a head molecule [(Figure 5.2(a)].[45] Unlike solid-phase synthesis, this conjugation reaction is straightforward in that it is a one step process with no added solvents or catalysis, and does not require timely protection/de-protection steps. In addition, this synthesis yields highly selective product formation in a predictable and efficient manner.[15,50] These features make the use of Michael addition chemistry favorable when creating large libraries of structurally-diverse lipid-like molecules.

In the first lipidoid library, Akinc *et al.*[15] varied four independent aspects of the lipid structure including the carbon chain length of the tail (between 10 and 18), the tail linker to the ester or amide, the substitution group on the amine, and the positive charge on the lipid. As shown in Figure 5.3, the authors carried out Michael addition chemistry using 17 different α, β unsaturated carbonyl hydrocarbon tails and 54 diverse amine core structures. The lipidoids were synthesized by simply mixing the head and tail structures at 90°C for one day (for acrylate) or seven days (for acrylamide) and mostly did not require an additional purification step.[15] Using this simple synthesis method, a first generation library of 700 structures as well as a second library of 500 structures were created; these diverse libraries are too large to be synthesized in a timely manner by previous synthesis methods. Evaluation of the first generation library identified 56 new cationic lipid-like structures capable of equal or better transfection efficacy compared to the control, Lipofectamine 2000. Furthermore, after second generation library synthesis, rigorous *in vitro* screening,

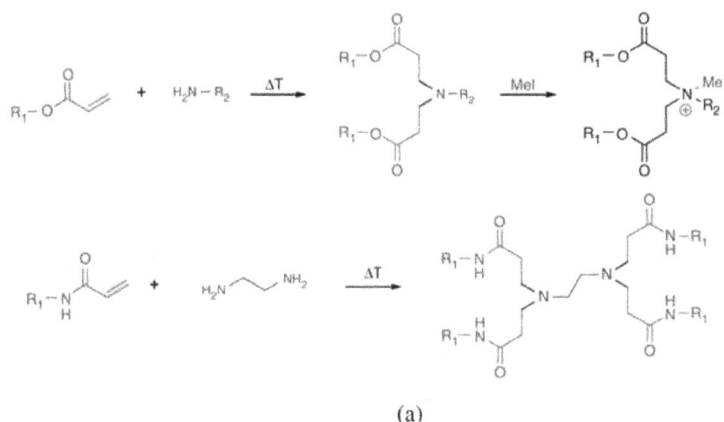

(a)

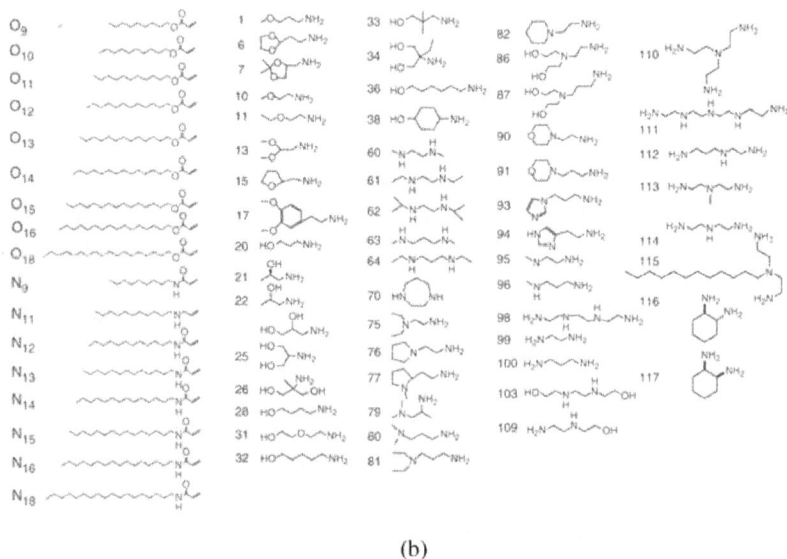

(b)

Figure 5.3: Synthesis of a combinatorial library. (a) Michael addition reaction used to create lipidoid structures. (b) Tail (alkyl-acrylates and alkyl-acrylamides) and amine core structures used in library synthesis. Figure used with permission of Akinc et al.[15] with modification.

and mouse, rat and non-human primate studies, one lipidoid structure ($98N_{12}$-5) stood out above the rest and showed the ability to deliver siRNA or anti-miRNA in both localized and systemic delivery models.[15]

Following the first published use of Michael addition for lipidoid library synthesis, several other lipidoid libraries have been generated using a similar approach.[34,37,38,51,53,53] For example, building on top of the original lipidoid library, Mahon *et al.*[34] introduced diversity to the two top performing amine core structures [core 98 and core 100, Figure 5.3(b)] by functionalizing the lipidoids with side chains varying in the amounts of hydrogen bonding, hydrophobic interactions and protonation state. From this, it was shown that small changes in tail functionalization can greatly impact the delivery performance of the structure, as tested *in vitro* and *in vivo*. The results also set up some general structure-performance patterns, such as an increase in siRNA delivery efficacy with added moieties such as hydroxyl, ether and carbamate, as well as greater *in vivo* tolerance using tails consisting of hydrocarbon with ether group structures.[34]

5.3.3 *Epoxide-Amine Addition*

Expanding upon the use of simple chemistries for the synthesis of combinatorial libraries, Love *et al.*[35] adapted a ring-opening reaction between epoxide and amine to develop a new lipidoid library. Similar to the Michael addition reaction, the epoxide-amine addition is an efficient, straightforward and fast reaction. Here, the epoxide ring on the alkyl tail will easily open and conjugate with the electron-rich amine group on the core structure, as shown in Figure 5.2(b). The synthesis is carried out at 90°C without the addition of solvent, thus making the product directly usable in the next screening step.[35] A key advantage to using this synthesis method is that the reaction chemistry is rapid, taking just two and half days to complete. In addition, utilizing the epoxide group instead of an acrylate or acrylamide also adds more diversity to the type of lipidoid structures that can be synthesized.

Using this method, Love *et al.*[35] produced 126 lipid-like compounds within a short three-day time span. The library was built from nine different epoxide terminated alkyl chains and 14 different amine-containing monomers. Through the synthesis of the library and the subsequent screening process, a newly created lipidoid (C12-200) was identified as being able to far surpass previously reported delivery vehicles for siRNA specific knockdown in non-human primates, with efficacious results down to 0.03 mg/kg. The group hypothesized that this type of lipidoid structure might have enhanced delivery properties due to cellular uptake via macropinocytosis instead of the typical endocytic pathway, which often leads to liposomal degradation.[35]

Recently, additional combinatorial libraries have been created using epoxideamine addition to evaluate the use of lipidoid materials for the delivery of various types of cargo. For example, Wang *et al.*[53] built a lipidoid library by synthesizing a 1,2-epoxyhexadecane tail structure on 14 different aliphatic amine core structures. This library of lipidoids was tested *in vitro* for its ability to deliver protein cargo. The authors identified one of the newly synthesized structures (EC16-1) as promising for protein delivery after it demonstrated efficient delivery of anti-cancer proteins that suppressed tumor growth in an *in vivo* breast cancer mouse model.[53]

Using a similar synthesis method, Dong *et al.*[54] developed and evaluated delivery vehicles capable of co-delivering plasmid DNA and siRNA. Here, a 1,3,5-triazinane-2,4,6-trione derivative featuring three epoxide groups was irradiated in a microwave oven for five hours at 150°C, with the irradiation and heat contributing to quicker synthesis. From this library, a new candidate lipid (TNT-4) was found to efficiently co-deliver both plasmid DNA and siRNA in a mouse model.[54]

5.3.4 *Click Chemistry*

Click chemistry was first introduced in 2001 by K. B. Sharpless as a method to generate stuctures built from smaller units and joined together in a powerful and selective way with heteroatom links.[55] For a reaction to be considered click chemistry it must conform to the following criteria: it must be modular, stereospecific, wide in scope, and produce high yields. In addition, the overall process should have simple reaction conditions, including readily available materials and easily removable solvents, if any.[55] Because this type of chemistry is so quick and easy and results in a well-defined product, many material libraries have been created using this synthesis method.

Li *et al.*[56] were the first to apply thiol-yne click chemistry for the parallel synthesis of cationic lipids. The authors incorporated an alkyne linker to alkyl thiol tails by exposing the reaction to UV light for one hour in the presence of a photocuring agent. [Figure 5.2(c)]. Following the thiol-yne click reaction, amide coupling was utilized to add an amine head group to the thioether structure, by stirring the mixture for 16 h in the presence of solvents. A library of over 100 cationic lipids was synthesized from eight thiol tail structures varying in carbon length from six to 16, two different alkyne linkers, and a diverse selection of amine head groups. From this library, Li *et al.* were able to identify more than 10 lipid structures that outperformed Lipofectamine 2000.[56]

Using a similar method, Alabi *et al.*[57] produced a lipid library of 32 unique structures. Michael addition was first used to attach one of 32 possible amine head structures to propargyl acrylate followed by a thiol-yne click chemistry step. Combining these two fast and efficient synthesis methods, the entire library was created within a two-day time span. This lipid library was then used to evaluate a newly developed multiparametric approach to screening nanoparticles, which included testing for key delivery features such as entrapment, pKa, cellular uptake, stability in and outside the cell, and hemolysis compared to the measured *in vitro* effect.[57]

On a smaller scale, Sheng *et al.*[58] used another type of click chemistry to create four sterol-based cationic lipids. Here, synthesis was completed using copper-mediated azide-alkyne cycloaddition (CuAAC) click chemistry, where an alkynyl-bearing cholesterol derivative is coupled with azide-bearing moieties using copper to catalyze the cycloaddition reaction [Figure 5.2(d)]. This study explored the use of CuAAC click chemistry to build a series of cationic lipids for gene delivery.[58]

5.3.5 *Alkylation of Amines*

A lipidoid library was recently produced by Li *et al.*[59] from the simple alkylation of amines. Alkylation addition synthesis is effortless and straightforward, similar to Michael addition synthesis and epoxide-amine addition. Here, various alkyl halides are conjugated to an amine structure by simple shaking at 40°C for 20 h in the presence of cesium [Figure 5.2(e)], followed by short purification and solvent exchange steps.[59,60] A library of more than 200 lipidoid structures was synthesized from 20 different amine head groups and 10 different alkyl halide tails functionalized with bromine or iodine. From this library, 12 newly synthesized structures were shown to outperform Lipofectamine 2000 in plasmid DNA delivery.[59]

5.3.6 *Altering Tail Structure*

One way to add diversity to a lipidoid library is by varying or adding features to the tail structures.

5.3.6.1 *Unsaturated tail structures*

The degree of saturation on a cationic lipid structure can influence delivery efficacy. Previously, it was shown in cationic lipid delivery systems that a lower level of saturation is correlated with an increase in delivery activity.[61] Recently, Wang *et al.*[39] investigated this correlation in lipidoid structures. Utilizing the Michael addition reaction, the authors synthesized a lipidoid library from 16 diverse amine functionalized cores and two acrylamide tails, equivalent in structure except for the degree of saturation. A thorough investigation of delivery efficacy based on saturation was conducted *in vivo* by delivering plasmid DNA or mRNA to HeLa cells. Results showed a clear trend of decreasing delivery efficacy with increasing tail saturation.[39]

5.3.6.2 *Responsive tail structures*

Smart delivery vehicles have been designed to alleviate the difficulty in releasing siRNA or related cargo upon delivery into the cell because of strong binding between negatively-charged cargo and positive lipid tails.[53,62,64] A disulfide bond can be incorporated into drug delivery vehicles to enhance drug release upon cell entry while also decreasing cytotoxicity.[44,65] Recently, Wang *et al.*[37] incorporated a bioreducible disulfide bond into the tail structure of a lipidoid. A small lipidoid library was produced with 12 different structures using two head groups and three tails, with or without the disulfide bond. Evaluation of this library showed the incorporation of the bioreducible disulfide bond facilitated lipidoid degradation and enhanced siRNA delivery compared to the same lipid structures that were not bioreducible.[37]

5.3.7   *Adding Diversity to Library Synthesis*

In addition to using different synthesis methods to build lipidoid libraries, diversity can also be enhanced by using multiple lipidoid structures or chemically modifying the lipidoid charge.

5.3.7.1   *Synergistic coupling of lipidoid structures*

From screening these large libraries, most material structures are labeled as not suitable for delivery and discarded. For example, in the first lipidoid library (created by Akinc *et al.* and detailed in section 5.3.2 of this chapter), over 700 lipidoids were synthesized but after screening for siRNA delivery, only 56 were identified as having similar or better delivery efficacy than Lipofectamine 2000.[15] Whitehead *et al.*[66] hypothesized that by only screening for overall delivery efficacy, lipidoids that could aid in specific parts of the delivery process, such as cellular entry or endosomal escape, were wrongfully discarded.[66] To investigate this, binary combinations of ineffective lipidoids were evaluated for synergistic delivery efficacy. Taking 36 of the poor performing lipidoid structures from the library, Akinc *et al.* and Whitehead *et al.*[66] formulated 3,780 unique pairs of lipidoids, at six different weight fractions. From this group, a binary combination was identified that was capable of up to 85% gene silencing in a mouse model. As predicted, further investigation indicated one lipidoid structure aided more in cellular entry while the other aided more in intracellular compartmental escape.[66] In another example of a synergistic effect, Li *et al.*[59] selected two very different lipidoid structures from their alkylation-synthesized library to test for enhanced delivery when formulated in a combination approach. Both lipidoid structures consisted of the same head and tail groups except one, 1C16-S, which had only one tail structure while the other, 1C16-D, had two. The structures also varied in performance: 1C16-S had a low level of efficacy and 1C16-D had an efficacy almost equivalent to the positive control of Lipofectamine 2000. When mixed together at a ratio of 1:2, the transfection efficacy of these structures outperformed that of Lipofectamine 2000 by two-fold.[59]

5.3.7.2 *Quaternization of lipidoid structures*

Quaternization of lipids has an impact on delivery efficacy; in fact, many common delivery systems already utilize quaternized cationic lipids.[11,12] In the lipidoid library synthesis by Akinc *et al.* described earlier, some lipidoid amine cores were quaternized but showed poor transfection efficacy.[15] To investigate on a larger scale if quaternized lipidoids are suitable for plasmid DNA delivery, Sun *et al.*[67] synthesized a lipidoid library from 18 different amine cores and three different tail structures using epoxide-amine addition followed by methyl iodide exposure for quaternization. Although quaternized lipidoids alone showed poor transfection efficacy, the use of DOPE helper lipids in formulations with some of the quaternized lipidoid structures showed an even higher transfection efficacy than that of Lipofectamine 2000.[67]

5.4   Library Screening for Desired Delivery Properties

Besides the design of lipid-like libraries, the testing and evaluation of each material in a high-throughput manner, including crude screening, purification, optimization and *in vivo* formulation, are just as important (Figure 5.4).[45] A crude screen can quickly eliminate poor delivery structures and highlight those with potential, while purification followed by delivery optimization can highlight lipid-like structures and doses relevant to the intended application. In the final step, *in vivo* formulation and screening can pinpoint clinically relevant formulations. This section will focus specifically on high-throughput screening of lipidoid structures for therapeutic delivery.

5.4.1   *Crude Screening*

Crude screening is used to eliminate poorly performing lipidoid structures on a large scale and in a rapid manner. There are two key aspects that set crude

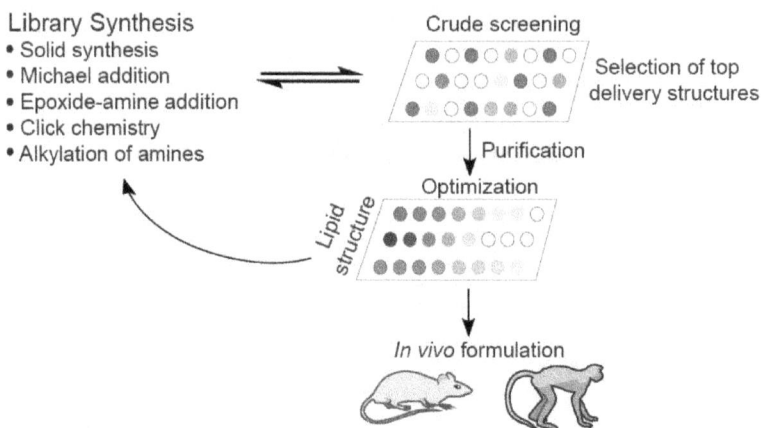

Figure 5.4: Schematic of high-throughput screening for the best lipidoid structure for therapeutic delivery.[45]

high-throughput screening apart from other downstream screening methods; firstly, the lipidoids are used in an unpurified state to eliminate time-consuming purification steps with such large sample sizes, and secondly, the intended medical application (i.e., the specific therapeutic cargo and delivery to a target cell type) is not taken into account. Here, the lipidoid is formulated with reporter cargo, delivered to a general cell line, and rapidly assayed for delivery efficacy.

Selecting the appropriate reporter cargo is important for achieving both high-throughputness and relevant results. Here, a payload that is similar to the therapeutic cargo and that can be analyzed in a reliable and quick manner for delivery to cells should be selected. For example, if siRNA is the intended therapeutic molecule, a common reporter gene is firefly luciferase or green fluorescent protein (GFP) siRNA using a FF luciferase or GFP-expressing cell line.[34,35,57,66] Both these siRNAs are easily analyzed in a high-throughput 96 well plate format by a simple luciferase activity assay or flow cytometry analysis.[39,54]

In addition to testing for cargo delivery efficacy, lipidoid toxicity should also be evaluated to remove structures that cause cytotoxicity issues downstream. Here, the lipidoid alone is added to the cells and a simple cell viability test, such as an MTT assay, can be used to quantify toxicity.[15,34]

Formulation of the lipidoids with a reporter gene for crude screening is designed specifically for simplicity and speed. The use of helper lipids or polyethylene glycol (PEG) is usually avoided until the next screening step. Here, the newly developed lipidoids are simply mixed with the cargo of interest to form a nanocomplex of cargo and lipidoid.[15,39] The ratio of genetic material to lipidoid is an important parameter and will be discussed in the section on optimization. After formulation, the lipidoid complex is co-incubated with easily-transfected or target cells.

After measuring the delivery efficacy of the lipidoid structures, the results can be used two different ways: to eliminate structures before the next screening step, or to determine SAR for second generation library construction. Interpreting the results to determine which structural features aid in top delivery performance is an important step and can be a daunting task when hundreds of materials are being screened. Some groups have utilized a heat map to help visualize results.[15,59]

For example, Li *et al.*[59] used a heat map of cationic head groups versus hydrophobic tail length, making it easy to see a general trend of increasing transfection efficiency with increasing tail length.[59] SAR found from initial screening often leads to a second generation library that is built by expanding on the key features identified.[15,38,51] This secondary library can then be put through crude screening again.

5.4.2 *Optimization*

To ensure that the lipidoid structure selected is relevant for formulation and testing *in vivo*, lipidoid and cargo complexes need to be optimized for the intended therapeutic application. This evaluation consists of delivery to multiple cell lines including the targeted cell type, as well as screening for the best cargo to lipid ratio, both of which can have a significant impact on transfection efficacy.

The ratio of lipid to cargo affects many aspects of the resulting complex, such as overall charge, particle size heterogeneity and stability. Because cationic delivery systems depend heavily on the interaction between the positively-charged particle and the negative cell membrane for delivery, the overall charge of the particle, and therefore the lipid to cargo ratio, can greatly affect delivery efficiency.[36,68] To optimize this lipid to cargo ratio, a range of several different weight ratios are typically dosed and evaluated similar to previously mentioned methods.[15,34,38]

During optimization, typically all but the top few performing lipidoid structures will be eliminated due to either an inability to show high efficacy transfection in the intended cell line, or due to poor delivery in the projected dose range. Results from this screening step can demonstrate lipidoid efficacy in the intended downstream application and therefore lead to improved, more apparent SAR and possibly the synthesis of another generation lipidoid library.

5.4.3 *In Vivo Screening*

While crude screening and optimization *in vitro* can rapidly eliminate poor delivery lipid candidates, there is not always a direct correlation between *in vitro* and *in vivo* results.[32] For this reason, several of the top performing structures must next be subjected to *in vivo* screening, which introduces additional elements into the evaluation, including exposure of the delivery material to the physiological environment as well as off-target cell types. Lipidoids alone have rather low *in vivo* tolerance due to poor serum stability and their tendency to form aggregates in physiological fluids, and must therefore be formulated with supplementary components before delivery *in vivo*.[15,42,69] A typical *in vivo* formulation includes cholesterol, which can be incorporated into the lipid bilayer structure to facilitate intracellular gene material release, and PEG, which can prevent colloidal aggregation and increase circulation time.[15,34,35,70] This *in vivo* formulation is now ready for administration in an animal model, where it will be evaluated for therapeutic efficacy, off-target delivery and tolerance.

5.5   Applications in Medicine

Delivery materials produced from combinatorial library synthesis and high-throughput screening have many potential applications in medicine. Specifically, lipid-like materials have been used to deliver multiple types of cargo for various therapeutic applications, as shown in Table 5.1.[45] This section will highlight some of the ways these lipidoid structures have been used.

5.5.1   *RNA Interference*

RNA interference (RNAi) is a post-transcriptional gene regulation process that is necessary for the regulation of cell functions such as development, apoptosis, differentiation and routine maintenance.[71] One of the best studied types of RNAi is short interfering RNAs (siRNA). When siRNA binds to a complementary mRNA sequence, the mRNA will immediately be cleaved.[72] The ability to knock down genes with the delivery of siRNA could have a significant impact on therapeutic medicine; however, there are still many challenges to delivering siRNA. The greatest barrier to overcome is endosomal escape. Once the delivery vehicle is taken up by the cell, the siRNA must be released into the cytoplasm for it to cause knockdown.[72] There are several good reviews on siRNA delivery[54,73–76] and specifically the use of cationic lipids for siRNA delivery.[10,42,70,72,77–79]

Lipidoid materials built from combinatorial lipidoid synthesis have shown great potential for the delivery of siRNA. One of the most promising siRNA delivery materials was synthesized by Akinc *et al.*[15] Structure 98N$_{12}$-5 [Figure 5.5(a)] stood out from more than 1,200 other structures after *in vivo* and *in vitro* screening.

Table 5.1:   Lipid-like materials used in therapeutic applications.

Year	Library Synthesis	Cargo Delivered	Model for Delivery	Novelty	Ref.
2002	SP	pDNA	*in vitro*, MDA-MB-468, MCF-7, MDCK-C7, COS-7 primary dendritic cells, KL-1-14	Development of a synthesis-screening approach for cationic lipids intended for gene-transfer	41
2008	MN-N, MN-O	siRNA, anti-miRNA	*in vitro*, HeLa, HepG2, primary macrophages *in vivo*, C57BL/6 mice, C57Bl/6J mice, BALB/c mice, C57BL/6NCRL mice, Sprague-Dawley rats, cynomolgus monkeys	Development of the first combinatorial lipidoid library using Michael addition chemistry and high throughput screening	15
2009	MN-N	siRNA	*in vitro*, HUVEC	siRNA delivery using lipidoid to human endothelial cells	52
2010	MN-N, MN-O	siRNA	*in vitro*, HeLa *in vivo*, C57BL/6 mice	Further diversified lipidoid structures by adding various functional groups	34
2010	EA	siRNA	*in vitro*, HeLa *in vivo*, C57BL/6 mice or cynomolgus monkeys	Development of epoxide-amine addition for lipidoid library synthesis	35
2011	MN-N, MN-O, EA	pDNA	*in vitro*, HeLa	Plasmid-DNA delivery using lipidoid structures	38
2011	MN-N, MN-O	siRNA	*in vitro*, HeLa *in vivo*, C57BL/6 mice and SKH1 hairless mice	Synergistic siRNA delivery using a combination of poorly performing lipidoids	66

(Continued)

Table 5.1: (*Continued*)

Year	Library Synthesis	Cargo Delivered	Model for Delivery	Novelty	Ref.
2012	MN-N	pDNA, and mRNA	*in vitro*, HeLa, MCF-7, HepG2, MDA-MB-231, NIH-3T3, BJ	A comparison between saturated and unsaturated tail structures for delivery of pDNA and mRNA	39
2012	Thiol-yne click	pDNA, and siRNA	*in vitro*, HEK293T, HeLa, mouse embryonic fibroblasts, Mouse D3 embryonic stem cells	Development of thiol-yne click chemistry for lipidoid library synthesis	56
2012	MN-N, MN-O	isRNA	*in vitro*, human PBMCs *in vivo*, 129sv mice, BALB/c mice, C57Bl/6 mice and MyD88−/− C57Bl/6 mice	Lipidoid library synthesis for delivery of isRNA for activation of immune response	51
2013	CuAA click	pDNA	*in vitro*, A549, HeLa, HEK193T, COS-7	Development of CuAA click chemistry for lipidoid library synthesis	58
2013	EA	pDNA	*in vitro*, HeLa	Comparison of quaternized and un-quaternized lipidoid structures for pDNA delivery	67
2013	MN-O, thiol-yne click	siRNA	*in vitro*, HeLa *in vivo*, C57BL/6 mice	Development of multiparametric screening of lipidoid siRNA delivery	57
2013	Alkylation of amines	pDNA	*in vitro*, HEK293	Development of alkylation of amines for lipidoid library synthesis	59
2014	MN-O	siRNA	*in vitro*, MDA-MB-231, HeLa, 4T1	Incorporation of a bio-reducible, disulfide bond into the lipidoid structure	37

(*Continued*)

Table 5.1:   (*Continued*)

Year	Library Synthesis	Cargo Delivered	Model for Delivery	Novelty	Ref.
2014	EA	Co delivery: pDNA, and siRNA	*in vitro*, HeLa *in vivo*, C57BL/6 mice	Co-delivery of pDNA and siRNA using lipidoid structures	54
2014	EA	Protein	*in vitro*, B16F10, MCF-7, MDA-MB-231, HepG 2, PC-3, LNCaP, HeLa, 4T1 *in vivo*, Balb/c	Modification and delivery of proteins using lipidoids	53

Abbreviations: SP: Solid-phase synthesis, MN-N: Michael addition amine to acrylamide, MN-O: Michael addition amine to acrylate, EA: Epoxide-Amine addition, pDNA: plasmid DNA, siRNA: short interfering RNA, miRNA: micro RNA, mRNA: messenger RNA, isRNA: immunostimulatory RNA.

Using $98N_{12}$-5, Factor VII siRNA was successfully delivered in mice and rats, and showed greater than 90% silencing at 2.5 mg/kg and 5 mg/kg doses, respectively.[15] In addition, $98N_{12}$-5, which was also tested in cynomolgus monkeys for the delivery of ApoB siRNA, resulted in a 75% reduction in ApoB levels at a dose of 6.25 mg/kg. Importantly, after only a single intravenous injection, this silencing was sustained for two weeks followed by a full recovery after 30 days, and no toxicity or side effects were observed.[15]

In 2009, Nguyen *et al.*[80] used the lipidoid structure $98N_{12}$-5 to deliver immunostimulatory RNAs (isRNAs). isRNAs are unmodified siRNAs that are capable of activating an antiviral immune response. $98N_{12}$-5 was shown to effectively deliver siRNA targeting an influenza gene (NP-1496 or PR8) to Vero cells. Unfortunately, no efficacy was seen in a mouse model.[80] However, in 2012, Nguyen *et al.*[51] synthesized a new lipidoid library specifically for isRNA delivery and were able to identify two structures that outperformed $98N_{12}$-5. These new structures, LRNP-I and LRNP-II, were able to stimulate antiviral activity that suppressed influenza virus replication in an *in vivo* mouse model.[51] This is a prime example of how combinatorial library synthesis and high-throughput screening can lead to delivery vehicles specific for the intended therapeutic application.

Love *et al.*[35] synthesized lipidoid C12-200 [Figure 5.5(b)][35] for siRNA delivery via the route of epoxide-amine ring-opening reactions (as detailed in section 5.3.3).

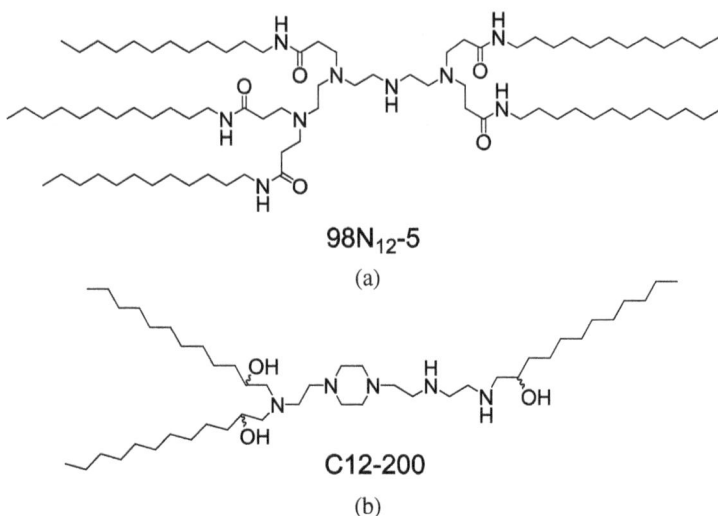

Figure 5.5:   Chemical structures of (a) $98N_{12}-5$[15] and (b) C12-200.[35]

Using *in vitro* and *in vivo* screening for siRNA delivery, the authors were able to narrow the library down to three top performing structures, C16-96, C14-110 and C12-200. These three structures showed almost 100% gene silencing in a mouse model, which led the authors to investigate if these lipids were still effective over a range of low doses. Their results showed that C12-200 was a far superior structure; to attain 80% gene silencing, a dose of 2 mg/kg or 0.5 mg/kg was needed with C16-96 or C14-110, respectively, while a dose of only 0.03 mg/kg was needed with C12-200.[35] In addition, this structure showed orders of magnitude higher potency compared to structure $98N_{12}-5$ for liver delivery in a mouse model[36] and non-human primate model.[15,35] The main advantage of using such low doses for human medical application is the higher tolerance associated with using lower levels of both siRNA and delivery material for silencing.

5.5.2 *DNA*

There are many differences to consider when it comes to delivering plasmid DNA (pDNA) instead of siRNA for gene therapy. In terms of the mechanism of therapy, siRNA silences a gene, while pDNA delivers a therapeutic gene that is expressed by the cell as a protein or molecule of interest. Therefore, to be functional, pDNA must be delivered into the nucleus for it to be expressed, compared to siRNA which can silence mRNA in the cytoplasm.[80] Structurally, both pDNA and siRNA are negatively-charged nucleic acids; however, pDNA is on a much larger size scale (thousands of base pairs versus 21 to 23 for siRNA) and therefore can more strongly interact with positively-charged delivery materials.[81] Cationic lipids have shown great promise for the delivery of pDNA, as reviewed in these articles.[16,68,82] However, the use of lipidoid materials for pDNA delivery is still relatively new.

Sun et al.[38] were the first to create and screen a lipidoid library specifically for pDNA delivery. Here, the authors utilized two different synthesis methods, Michael addition reaction and epoxide-amine addition, to create a library of 23 unique structures that had variations in tail length (8–16 carbon), functional group on the tail (acrylamide acrylate and epoxide), and amine core structure. By screening for β-gal pDNA delivery, five of the lipidoids were found capable of delivering pDNA at a higher efficacy than Lipofectamine 2000.[38] These lipidoids were also shown to deliver to several different cell lines, proving that lipidoids are well suited as materials for pDNA delivery.[39]

Other articles have published results from library synthesis and screening for the delivery of pDNA.[56,58,59,67] However, it was only recently in 2012 that Dong et al[54] demonstrated in vivo delivery of pDNA after screening. Using a mouse model, lipidoid-encapsulated luciferase pDNA was administered via tail vein injection. Eight hours after injection, successful and efficient pDNA delivery was observed in the lungs and spleen.[54] This result is seen as a positive confirmation that lipidoid structures made from combinatorial synthesis and screening have the potential to be therapeutic pDNA delivery vehicles.

5.5.3 *Protein*

Therapeutic proteins can be used to treat a wide range of illnesses such as infectious diseases (Fuzeon, used to treat HIV),[83] cancer (Herceptin, used to treat breast cancer)[84] and pulmonary disorders (Xolair, used to treat moderate to severe asthma).[85,86] The main challenges involved with delivering proteins are their high molecular weight, often short half-lives, instability, and immunogenicity upon multiple doses.[1] Protein modifications such as PEG and polysialic acid (PSA) can improve stability and pharmacokinetics,[1,86] but can also lower protein activity. Using a carrier to deliver the protein safely into the cell cytoplasm alleviates most of these problems, while also reducing the need for inactivating protein modifications.[1] Wang et al.[53] were the first to investigate the use of a combinatorial library of lipidoids for protein delivery. They also investigated a protein modification strategy to increase the typically low electrostatic interactions between proteins and lipidoid material, which can lead to poor loading. Using two different cytotoxic proteins, RNase-A and saporin, a *cis*-aconitic anhydride modification was completed by mixing, followed by dialysis purification. This modification created proteins with improved electrostatic interactions, but that is also reversible in a slightly acidic environment (pH 5.2) similar to that of an endosome or lysosome.[53] To identify lipidoids that provide efficient protein delivery, a combinatorial library of 14 lipidoid structures was synthesized and screened. *In vitro* results showed a significant increase in delivery efficacy using modified RNase-A compared with unmodified RNase-A. However, for saporin, the modification did not make a significant difference in delivery efficacy. One lipidoid structure, EC16-1, showed great potential for protein delivery, with cell viability below 30%, when complexed with either unmodified saporin, modified saporin or modified RNase-A. The structure was also tested *in vivo* for its ability to deliver saporin in a 4T1 breast cancer mouse model. EC16-1/saporin reduced tumor sizes by 80% compared with controls, including free saporin.[53] In addition, modified EC16-80 was shown *in vitro* to efficiently deliver a reactive oxygen species (ROS)-modified RNase-A protein to target cancer cells.[87] Together, these results provide a reversible protein modification scheme to increase protein delivery as well as show the potential of lipidoids for protein delivery.

5.6　Conclusions and Future Directions

In this chapter, we have summarized recent progress in the synthesis of combinatorial cationic lipid-like nanoparticles and the high-throughput screening of these lipidoids for gene and protein delivery. The use of simple, quick and convenient chemistries such as Michael addition and epoxide-amine addition has facilitated the parallel synthesis of large libraries of lipidoids with diversified chemical structures. Rapid high-throughput screening identifies the most promising structure for specific medical applications, making it facile to develop, analyze and select highly-effective delivery structures. Moreover, the combinatorial library strategy simplifies the identification of SAR of lipidoid-facilitated delivery, and provides data that can be used to optimize and develop the next generation of lipidoids to further improve delivery efficiency.

The future of combinatorial synthesis of cationic lipids lies in the development of novel chemistry principles, such that even faster synthesis routes than those reviewed here can be achieved. Furthermore, the key to success will be the development of even greater structural diversity. In particular, identifying biodegradable lipid candidates with enhanced systemic clearance mechanisms will lead to safer and more effective delivery systems. Also important are strategies for internalizing and delivering larger biotherapeutics, such as antibodies, which would broaden the spectrum of therapeutic applications based on lipidoids.

Acknowledgements

Q. Xu acknowledges support from the Pew Scholar for Biomedical Sciences program of the Pew Charitable Trusts.

References

1. Pisal, D.S. *et al.* Delivery of therapeutic proteins. *Journal of Pharmaceutical Sciences* 99(6), 2557–2575 (2010).
2. Pouton, C.W., and Seymour, L.W. Key issues in non-viral gene delivery. *Advanced Drug Delivery Reviews* 46(1–3), 187–203 (2001).
3. Kostarelos, K., and Miller, A.D. What role can chemistry play in cationic liposome-based gene therapy research today? *Non-Viral Vectors for Gene Therapy, 2nd Edition: Part 1*, 71–118 (2005).
4. Namiki, Y. *et al.* Nanomedicine for cancer: lipid-based nanostructures for drug delivery and monitoring. *Accounts of Chemical Research* 44(10), 1080–1093 (2011).
5. Donkuru, M. *et al.* Advancing nonviral gene delivery: lipid- and surfactant-based nanoparticle design strategies. *Nanomedicine* 5(7), 1103–1127 (2010).
6. Lin, Q., Chen, J., Zhang, Z., and Zheng, G. Lipid-based nanoparticles in the systemic delivery of siRNA. *Nanomedicine* 9(1), 105–120 (2014).
7. Zhang, S., Zhi, D., and Huang, L. Lipid-based vectors for siRNA delivery. *Journal of Drug Targeting* 20(9), 724–735 (2012).
8. Li, W., and Szoka, F., Jr. Lipid-based nanoparticles for nucleic acid delivery. *Pharmaceutical Research* 24(3), 438–449 (2007).
9. Allen T., and Cullis, P. Liposomal drug delivery systems: From concept to clinical applications. *Advanced Drug Delivery Reviews* 65(1), 36–48 (2013).
10. Schroeder, A., Levins Cg and Cortez, C., Langer, R., and Anderson, D. Lipid-based nanotherapeutics for siRNA delivery. *Journal of Internal Medicine* 267(1), 9–21 (2010).
11. Felgner, P., and Ringold, G. Cationic liposome-mediated transfection. *Nature* 337(6205), 387–388 (1989).
12. Malone, R., Felgner, P., and Verma, I. Cationic liposome-mediated RNA transfection. *Proceedings of the National Academy of Sciences* 86(16), 6077–6081 (1989).
13. Pierrat, P., Creusat, G., Laverny, G., Pons, F., Zuber, G., and Lebeau, L. A cationic phospholipid–detergent conjugate as a new efficient carrier for siRNA delivery. *Chemistry — A European Journal* 18(13), 3835–3839 (2012).
14. Zhao, Y.-N., Qureshi, F., Zhang, S.-B. *et al.* Novel Gemini cationic lipids with carbamate groups for gene delivery. *Journal of Materials Chemistry B* 2(19), 2920–2928 (2014).
15. Akinc, A., Zumbuehl, A., Goldberg, M. *et al.* A combinatorial library of lipid-like materials for delivery of RNAi therapeutics. *Nat Biotech.* 26(5), 561–569 (2008).

16. De Lima, M., Simoes, S., Pires, P., Faneca, H., and Duzgunes, N. Cationic lipid-DNA complexes in gene delivery: From biophysics to biological applications. *Advanced Drug Delivery Reviews* 47(2–3), 277–294 (2001).

17. Rehman, Z., Zuhorn, I., and Hoekstra, D. How cationic lipids transfer nucleic acids into cells and across cellular membranes: Recent advances. *Journal of Controlled Release* 166(1), 46–56 (2013).

18. Rejman, J., Oberle, V., Zuhorn, I., and Hoekstra, D. Size-dependent internalization of particles via the pathways of clathrin- and caveolae-mediated endocytosis. *Biochem. J.* 377(1), 159–169 (2004).

19. Saffari, M., Shirazi, F., Oghabian, M., and Moghimi, H. Preparation and *in-vitro* evaluation of an antisense-containing cationic liposome against non-small cell lung cancer: a comparative preparation study. *Iran. J. Pharm. Res.* 12, 1–8 (2013).

20. Zhang, H., Zhang, L., Sun, X., Diao, S., and Zhang, Zr. Assembly of plasmid DNA into liposomes after condensation by cationic lipid in anionic detergent solution. *Biotechnology Letters* 27(21), 1701–1705 (2005).

21. Fenske, D.B., and Cullis, P.R. Entrapment of small molecules and nucleic acid-based drugs in liposomes. In N. Duzgunes (Ed.), *Liposomes, Part E*. Elsevier Academic Press Inc, San Diego, 7–40 (2005).

22. Maulucci, G., De Spirito, M., Arcovito, G., Boffi, F., Castellano, A., and Briganti, G. Particle size distribution in DMPC vesicles solutions undergoing different sonication times. *Biophysical Journal* 88(5), 3545–3550 (2005).

23. Balbino, T., Aoki, N., Gasperini, A. *et al.* Continuous flow production of cationic liposomes at high lipid concentration in microfluidic devices for gene delivery applications. *Chemical Engineering Journal* 226(0), 423–433 (2013).

24. Lapinski, M.M. *et al.* Comparison of liposomes formed by sonication and extrusion: Rotational and translational diffusion of an embedded chromophore. *Langmuir* 23(23), 11677–11683 (2007).

25. Koh, C., Zhang, X., Liu, S. *et al.* Delivery of antisense oligodeoxyribonucleotide lipopolyplex nanoparticles assembled by microfluidic hydrodynamic focusing. *Journal of Controlled Release* 141(1), 62–69 (2010).

26. Torchilin, V. Intracellular delivery of protein and peptide therapeutics. *Drug Discovery Today: Technologies* 5(2–3), e95–e103 (2008).

27. Mishra, G., Bagui, M., Tamboli, V., and Mitra, A. Recent applications of liposomes in ophthalmic drug delivery. *Journal of Drug Delivery* 2011, 863734 (2011).

28. Gujrati, M., Malamas, A., Shin, T., Jin, E., Sun, Y., and Lu, Z.-R. Multifunctional cationic lipid-based nanoparticles facilitate endosomal escape and reduction-triggered cytosolic siRNA release. *Molecular Pharmaceutics* 11(8), 2734–2744 (2014).

29. Hess, G., Humphries, W., Fay, N., and Payne, C. Cellular binding, motion, and internalization of synthetic gene delivery polymers. *Biochimica Et Biophysica Acta-Molecular Cell Research* 1773(10), 1583–1588 (2007).

30. Hoekstra, D., Rejman, J., Wasungu, L., Shi, F., and Zuhorn, I. Gene delivery by cationic lipids: in and out of an endosome. *Biochemical Society Transactions* 35, 68–71 (2007).

31. Lazo, J., and Wipf, P. Combinatorial chemistry and contemporary pharmacology. *Journal of Pharmacology and Experimental Therapeutics* 293(3), 705–709 (2000).

32. Whitehead, K., Matthews, J., Chang, P., *et al.* *In vitro–in vivo* translation of lipid nanoparticles for hepatocellular siRNA delivery. *ACS Nano* 6(8), 6922–6929 (2012).

33. Whitehead, K., Dorkin, J., Vegas, A., *et al.* Degradable lipid nanoparticles with pre-dictable *in vivo* siRNA delivery activity. *Nat Commun* 5, (2014).
34. Mahon, K., Love, K., Whitehead, K., *et al.* Combinatorial approach to determine functional group effects on lipidoid-mediated siRNA delivery. *Bioconjugate Chemistry* 21(8), 1448–1454 (2010).
35. Love, K., Mahon, K., Levins, C. *et al.* Lipid-like materials for low-dose, *in vivo* gene silencing. *Proceedings of the National Academy of Sciences of the United States of America* 107(5), 1864–1869 (2010).
36. Akinc, A., Goldberg, M., Qin, J., *et al.* Development of lipidoid-siRNA formulations for systemic delivery to the liver. *Mol Ther* 17(5), 872–879 (2009).
37. Wang, M. *et al.* Enhanced intracellular siRNA delivery using bioreducible lipid-like nanoparticles. *Advanced Healthcare Materials* 3(9), 1398–1403 (2014).
38. Sun, S., Wang, M., Knupp, S. *et al.* Combinatorial library of lipidoids for *in vitro* DNA delivery. *Bioconjugate Chemistry* 23(1), 135–140 (2012).
39. Wang, M., Sun, S., Alberti, K., and Xu, Q. A combinatorial library of unsaturated lipidoids for efficient intracellular gene delivery. *ACS Synthetic Biology* 1(9), 403–407 (2012).
40. Xue, W., Dahlman, Je, Tammela, T., *et al.* Small RNA combination therapy for lung cancer. *Proceedings of the National Academy of Sciences of the United States of America* 111(34), E3553–E3561 (2014).
41. Lenssen, K., Jantscheff, P., Von Kiedrowski, G., and Massing, U. Combinatorial synthe-sis of new cationic lipids and high-throughput screening of their transfection properties. *ChemBioChem* 3(9), 852–858 (2002).
42. Howard, K.A. and Goldberg, M. Lipidoids: A combinatorial approach to siRNA deliv-ery. In: *RNA Interference from Biology to Therapeutics*, (Ed. K.A. Howard). Springer US, 143–160 (2013).
43. Behr, J., Demeneix, B., Loeffler, J., and Mutul, J. Efficient gene-transfer into mammalian primary endocrine-cells with lipopolyamine-coated DNA. *Proceedings of the National Academy of Sciences of the United States of America* 86(18), 6982–6986 (1989).
44. Shirazi, R., Ewert, K., Leal, C., Majzoub, R., Bouxsein, N., and Safinya, C. Synthesis and characterization of degradable multivalent cationic lipids with disulfide-bond spac-ers for gene delivery. *Biochimica et Biophysica Acta (BBA) — Biomembranes* 1808(9), 2156–2166 (2011).
45. Altinoglu, S., Wang, M., and Xu, Q. Combinatorial library strategies for synthesis of cationic lipid-like nanoparticles and their potential medical applications. *Nanomedicine* In Press, (2015).
46. Merrifield, B. Concept and early development of solid-phase peptide synthesis. In: *Methods in Enzymology* (Ed. B. Gregg). Academic Press, 3–13 (1997).
47. Yingyongnarongkul, B.-E., Howarth, M., Elliott, T., and Bradley, M. Solid-phase syn-thesis of 89 polyamine-based cationic lipids for DNA delivery to mammalian cells. *Chemistry — A European Journal* 10(2), 463–473 (2004).
48. Radchatawedchakoon, W., Watanapokasin, R., Krajarng, A., and Yingyongnarongkul, B.-E. Solid phase synthesis of novel asymmetric hydrophilic head cholesterol-based cationic lipids with potential DNA delivery. *Bioorganic & Medicinal Chemistry* 18(1), 330–342 (2010).

49. Liberska, A., Lilienkampf, A., Unciti-Broceta, A., and Bradley, M. Solid-phase synthesis of arginine-based double-tailed cationic lipopeptides: potent nucleic acid carriers. *Chemical Communications* 47(48), 12774–12776 (2011).
50. Nair, D., Podgórski, M., Chatani, S., *et al.* The thiol-Michael addition click reaction: A powerful and widely used tool in materials chemistry. *Chemistry of Materials* 26(1), 724–744 (2013).
51. Nguyen, D., Mahon, K., Chikh, G., *et al.* Lipid-derived nanoparticles for immunostimulatory RNA adjuvant delivery. *Proceedings of the National Academy of Sciences,* (2012).
52. Cho, S., Goldberg, M., Son, S., *et al.* Lipid-like nanoparticles for small interfering RNA delivery to endothelial cells. *Advanced Functional Materials* 19(19), 3112–3118 (2009).
53. Wang, M., Alberti, K., Sun, S., Arellano, C., and Xu, Q. Combinatorially designed lipid-like nanoparticles for intracellular delivery of cytotoxic protein for cancer therapy. *Angewandte Chemie-International Edition* 53(11), 2893–2898 (2014).
54. Dong, Y., *et al.* Lipid-like nanomaterials for simultaneous gene expression and silencing *in vivo*. *Advanced Healthcare Materials* 3, 9 (2014).
55. Kolb, H., Finn, M., and Sharpless, K. Click chemistry: Diverse chemical function from a few good reactions. *Angewandte Chemie-International Edition* 40(11), 2004-2021 (2001).
56. Li, L., Zahner, D., Su, Y., Gruen, C., Davidson, G., and Levkin, P. A biomimetic lipid library for gene delivery through thiol-yne click chemistry. *Biomaterials* 33(32), 8160–8166 (2012).
57. Alabi, C., Love, K., Sahay, G., *et al.* Multiparametric approach for the evaluation of lipid nanoparticles for siRNA delivery. *Proceedings of the National Academy of Sciences,* (2013).
58. Sheng, R., Luo, T., Li, H., Sun, J., Wang, Z., and Cao, A. 'Click' synthesized sterolbased cationic lipids as gene carriers, and the effect of skeletons and headgroups on gene delivery. *Bioorganic & Medicinal Chemistry* 21(21), 6366–6377 (2013).
59. Li, L., Wang, F., Wu, Y., Davidson, G., and Levkin, P. Combinatorial synthesis and high-throughput screening of alkyl amines for conviral gene delivery. *Bioconjugate Chemistry* 24(9), 1543–1551 (2013).
60. Salvatore, R., Nagle, A., and Jung, K. Cesium effect: high chemoselectivity in direct N-alkylation of amines. *The Journal of Organic Chemistry* 67(3), 674–683 (2002).
61. Felgner, J., Kumar, R., Sridhar, C., *et al.* Enhanced gene delivery and mechanism studies with a novel series of cationic lipid formulations. *Journal of Biological Chemistry* 269(4), 2550–2561 (1994).
62. Suma, T., Miyata, K., Anraku, Y., *et al.* Smart multilayered assembly for biocompatible siRNA delivery featuring dissolvable silica, endosome-disrupting polycation, and detachable PEG. *ACS Nano* 6(8), 6693–6705 (2012).
63. Gehin, C., Montenegro, J., Bang, E.-K., *et al.* Dynamic amphiphile libraries to screen for the "fragrant" delivery of siRNA into HeLa cells and human primary fibroblasts. *Journal of the American Chemical Society* 135(25), 9295–9298 (2013).
64. Candiani, G., Pezzoli, D., Ciani, L., Chiesa, R., and Ristori, S. Bioreducible liposomes for gene delivery: From the formulation to the mechanism of action. *PLoS ONE* 5(10), e13430 (2010).

65. Saito, G., Swanson, J., and Lee, K.-D. Drug delivery strategy utilizing conjugation via reversible disulfide linkages: role and site of cellular reducing activities. *Advanced Drug Delivery Reviews* 55(2), 199–215 (2003).

66. Whitehead, K., Sahay, G., Li, G., *et al.* Synergistic silencing: combinations of lipid-like materials for efficacious siRNA delivery. *Mol Ther* 19(9), 1688–1694 (2011).

67. Sun, S., Wang, M., Alberti, K., Choy, A., and Xu, Q. DOPE facilitates quaternized lipidoids (QLDs) for *in vitro* DNA delivery. *Nanomedicine: Nanotechnology, Biology and Medicine* 9(7), 849–854 (2013).

68. Karmali, P., and Chaudhuri, A. Cationic liposomes as non-viral carriers of gene medicines: Resolved issues, open questions, and future promises. *Medicinal Research Reviews* 27(5), 696–722 (2007).

69. Wheeler, J., Palmer, L., Ossanlou, M., *et al.* Stabilized plasmid-lipid particles: Construction and characterization. *Gene Therapy* 6(2), 271–281 (1999).

70. Shim, G., Kim, M.-G., Park, J., and Oh, Y.-K. Application of cationic liposomes for delivery of nucleic acids. *Asian Journal of Pharmaceutical Sciences* 8(2), 72–80 (2013).

71. Röther, S., Meister, G. Small RNAs derived from longer non-coding RNAs. *Biochimie* 93(11), 1905–1915 (2011).

72. Whitehead, K., Langer, R., and Anderson, D. Knocking down barriers: advances in siRNA delivery. *Nat Rev Drug Discov* 8(2), 129–138 (2009).

73. Lin, Q., Chen, J., Zhang, Z., Zheng, G. Lipid-based nanoparticles in the systemic delivery of siRNA. *Nanomedicine* 9(1), 105–120 (2014).

74. Ku, S., Kim, K., Choi, K., Kim, S., and Kwon, I. Tumor-targeting multifunctional nanoparticles for siRNA delivery: Recent advances in cancer therapy. *Advanced Healthcare Materials* 3(8), 1182–1193 (2014).

75. Williford, J.-M., Wu, J., Ren, Y., Archang, M., Leong, K., and Mao, H.-Q. Recent advances in nanoparticle-mediated siRNA delivery. *Annual Review of Biomedical Engineering* 16(1), 347–370 (2014).

76. Yin, H., Kanasty, R., Eltoukhy, A., Vegas, A., Dorkin, J., and Anderson, D. Non-viral vectors for gene-based therapy. *Nat Rev Genet* 15(8), 541–555 (2014).

77. Zhang, S., Zhao, Y., Zhi, D., and Zhang, S. Non-viral vectors for the mediation of RNAi. *Bioorganic Chemistry* 40(0), 10–18 (2012).

78. Shu, Y., Pi, F., and Sharma, A., *et al.* Stable RNA nanoparticles as potential new generation drugs for cancer therapy. *Advanced Drug Delivery Reviews* 66(0), 74–89 (2014).

79. Wasungu, L., and Hoekstra, D. Cationic lipids, lipoplexes and intracellular delivery of genes. *Journal of Controlled Release* 116(2), 255–264 (2006).

80. Nguyen, D., Chen, S., Lu, J., *et al.* Drug delivery-mediated control of RNA immunostimulation. *Mol Ther* 17(9), 1555–1562 (2009).

81. Scholz, C., and Wagner, E. Therapeutic plasmid DNA versus siRNA delivery: Common and different tasks for synthetic carriers. *Journal of Controlled Release* 161(2), 554–565 (2012).

82. Immordino, M., Dosio, F., Cattel, L. Stealth liposomes: review of the basic science, rationale, and clinical applications, existing and potential. *Int. J. Nanomed.* 1(3), 297–315 (2006).

83. Matthews, T., Salgo, M., Greenberg, M., Chung, J., Demasi, R., and Bolognesi, D. Enfuvirtide: the first therapy to inhibit the entry of HIV-1 into host CD4 lymphocytes. *Nat Rev Drug Discov* 3(3), 215–225 (2004).

84. Vogel, C., Cobleigh, M., Tripathy, D., *et al.* Efficacy and safety of trastuzumab as a single agent in first-line treatment of HER2-overexpressing metastatic breast cancer. *Journal of Clinical Oncology* 20(3), 719–726 (2002).

85. Silkoff, P., Romero, F., Gupta, N., Townley, R., and Milgrom, H. Exhaled nitric oxide in children with asthma receiving xolair (Omalizumab), a monoclonal anti-immunoglobulin E antibody. *Pediatrics* 113(4), e308–e312 (2004).

86. Leader, B., Baca, Q., and Golan, D. Protein therapeutics: a summary and pharmacological classification. *Nat Rev Drug Discov* 7(1), 21–39 (2008).

87. Wang, M., Sun, S., Neufeld, C., Perez-Ramirez B., and Xu, Q. Reactive oxygen species-responsive protein modification and its intracellular delivery for targeted cancer therapy. *Angewandte Chemie (International ed. in English)* 53(49), 13444–13448 (2014).

Biodegradable Polymeric Nanoparticles for Gene Delivery

6

Jayoung Kim*, Kristen L. Kozielski*, David R. Wilson* and Jordan J. Green*,†

*Department of Biomedical Engineering, Translational Tissue Engineering Center, and Institute for Nanobiotechnology, Johns Hopkins University School of Medicine, 733 N Broadway, Baltimore, MD 21205, USA
†Departments of Materials Science and Engineering, Ophthalmology, and Neurosurgery, Johns Hopkins University School of Medicine, 733 N Broadway, Baltimore, MD 21205, USA

Abstract

Increasingly, scientists and clinicians are discovering the genetic basis of diseases, ranging from monogenic disorders to multigenic diseases such as various cancers. Moreover, human diseases are increasingly being understood on the molecular level, rather than as a phenotype, which is moving treatments towards stratified and personalized medicine. Gene therapy holds the promise of a technology that could address this growing need for genetic medicine as it can potentially tune individual cell gene expression on or off in a targeted and precise manner. The technology could, theoretically, also be applied to almost any human disease. The central challenge is that in practice, safe and effective delivery of desired nucleic acids to targeted human cells is very difficult. This chapter outlines these challenges in gene delivery and discusses state-of-the art approaches using biodegradable polymers to overcome these obstacles and obtain successful gene delivery *in vitro* and *in vivo*. Biodegradable gene delivery polymers, or plastics that are designed to safely deliver a biological cargo inside cells and then degrade, have certain advantages over other materials and viruses for the delivery of genes. This chapter elucidates the diverse types of biodegradable polymers used for gene delivery, the related nanoparticulate systems they form with nucleic acids, and the structural properties that increase their efficacy and safety.

6.1 Introduction

The delivery of nucleic acids to manipulate gene regulation can be both a therapeutic and scientific tool. Diseases caused by missing or defective genes can potentially be cured by replacing these genes, such as upregulating tumor suppressor genes in cancer.[1–3] The immune system can be modulated by the introduction of DNA-based vaccines,[4,5] or genes that allow the immune system to better recognize or fight cancer.[6,7] Additionally, suicide genes can be introduced to kill cancer cells.[8] *Ex vivo*, gene therapy can be used to manipulate stem cells for targeted differentiation,[9] or reprogram induced pluripotent stem cells from differentiated cells.[10] Turning off or down regulating genes can be used to treat diseases caused by gene overexpression.[11,12] Technology to selectively turn genes off is also a valuable biological tool to elucidate the function of genes within a cell and in the context of a disease.[13]

Viral gene delivery vectors, although effective, come with risks such as tumorigenicity and immunogenicity.[14] Adenovirus-mediated gene delivery studies have shown dosage repeatability to be challenging and concentrations to be limited by toxicity and humoral immune response.[15] Although non-viral nucleic acid delivery can avoid these issues, it is typically less effective.[16]

Lipid-based and inorganic delivery vehicles have previously been examined for their potential to deliver nucleic acids. Lipid-based delivery is well characterized,[17–19] and lipid-based delivery vehicles are commercially available for the *in vitro* delivery of DNA[20,21] and siRNA.[22] Lipid-based nanoparticles can potentially generate off-target and immunogenic effects,[20] but there are strategies to attenuate these unwanted interactions, such as the introduction of poly(ethylene glycol) (PEG) shielding.[23,24] Calcium phosphate crystals,[25–27] gold nanoparticles, quantum-dots and other inorganic materials have also been employed for non-viral gene delivery. Gold is advantageous because it is biocompatible, easy to functionalize, and has malleable physical properties.[28–32] Quantum dots are useful for fluorescent imaging as they are brighter and less prone to photobleaching than typical fluorophores.[33,34] Several lipid-based and inorganic nanoparticle systems have also been combined with polymeric materials for enhanced gene delivery, particularly through the incorporation of PEG coatings[24,35,36] for nanoparticulate shielding or polyamines for improved interaction with DNA and intracellular delivery.[37,38]

Biodegradable polymeric gene delivery systems are a relatively newer class of materials for non-viral gene therapy. They are promising due to key features such as safety mediated by their biodegradability, design flexibility due to their tunable structure, large cargo capacity, and relative ease of manufacture. This chapter will focus on biodegradable polymeric nanoparticles for gene delivery. They will be discussed in the context of systemic and intracellular barriers to gene delivery and how polymer design can be utilized to overcome these barriers. New developments in the field of biodegradable polymeric gene delivery and an outlook for the future will be highlighted.

6.2   Obstacles to Gene Delivery

The central limitation of non-viral vectors for gene delivery is inefficient gene trans-fection arising from natural mechanisms of the human body to protect itself against foreign substances.[39] These barriers to biomaterial-mediated gene transfection span a spectrum from the systemic level to the cellular level (Figure 6.1).[40] Different bio-material properties and modifications to gene carriers are important for each step of the delivery process leading to successful expression of the exogenously deliv-ered nucleic acid. We will discuss seven major biological barriers to gene transfer using non-viral vectors and strategies to overcome each of these barriers. It is also important to note that design properties that seemingly overcome one of the delivery obstacles could pose a challenge to other obstacles; hence, further effort is needed to globally optimize polymeric nanoparticles and balance potential trade-offs.

6.2.1   *Nucleic Acid Binding/Encapsulation*

An efficient gene delivery vector must condense or encapsulate the nucleic acid to prevent enzymatic degradation and facilitate its cellular entry.[41] As DNA is a strongly negatively-charged material, early work in the field of non-viral gene delivery focused on the use of naturally occurring biological materials with signif-icant positive charge that could electrostatically bind to the DNA. In this manner, DNA could be condensed into a smaller size, be made more resistant to poten-tial enzymatic degradation, and have improved ability to enter cells. Polycation poly(*L*-lysine) (PLL) was observed to bind to nucleohistones,[42,43] and was later investigated as one of the earliest polymers to form nanocomplexes with DNA.[44–46] PLL is capable of complexing with DNA to form nanoparticles that successfully undergo cellular uptake but fail to escape from the endosome.[47] To overcome this challenge, PLL has been used in combination with other materials that aid in endo-somal escape, including other peptide molecules and pH-sensitive moieties that make use of the proton sponge effect.[48–50] To improve delivery in comparison to PLL, alternative gene delivery materials had to be discovered. An off-the-shelf commercially produced polymer with very high charge density, polyethylenimine (PEI), was first reported for use in transfection in 1995 by Boussif *et al.*, who attributed its high transfection efficiency to its ability to undergo endosomal escape via the proton sponge effect.[51] Unfortunately, while transfection efficacy is corre-lated with the molecular weight of PEI, cytotoxicity is similarly correlated, mak-ing unmodified high MW PEI largely unsuitable for *in vivo* applications. PEI is a non-biodegradable polycation that requires excess polymer to effectively transfect

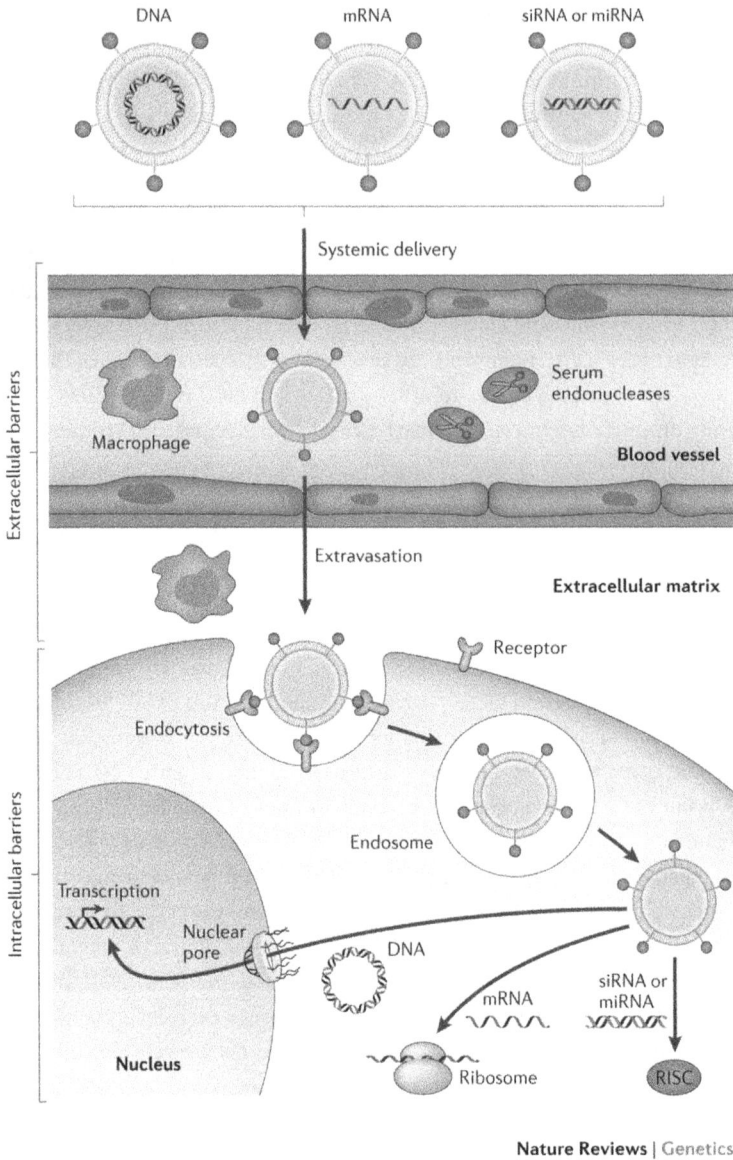

Figure 6.1:　Major barriers to nucleic acid delivery using nanoparticles include stable particle formation, systemic circulation, tissue and cell targeting, cellular uptake, endosomal escape, and release of nucleic acid. Reprinted with permission from Macmillan Publishers Ltd.[40]

cells and can lead to accumulation upon repeated administration.[52,53] To over-come this first obstacle to non-viral gene delivery, other biomaterials can be used to encapsulate nucleic acids into particulates. Amphiphilic lipids, such as N-[1-(2,3-dioleyloxy)propyl-N,N,N-trimethylammonium chloride (DOTMA), and relatively hydrophobic polymers, including poly(lactide-*co*-glycolide) polymer (PLG), form DNA-encapsulated nanoparticles by either phase separation alone or in combination with electrostatic interaction.[54,55]

Polymeric vectors are able to deliver different types of nucleic acids. For DNA, standard double-stranded plasmids as well as minicircles, which are plasmids that have had prokaryote sequences such as the CpG islands removed, are widely used to introduce exogenous genes that encode for proteins of interest.[56] Because of the large size of DNA molecules, they are able to bind to cationic polymers such as PEI and poly(β-amino ester)s (PBAE) and form stable nanoparticles.[57,58] Small interfering RNA (siRNA), through the RNA interference pathway, causes mRNA to be degraded and translation to be inhibited.[59] As an siRNA molecule is dram-atically shorter ($\sim$20 bp vs. $>$1,000 bp) and stiffer than a plasmid DNA molecule, stable particle formulation through electrostatic interactions with cationic polymers is more difficult.[60,61] Therefore, siRNA-delivering nanoparticles can require more complex engineering solutions to form effective particles, such as making siRNA more multivalent by introducing short complementary overhangs[62] or multimeriz-ing siRNA molecules with cleavable disulfide linkages.[63] More recently, an enzy-matic RNA polymerization technique has been used to condense RNA structures into self-assembled RNAi-microsponges.[64] While siRNA binding and encapsula-tion can be challenging, siRNA delivery overall is not necessarily more difficult than DNA delivery, as unlike DNA, siRNA does not require nuclear import to function.[65]

6.2.2 *Systemic Circulation*

Both viral-based and non-viral-based vectors face the problem of rapid clearance from the systemic circulation on the order of minutes.[66,67] While viruses often suffer from specific antibody-mediated immune response, non-viral platforms are quickly cleared by several non-specific mechanisms. Most polymeric gene delivery materials are positively charged as electrostatic interaction is the prevalent driving force in forming many types of nanoparticles for gene delivery. The resulting positive surface charge of nanoparticles provides colloidal stability in aqueous solutions and facilitates interaction with cellular membranes. However, it also attracts anionic counter ions in physiological salt and serum proteins that cause opsonization and aggregation, leading to increased clearance by the mononuclear phagocyte system (MPS) and the reticuloendothelial system (RES).[68–70]

Several strategies have been employed to minimize clearance from the systemic circulation, including charge shielding and shape control. One common "stealth" technology involves coating nanoparticles with PEG, which provides a relatively inert surface due to its neutral and hydrophilic structure.[71,72] Conjugation of PEG or "PEGylation" to gene delivery particles composed of cationic polymers,[73,74] lipids,[75,76] dextran-spermine,[77] and other materials has been reported to show beneficial effects. An alternative approach is to neutralize excess positive surface charges by coating the particles with anionic biomacromolecules such as negatively-charged polypeptides.[78] More recently, engineering of particle shape has been shown to be an important parameter to extending circulation time. Nanoparticles with higher aspect ratios were shown to have longer circulation half-lives than spherical particles,[79] giving rise to a strategy that researchers have begun to use for gene delivery.[80]

6.2.3 *Tissue and Cell Targeting*

Tissue targeting can be accomplished through the design of a targeted systemically-administered nanoparticle, through a tissue-specific promoter, or through local injection. Nanoparticles can be injected into anatomically-accessible sites to enhance delivery to a local region of interest while reducing non-specific transfection at other sites. For example, intracerebroventricular delivery grants direct access to the brain, retrograde intrabiliary infusion to the liver, and intratumoral injection to tumors.[81–83]

Nanoparticles that are administered intravenously must have mechanisms to exit the circulation at the target tissue. The application of nanoparticles to solid tumors often benefits from passive targeting. The formation of new blood vessels near rapidly growing solid tumors allows nanoparticles with diameters of 400 nm or less to passively leak out of the neovasculature and distribute to the tumor tissue. Therefore, prolonging systemic circulation by the aforementioned strategies, such as PEGylation, can increase the accumulation of nanoparticles at a tumor site, and this enhanced permeability and retention (EPR) effect has become an important tool in nanoparticle-mediated gene delivery for cancer therapy.[84] On the other hand, molecular ligands and chemical moieties that bind specifically to overexpressed receptors on the surface of vascular endothelial cells near a solid tumor have been conjugated to various nanoparticles as an active targeting mechanism. For example, arginine-glycine-aspartic acid (RGD) peptide sequences and other chemical antagonists to various integrin isoforms have been covalently conjugated or electrostatically bound via negatively-charged polypeptides to biomaterials for tumor vasculature targeting.[85–89] Another emerging method of conferring tissue specificity to nanoparticles utilizes aptamers as targeting ligands.[90]

Nanoparticles with targeting ligands can be used to target many additional cell types as well. When cell-specific RNA interference or therapeutic exogenous DNA expression is critical, nucleic acid delivery nanoparticles with cell-specific targeting can increase efficacy and reduce potential off-target side effects. Overexpressed receptors on the surface of specific cell types of interest are good candidates for ligand targeting. The literature on gene delivery and nanomedicine shows that modifying nanoparticles with different types of ligands, including galactosylated PEI targeting hepatocytes, antibodies specific to the insulin receptor to target cancer cells in brain, synthetic peptides that bind to integrin $\alpha 5\beta 1$ on neuroblastoma cells, small molecules targeting CD40 on ovarian cancer cells, and leukocyte function-associated antigen-1 to bind to melanoma cells, can lead to effective cell-specific gene delivery.[91–95]

Other methods for cell-specific gene delivery are also possible. Following a polymer library approach, a recent study by Guerrero-Cázares et al. showed that the specific chemical structure of the polymer comprising a polymer/DNA gene delivery nanoparticle can confer specificity to one type of cell over another, such as primary brain tumor initiating cells over healthy neural progenitor cells.[96] Similar results have also been shown for polymeric nanoparticles that can target liver cancer cells[97] and endothelial cells.[98,99] A final approach is to include a cell type-specific promoter in the plasmid to promote the targeting of specific cells such as cancer cells.[100] In this manner, even if some polymeric nanoparticles are delivered to off-target cells, the exogenous gene is only successfully expressed in target cells where the specific promoter is active.

6.2.4 *Cellular Uptake*

Once the nanoparticles reach the cells of interest, they must overcome several barriers at the cellular level before successful transfection can be achieved. First, gene carriers need to cross the cellular membrane, for which exist both non-specific and specific mechanisms. Macropinocytosis is a non-specific cellular uptake mechanism where cells engulf extracellular fluid through actin-driven evagination.[101] However, gene vectors that enter via macropinocytosis result in poor transfection efficacy due to the high rate of recycling.[102] Also, the positive surface charge on nanoparticles formulated with cationic polymers or lipids promotes electrostatic interaction with the negatively-charged cell surface, which in turn triggers another non-specific pathway called adsorptive endocytosis.[103] It should be noted that the positive surface charge can be shielded by approaches designed to increase systemic circulation time, such as PEGylation. These coatings minimize electrostatic interactions between the nanoparticle and cellular membrane.

On the other hand, specific uptake mechanisms are mediated by receptors on the cellular membrane, which can recognize various molecular ligands as well as chemical moieties on nanoparticles. In the case of PEGylated nanoparticles, ligands can be conjugated to the terminal ends of the PEG chain, and chemical moieties can be exposed upon environment-stimulated PEG cleavage.[104,105] There are two major specific uptake routes for gene carriers. Clathrin-mediated endocytosis is initiated within clathrin-coated pits of approximately 100–150 nm in size that pinch off from the plasma membrane to form endosomes.[106] Nanoparticles endocytosed via this pathway include those modified with MC1SP-peptide and transferrin that target the melanocortin receptor-1 and the transferrin receptor, respectively, as well as unmodified lipoplexes and liposomes.[107–109] In comparison, caveolae-mediated endocytosis is characterized by flask-shaped invaginations of about 50–100 nm in diameter.[110] Folic acid ligands that bind with folate receptors, as well as unmodified polymeric nanoparticles, can be directed to caveolae-mediated uptake.[109,111]

Different uptake pathways lead to different intracellular fates, which underscores the importance of cellular uptake on successful transfection.[112] For example, the major route of uptake for PBAE nanoparticles does not necessarily lead to high transfection efficiency.[113] The surface of nanoparticles can be modified to direct their uptake pathway, improve the intracellular fate of the plasmids that they contain, and increase subsequent transfection efficiency.

6.2.5 *Endosomal Escape*

Once endocytosed, nanoparticles must escape the endosomal compartment and reach the cytoplasm. One mechanism of endosomal escape for polymeric nanoparticles is through the proton sponge effect.[114] Reversibly protonated biomaterial vectors can act as buffers as the endosome gradually becomes acidic, thereby protecting the cargo. Subsequently, chloride ions enter the endosomes to neutralize the charge, creating osmotic pressure that eventually leads to endosomal burst and cargo release. PLL, owing to its primary amines that are easily protonated at pH 7, is unable to provide strong buffering capacity at endosomal pH. In order to neutralize the acidic pH, researchers have either co-delivered PLL with amphipathic amines, such as chloroquine, or substituted its lysine residues with histidine or arginine residues that have titratable amines.[48,115] PEI and PBAE, on the other hand, have weakly basic tertiary amines in their structure that allow for the proton sponge effect.[51,116] Although the proton sponge hypothesis is widely believed to be the mechanism of endosomal escape, it has been challenged and remains to be clearly elucidated.[117]

Nanoparticles can also escape the endosome by destabilizing endosomal membranes. For example, amphiphatic, fusogenic peptides such as GALA (repeats of Glu-Ala-Leu-Ala) and KALA (repeats of Lys-Ala-Leu-Ala) have been utilized as the primary component of non-viral vectors and have also been associated non-covalently with nanoparticles.[118,119] These fusogenic peptides form alpha-helical structures at endosomal pH that can destabilize the endosomal membranes. Other amphiphatic lipids, such as dioleoylphosphatidylethanolamine (DOPE), assume non-bilayer structures that can also facilitate endosomal membrane destabilization when associated with liposomes.[120,121]

6.2.6   *Release of Nucleic Acid and Nuclear Transport of DNA*

Although strong binding or encapsulation of nucleic acid is necessary to form stable particles, nucleic acids must be released from the particles once they enter the cytoplasm. This release typically occurs due to thermodynamics-driven disassociation of anionic DNA from cationic polymers, or due to degradation of the particle. Release is necessary for the nucleic acid to have a biological effect, as DNA that is not released from its carrier cannot be transcribed as efficiently.[122,123] This was illustrated in a recent study showing that the binding constant between polycations and DNA is biphasic with transfection efficiency.[124] Additionally, increased polymer degradability imparts decreased cytotoxicity, as polymer molecular weight has been shown to correlate positively with toxicity.[125] Lack of degradability contributes to the toxicity of conventional polymers such as 25 kDa branched PEI.[126,127] Below, we will discuss methods to chemically modify conventional polymers to enable them to degrade (Figure 6.2), as well as design polymers with degradable moieties inherent in their chemical structure.

6.2.6.1   *Hydrolysis*

Nanoparticle cargo can be released through polymer hydrolysis, via the cleavage of ester, urethane, imine and orthoester linkages. Poly[alpha-(4 aminobutyl)-L-glycolic acid] (PAGA) is a hydrolyzable analog of PLL in which the amides linking conventional PLL monomers are replaced with ester linkages. PAGA-based DNA delivery has been shown to lead to higher transfection efficacy and lower toxicity versus

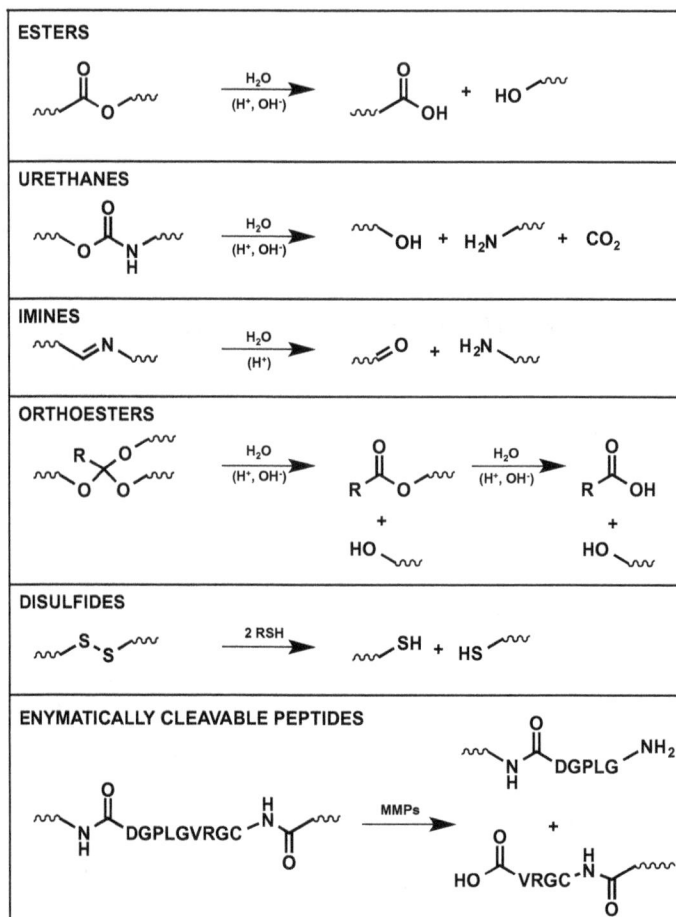

Figure 6.2: Chemical moieties that can enable the degradation of polymeric nanoparticles. Chemicals listed above the reaction scheme arrow indicate those necessary for the reaction, while chemicals listed below indicate reaction catalysts.

conventional PLL.[128] Hydrolytically-cleavable PEI can be synthesized by linking shorter PEI polymers together with ester-containing crosslinkers. Diacrylate monomers used to crosslink 800 Da PEI were shown to produce nanoparticles of the same size, shape, charge and DNA binding as 25 kDa PEI, while achieving a 16-fold enhancement in transfection efficacy with no measureable toxicity.[129] PBAE polymers are formed via Michael addition of amine-containing monomers with diacrylate monomers and therefore contain esters within the polymer backbone.[130]

Combining various amine and acrylate monomers enables the creation of PBAE libraries with various chemical properties,[131] and whose binding constants are affected by polymer molecular weight.[124] Other poly(amino ester)s have been synthesized to have a similar chemical structure and transfection efficacy as 25 kDa PEI, but with reduced toxicity.[132] Poly(lactic-*co*-glycolic) acid (PLGA)-based nanoparticles can encapsulate and successfully deliver nucleic acids, particularly when using the double emulsion method and polyamines.[133] PLGA chemically modified with amine-containing molecules grafted onto their polymer backbone has also been shown to effectively deliver DNA and siRNA.[134,135] Other hydrolytically-cleavable polymer linkages have been explored for biodegradable polymer design. Amine-containing polyurethanes can be designed to deliver DNA.[136] Polyimines, which specifically allow for acid-labile hydrolysis, can be used to link short chain PEI.[137] Additionally, polyorthoester polymers can form stable nanoparticles at neutral pH, but are acid-labile and release DNA at pH 5.[138,139]

6.2.6.2   *Reduction*

Polymers can be made bioreducible by incorporating disulfide bonds into them, a property that enables cargo release within the cytoplasm. Cytosolic reduction is due to the presence of reducing agents such as glutathione, a molecule present in concentrations roughly 1,000 times higher in the cytosol versus extracellular space.[140] This cellular compartment-specific degradation and release makes bioreducible polymers particularly useful for the delivery of siRNA, mRNA and miRNA, whose site of action is within the cytosol. In contrast, delivery of DNA via disulfide-containing polymers has sometimes been found to be less effective.[141] As with hydrolytic linkages, reducible linkages can improve delivery efficacy and reduce cytotoxicity of conventional polymers. PLL linked with disulfides has shown improved nucleic acid delivery.[142] Methods to crosslink PLL with disulfides, either by incorporating cysteines into the polypeptide backbone,[143] or by chemically modifying lysine side chains to contain thiols,[144,145] have led to improved siRNA and DNA delivery. Linear PEI linked with disulfides also showed improved siRNA delivery with lower toxicity than 25 kDa PEI.[146] The KALA fusogenic peptide has also been modified with cysteines to allow for crosslinking.[147]

Polymer bioreducibility can be imparted by synthesizing polymers from disulfide-containing monomers. Poly(amido amine)s (PAAs) are synthesized from diacrylamide and amine-containing monomers, and are therefore not biodegradable. However, disulfide-containing diacrylamides can also form PAAs for targeted cargo release into the cytoplasm. Disulfide-containing PAAs have been extensively studied for both DNA and siRNA delivery.[146,148–151] PBAEs can also be made bioreducible either by end-capping with disulfide-containing monomers[9,152] or by incorporating diacrylate monomers with disulfides into the polymer backbone.[153] Bioreducible PBAEs have been shown to be successful as siRNA delivery vehicles, achieving near-complete gene knockdown with little toxicity,[154] including at low siRNA doses.[141]

6.2.6.3 *Triggered release*

Another mode of degradability is the triggered release of polymeric nanoparticles in a tissue or environment-specific manner. As an example, enzyme-cleavable linkages, specifically matrix-metalloproteinase (MMP)-cleavable groups, can allow for release within tumor spaces.[155] Polymers can be degraded by enzymes and polymeric nanoparticles can become deshielded in the presence of specific enzymes. Nanoparticle degradation and nucleic acid release can also be triggered by an external source. For example, light-induced degradation could allow for user-controlled release and spatially-controlled cargo release.[156]

6.2.6.4 *Nuclear transport*

Lastly, the potential final delivery step following delivery of nucleic acids to the cytoplasm is their transport into a specific organelle, such as the nucleus. Unlike siRNA, DNA plasmids have to be transported across the nuclear membrane into the nucleus for their biological effects to occur. Following nuclear import, exogenous DNA expression requires transcription and translation. The simian virus 40 large T antigen nuclear localization signal (NLS), which is a peptide sequence rich in lysine amino acid, is known to facilitate nuclear transport.[157] Many vectors, including cationic peptides and lipids as well as DNA plasmids, have been modified with variants of the NLS or NLS-binding motif to enhance transfection.[158–160]

6.3 Biomaterials used for Polymeric Gene Delivery

6.3.1 *Peptides*

6.3.1.1 *Poly(L-lysine)*

While poly(L-lysine) (PLL), shown in Figure 6.3(a), has generally low gene delivery efficacy, multicomponent nanoparticles that incorporate PLL as a polycation to bind nucleic acids have been more successful. In one example, PEG-PLL-DMMAn-Mel nanoparticles have been shown to improve PLL-based siRNA delivery. In this system, Meyer *et al.* designed PEG-modified PLL nanoparticles that utilized the

(a) Poly(L-lysine)

(b) Polyethylenimine

(c) Poly(β-amino ester)

(d) Poly(amidoamine)

(e) Chitosan

(f) Hyaluronan

(g) β-cyclodextrin

(h) Dextran

(i) Spermine

(j) Nucleic Acid Origami

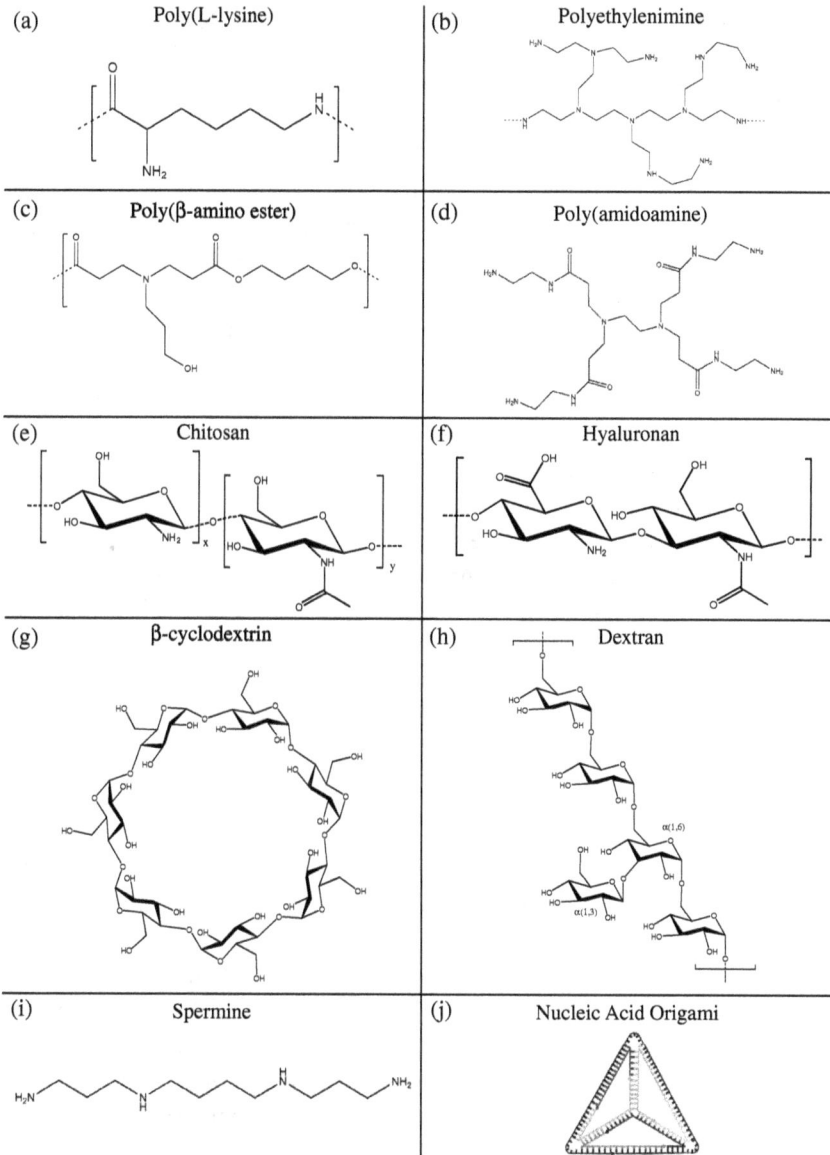

Figure 6.3: Characteristic chemical structures of polymers used for non–viral gene delivery.

lytic peptide melittin (Mel) shielded by pH-cleavable dimethylmaleic anhydride (DMMAn), so that the lytic peptides are only exposed for endosomal escape once a pH of 5 is reached.[50] In addition to pH-sensitive lytic peptide exposure, siRNA release was achieved by disulfide cleavage between the siRNA and the polymer.[50] The PEG-PLL-DMMAn-Mel nanoparticles achieved 90% knockdown *in vitro*, with a caveat being potential cytotoxicity (70% metabolic activity as measured by an MTT assay).[50] While the results of this nanoparticle formulation are promising for siRNA delivery *in vitro*, *in vivo* testing revealed high levels of toxicity in both healthy and tumor-bearing mice, requiring sacrifice of the animals shortly after application.[50] This study highlights the concerns of biomaterial-induced toxicity with synthetic gene delivery systems.

Other uses of PLL for nucleic acid delivery include pH-cleavable PEGylated PLL-cholic acid nanoparticles, which gave a nine-fold reduction in gene expression *in vitro* with cell viability over 90%.[161] *In vivo*, VEGF siRNA delivered using PEGylated PLL-cholic acid nanoparticles achieved 41% tumor size reduction and 70% knockdown of VEGF mRNA in treated mice, with no significant weight reduction observed.[161] A dendritic PLL nanoparticle formulation achieved significant knockdown of Apolipoprotein B *in vivo* leading to a 40% reduction in serum low-density lipoprotein levels.[162] PLL has been shown to be an effective polycation when modified to enable cellular targeting and endosomal escape, but its current use is limited as it lacks the versatility of many other polycations for nucleic acid delivery.

6.3.1.2 *Cell-penetrating peptides*

Peptides have been incorporated into many other nanoparticle designs as both the backbone structure and surface molecules. Amphipathic endosomal escape peptides, such as GALA and KALA as mentioned previously, are well documented for improving transfection among various cell types and with different nanoparticle formulations.[118,163] In the creation of nanoparticles for siRNA delivery, lysine residues are often used to increase the cationic nature of the peptides.[164] Peptides have also been incorporated into nanoparticles for the purposes of endosomal release. For example, sHGP, a 15 amino acid oligopeptide from HIV gp41, has been shown to improve endosomal release.[165] Peptide sequences from influenza, specifically INF7, have also been incorporated into nanoparticles to aid in endosomal escape. Nanoparticles containing INF7 showed improved transfection efficacy with siRNA.[166] Control over the enzymatic degradation rate of peptide-based nanoparticles has been achieved by Chu *et al.*, who designed nanoparticles utilizing both *D* and *L* amino acids for controlled cleavage by Cathepsin B.[167] This stereospecific enzymatic degradation strategy offered excellent extracellular stability with a controlled rate of intracellular degradation for nucleic acid release.[167]

CPP-based nanoparticles termed PF6 were created by Andaloussi *et al.* for siRNA delivery. PF6 nanoparticles were shown to knock down a reporter gene by up to 90% in serum-containing media with minimal cytotoxicity and inflammatory effects.[164] Importantly, PF6 nanoparticles were shown to be stable over a span of weeks in water as well as over a shorter term in serum-containing media. The nanoparticles have diameters between 125–200 nm and zeta potentials of approximately -10 mV.[164] Intravenous administration of PF6 with luc-siRNA in transgenic mice with bioluminescent liver cells showed effective knockdown, peaking on day 5 with a 75% reduction.[164] PF6 knockdown of the functional protein HPRT1 was observed to be greatest in the liver; in addition, there was silencing of greater than 60% in the kidneys.[164]

6.3.2 *Synthetic Polymers*

6.3.2.1 *Polyethylenimine*

High molecular weight 25 kDa Polyethylenimine (PEI), shown in Figure 6.3(b), has previously been shown to condense DNA into nanoparticles, undergo endosomal escape, and successfully deliver DNA.[51,168] Unfortunately, while transfection efficacy is correlated with the molecular weight of PEI, cytotoxicity is likewise correlated, making unmodified 25 kDa PEI largely unsuitable for *in vivo* applications.[168] Related to issues of immediate cytotoxicity, PEI is a non-biodegradable polycation that requires excess polymer to effectively transfect cells, which can lead to accumulation upon repeated administration.[52,53] PEI of approximately 25 kDa molecular weight was shown to have higher transfection efficiencies than higher molecular weight versions, such as 50 kDa and 800 kDa, but still suffered from *in vivo* cytotoxicity in mice.[51,169] To further minimize the cytotoxicity of non-biodegradable PEI, lower molecular weight versions of linear and branched PEI have been investigated, as have partially biodegradable cross-linked PEI and hyperbranched oligoethyleneimine.[170–173] Biodegradable linkages between low molecular weight PEI segments, such as bioreducible disulfides and hydrolyzable esters, have been shown to improve transfection efficacy as well as decrease cytotoxicity compared to 25 kDa PEI.[129,172] In 2003, Forrest *et al.* created hydrolyzable PEI polymers from 800 Da PEI and diol-diacrylate monomers that improved transfection efficiency and reduced cytotoxicity.[129] Using disulfide crosslinked 1.8 kDa low molecular weight PEI, Liu *et al.* were able to achieve greater than 60% transfection with 90% cell viability in serum-containing media.[174]

These improvements to the polymer structure have made PEI much less toxic *in vitro* but do not fully avoid the problems of polycation accumulation *in vivo* that make non-biodegradable polymeric delivery challenging. PEI-based nanoparticle formulations have been investigated in human clinical trials. One example is a phase I clinical trial for the delivery of a plasmid encoding interleukin-12, which was administered to 13 patients with recurrent ovarian cancer and indicated favorable safety results.[175] The nanoparticle was composed of a lipopolymer and PEG-PEI-cholesterol, and was administered intrapleurally in four escalating doses over four weeks.[175]

6.3.2.2 *Poly(beta-amino ester)s*

The development of poly(beta-amino ester)s (PBAEs), shown in Figure 6.3(c), as a material for transfection has greatly benefitted from the high throughput screening of PBAE polymer libraries, in which monomers that make up the backbone, side chains, and end capping groups are systematically varied.[131,176–178] This rapid screening technique has allowed for a large variety of PBAEs to be tested and patterns in transfection to be determined. Transfection by PBAEs has been shown in some cases to be cell type specific, with differences in cellular uptake and transfection between healthy cells and tumor cells due to polymer structure variations.[96,97,179]

Using polymer libraries and high throughput screening, PBAE nanoparticle formulations for the delivery of siRNA and DNA to human glioblastoma cells *in vitro* have achieved up to 85% and 90% knockdown, respectively, which are values greater than when Lipofectamine 2000 and other commercial transfection reagents are used (Figure 6.4).[153] While most PBAE nanoparticles for nucleic acid delivery use linear PBAEs for rapid intracellular degradation and DNA release, cross-linked PBAEs formed by Michael addition using triacrylate monomers and N,N-dimethylethylenediamine have also been shown to reduce the rate of degradation and DNA release.[180]

In vivo results of PBAE nanoparticle-mediated delivery of DNA in mice have demonstrated functional transfection for treating diseases such as ovarian cancer.[181] PBAE nanoparticles have been used to transfect brain tumor-initiating cells (BTICs) in 3D oncospheres with pDNA at up to 76% transfection efficiency.[96] In this work, PBAE nanoparticle transfection specificity for BTICs over fetal neural progenitor cells (fNPCs) was demonstrated both *in vitro* and *in vivo*, supporting the notion that *in vitro* monolayer culture screening of PBAE nanoparticles has relevance for *in vivo* efficacy. Transfection of ovarian tumor bearing mice via intratumoral injection of PBAE nanoparticles containing a plasmid encoding diphtheria toxin showed a mean tumor load reduction greater than that of dual chemotherapeutics.[181] The

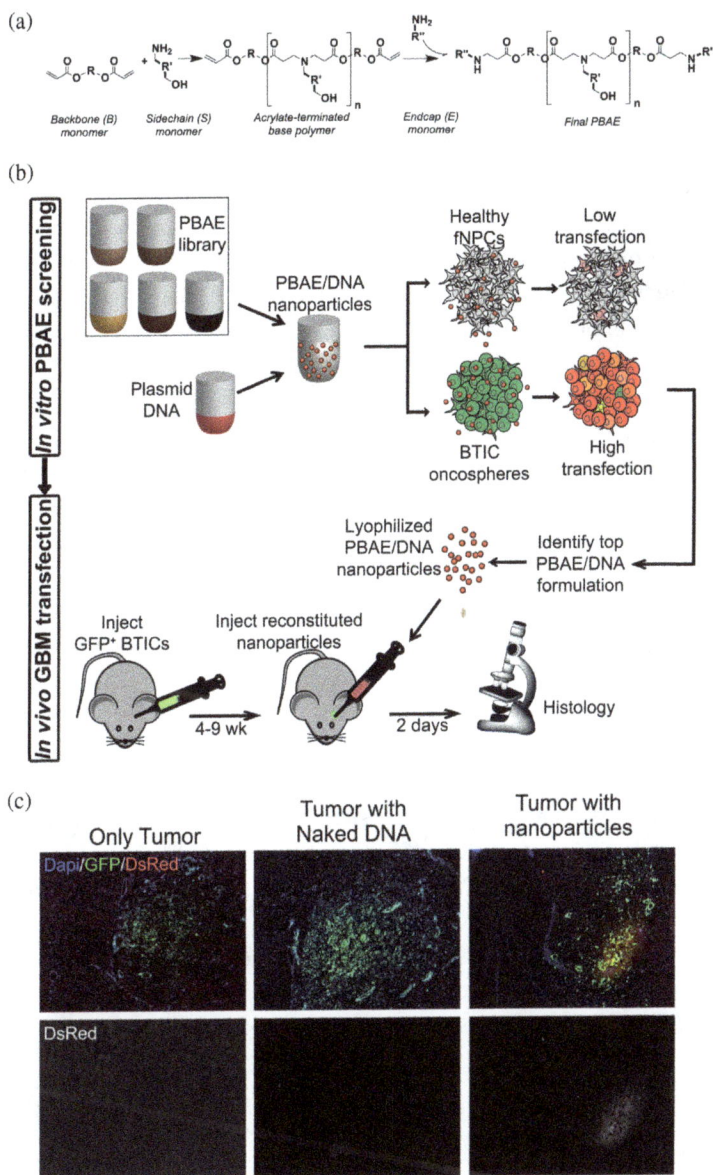

Figure 6.4: Poly(beta-amino ester)s (PBAE)s for gene delivery to brain cancer. (a) Libraries of PBAEs can be synthesized by reacting different acrylate and amine-containing monomers. (b) DNA-containing nanoparticles can be lyophilized and stored prior to *in vivo* administration. (c) Intracranially-administered PBAE/DNA nanoparticles selectively transfect (red) human brain cancer cells (green) while avoiding healthy tissue in a mouse model. Reprinted with permission from the American Chemical Society.[96]

intrapleural injection route used in this study mirrors the current injection route of chemotherapeutics for advanced ovarian cancer, supporting the clinical relevance of the work.[181] PBAE/DNA nanoparticles, although readily hydrolyzable, have been demonstrated to be stable when stored at $-20°C$ for up to two years after lyophilization with sucrose as a cryoprotectant, making them relevant for clinical drug delivery.[96]

PBAE cationic polymers have also been used to supplement other materials in the design of nanoparticles for siRNA delivery *in vivo*. Cohen *et al.* have created acetalated-dextran nanoparticles with 10 wt% PBAE that showed pH-sensitive degradation and release of DNA.[182] PBAE has been used as a cationic polymer for binding DNA in conjunction with PLGA to form microspheres capable of transfecting macrophages for the expression of a tumor specific antigen and the induction of an adaptive immune response in mice.[183]

For the localized delivery of DNA and siRNA amenable to tissue engineering applications, PBAEs have been used in multilayer polyelectrolyte films for contact-dependent transfection.[184] Multilayered films, such as those developed in the lab of David Lynn, rely on charge association between layers of polycations, in this case cationic PBAEs and anionic DNA, for localized transfection in a pH and temperature-dependent manner.[185] DNA release was largely dependent on multilayer film degradation, and in a relaxed conformation comparable to the typical supercoiled conformation resulting from nanoparticle delivery.[184] Multilayered polyelectrolyte films have been further developed for the localized delivery of siRNA with release due in large part to diffusion out of the film rather than from film degradation.[186] Notably, the multilayer film design allowed for sustained release of DNA over 30 h, while siRNA was released in a burst manner.[184,186] The Hammond and Irvine groups have developed PBAE-based layer-by-layer coatings of microneedles that can be used for DNA vaccination and delivery of immunostimulatory RNA through the skin.[187] Using this approach, the authors found potent cellular and humoral immunity *in vivo* in mice and enhanced gene delivery *ex vivo* in non-human primate skin.[187]

6.3.2.3 *Poly(amidoamine)s*

Dendrimers are symmetrically-branched polymer structures used as base units to encapsulate and deliver various materials through charge interactions or conjugation. Many dendrimers, including poly(amidoamine) (PAA or PAMAM) shown in Figure 6.3(d), are synthesized by a series of Michael addition reactions, which allow for great specificity in size and nitrogen content for complexation with nucleic acids in a fine-tuned N:P ratio. The exterior surface of dendrimers can also be modified with hydrophilic groups to improve water solubility, or with targeting ligands to improve active cellular uptake. The most frequently used dendrimer for nucleic acid delivery to date is PAMAM, although peptide dendrimers have also been used with some success.[188] Bioreducible PAMAM nanoparticles have been created to deliver DNA with very high cell viabilities and transfection efficiencies of up to 200 times that of branched PEI.[189] The degree to which the structure of these hyper-branched PAMAM particles could be reduced was fine-tuned by the changing the monomer molar ratios used in Michael addition reactions to create the polymer.[189] Beyond PAMAM, Barnard *et al.* developed an ester-hydrolyzable dendrimer with surface amine groups that was capable of *in vitro* transfection rates of up to 10 fold greater efficiency than PEI.[190]

6.3.2.4   *Poly(lactide-co-glycolide) and poly(caprolactone)*

PLGA microparticles containing 25 wt% PBAE were used to transfect macrophages with reporter gene DNA both *in vitro* and *in vivo*.[183] *In vivo* delivery of a plasmid expressing a tumor antigen induced the rejection of a transplanted tumor matching the antigen.[183] The resulting adaptive immune response was sufficient to cause a reduction in measured tumor growth by day 11 following transfection.[183] Another nanoparticle formulation, a copolymer hybrid poly(ester amine) of polycaprolactone and PEI, improved transfection of a number of cell lines compared to 25 kDa PEI.[191] Thus, biodegradable polymer blends are an appealing approach for nucleic acid delivery.

6.3.3   *Polysaccharides*

6.3.3.1   *Chitosan*

Chitosan, a natural linear polysaccharide derived from chitin and shown in Figure 6.3(e), has been used in the delivery of pDNA and siRNA.[192] Chitosan varies by the degree of deacetylation in chitin, expressed as a ratio of β-(1–4)-linked D-glucosamine to N-acetylated-D-glucosamine.[193] Highly deacetylated chitosan is used more frequently for nucleic acid delivery due to its greater cationic nature and corresponding ability to complex with the negatively-charged backbone of DNA or RNA. Mao *et al.* created PEGylated chitosan DNA nanoparticles with transferrin as the targeting molecule for transfection.[194] In another study, chitosan-DNA nanoparticles transfected intestinal epithelium *in vivo*, generated immunologic protection, and reduced allergen-induced anaphylaxis when administered orally to mice.[195] For siRNA delivery, chitosan thiamine pyrophosphate nanoparticles have been shown to achieve knockdown of up to 70% with cell viability above 90% in hepatocarcinoma cells, values greater than when Lipofectamine was used.[196] Trimethyl chitosan has also been used in conjunction with polysaccharide polysialic acid (PSA) to deliver transcription factor decoy oligonucleotides, reducing inflammation measured by excreted cytokines *in vitro*.[197]

6.3.3.2 *Hyaluronic acid*

Hyaluronic acid (HA), shown in Figure 6.3(f), has been utilized in nanoparticles for the delivery of nucleic acids and as a targeting molecule to the CD44 cell receptor, which is often overexpressed on the surface of tumor cells.[198,199] HA-chitosan-PEG nanoparticles synthesized for the delivery of pDNA and siRNA have been shown to achieve transfection efficiency equivalent to that of Lipofectamine 2000 *in vitro*.[198] Nanoparticles for siRNA delivery composed of HA-spermine and HA-PEI have achieved above 90% knockdown *in vitro* with specificity for the CD44 receptor.[199] When the HA-PEI particles were tested *in vivo* in a mouse model of metastatic lung cancer, knockdown of up to 55% was measured by qPCR.[199] HA, in combination with polycations such as PEI, have been utilized to improve the serum stability of nanoparticles, functioning in much the same way as the glycosylation of proteins *in vivo*.[200] HA has also been used in hydrogels for the controlled delivery of DNA in *in vivo* tissue engineering applications.[201]

6.3.3.3 *Cyclodextrin*

β-cyclodextrin, shown in Figure 6.3(g), is a three-dimensionally stable oligomer of glucose that forms cup-like structures with a hydrophobic core. Chemical modification of β-cyclodextrin with acetyl groups enables the polymer structure to complex with nucleic acids as a cationic polymer. Cyclodextrin-based nanoparticles developed in the lab of Mark Davis for intravenous delivery of siRNA have reached clinical trials. The nanoparticles, shown in Figure 6.5(a), are formulated from β-cyclodextrin, adamantine-PEG, and the targeting ligand transferrin, and have shown favorable characteristics for the delivery of siRNA, including small nanoparticle size between 60–80 nm, zeta potential of +10–20 mV, and the ability to protect siRNA from nuclease activity in the presence of serum for at least 4 h.[202] In 2009, following pre-clinical trials in monkeys, this nanoparticle formulation was tested in a phase I clinical trial for the delivery of RRM2-siRNA in cancer indications.[203] The phase I trial involving 24 patients concluded with favorable results for nanoparticle safety, including evidence for the lack of a complement response.[204] Additionally, intratumoral RRM2 mRNA levels were reduced by up to 77% and a 32% partial

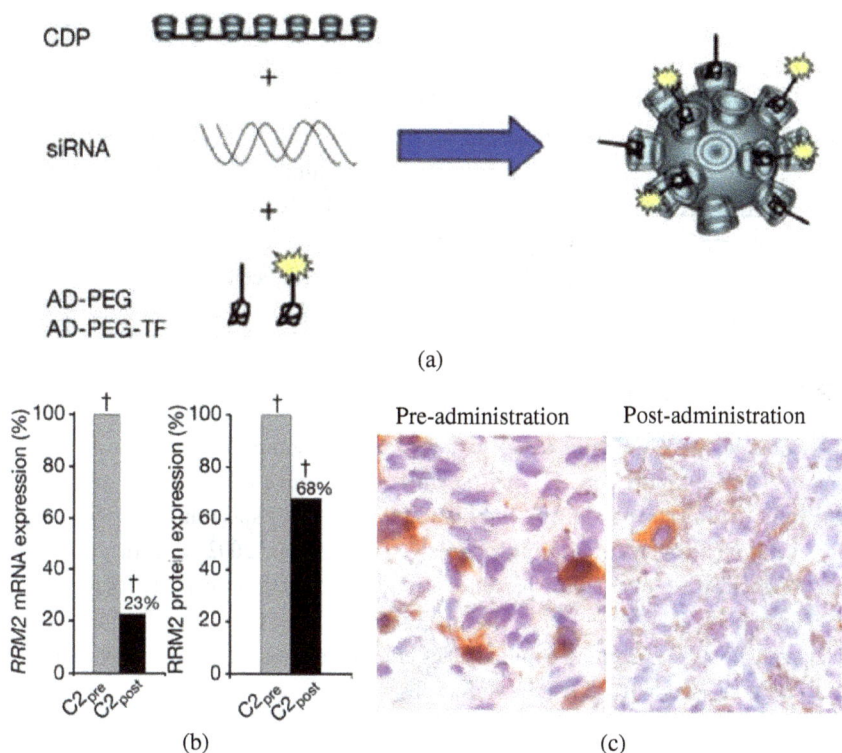

Figure 6.5: Cyclodextrins for siRNA delivery. (a) The three components that comprise CALAA-01, a cyclodextrin-containing nanoparticle for RRM2 RNAi tested in FDA phase I clinical trials. (b) mRNA and protein levels of RRM2 are both knocked down in the targeted tissue of one patient. (c) RRM2 staining (red) of human tumor tissue before and after systemic administration of the cyclodextrin-containing nanoparticles. Reprinted with permission from Nature Publishing Group.[205]

knockdown was measured for RRM2 protein levels in tumor tissues, as shown in Figure 6.5(b).[205] While this level of knockdown may not be effective as a monotherapy for cancer, it provides early evidence that with this non-viral system, siRNA can be targeted to tumor cells in human patients with measurable knockdown of a specific protein.

6.3.3.4 Dextran

Dextran, shown in Figure 6.3(h), is a branched polysaccharide of repeating glucose units and has been used in the formation of nanoparticles for the delivery of siRNA and pDNA. Dextran is often acetalated to improve its solubility in organic solvents and allow for pH-dependent degradation.[206] While the structure of unmodified dextran does not fit the requirements of an ideal biomaterial for nucleic acid delivery, its status as an easily-modified biocompatible and biodegradable polymer allows for it to be utilized with other materials in the formulation of nanoparticles for nucleic acid delivery. Acetalated dextran has been used in conjunction with PBAE and spermine for the successful delivery of both siRNA and DNA.[182,207] Ac-DEX/PBAE particles created for the delivery of DNA have been shown to undergo endosomal pH-dependent degradation and have also been coated with cell penetrating peptides for improved endosomal escape.[182]

6.3.3.5 *Spermine*

Spermine, shown in Figure 6.3(i), is a natural oligoamine used primarily as an oligomer grafted onto non-cationic polymers to improve charge association with DNA. Biodegradable polysaccharide-based particles using spermine as a polycation have been explored with varying degrees of success. In 2002, a library of over 300 polysaccharide-oligoamine particles were created, with some particles reaching transfection efficiencies equal to that of Transfast cationic lipids.[208] Since then, acetalated-dextran spermine nanoparticles created for siRNA delivery have caused up to 60% knockdown of GFP in HeLa cells.[207]

6.3.3.6 *Nucleic acid based particles*

Nanoparticles composed entirely of nucleic acids have been created for the delivery of siRNA. In this case, the natural biodegradable polymer is the nucleic acid itself, which functions as both the structure for the particle as well as the cargo. Lee *et al.* used six 30 bp segments of DNA with complementary overhanging siRNA segments to create self-assembling tetrahedral oligonucleotide nanoparticles from DNA and siRNA complementation, as shown in Figure 6.3(j).[209] These particles, each carrying six siRNA molecules, were shown to have a circulation time that was four times longer than unprotected siRNA.[209] Oligonucleotide nanoparticles improved siRNA delivery for the knockdown of luciferase both *in vitro* and *in vivo*, with a 60% reduction in the bioluminescence of luciferase-expressing tumors in a rat model two days after treatment.[209] These initial studies, particularly for *in vivo* siRNA delivery, indicate that nucleic acid origami particles are a high capacity siRNA delivery method that deserves further study.[209] Paula Hammond and colleagues have used RNA polymerase to form long strands of RNA that can self-assemble into nanostructures and microstructures.[64] These structures, termed RNA microsponges, were able to generate *in vivo* knockdown when administered intratumorally in mice, with or without PEI.[64] In an alternative approach, Chad Mirkin and colleagues have developed spherical nucleic acids, which are densely packed nucleic acids arranged in a spherical geometry, with or without a core. These materials are promising for nucleic acid delivery in a variety of applications, including cancer therapy. In one example, gold core, siRNA-based spherical nucleic acid nanoparticles were found to cross the blood brain barrier, knock down Bcl2L12, and induce apoptosis in brain cancer cells, increasing *in vivo* survival.[210]

6.4 Conclusion

Gene therapy holds great promise in treating various diseases of genetic origin by introducing exogenous nucleic acids to express desired proteins, and by knocking down the expression of undesirable genes. A key challenge in gene therapy is effective delivery, and significant effort has been invested into developing biomaterials that can form nanoparticles to deliver genes to specific targets safely and efficiently. As highlighted in this chapter, a number of non-viral, biodegradable polymers have been developed to form biodegradable nanoparticles for gene delivery and are promising due to their ease of synthesis, low toxicity, and transfection efficacy. Importantly, strategies for polymer modification have been identified to overcome the major biological barriers to gene delivery. While polymeric nanoparticle systems for human gene therapy have yet to be FDA-approved, numerous systems for polymeric DNA and siRNA delivery are in preclinical and clinical trials. These systems, or their future derivatives, may be able to achieve the promise of genetic medicine.

Acknowledgements

This work was supported in part by the NIH (1R01EB016721). J.K. thanks Samsung for fellowship support. K.L.K. thanks the NIH Cancer Nanotechnology Training Center (R25CA153952) at the JHU Institute for Nanobiotechnology for fellowship support. D.R.W. thanks the NSF for graduate student fellowship support.

References

1. Lang, F.F. *et al*. Phase I trial of adenovirus-mediated p53 gene therapy for recurrent glioma: Biological and clinical results. *Journal of Clinical Oncology* 21, 2508–2518 (2003).
2. Tolcher, A.W. *et al*. Phase I, pharmacokinetic, and pharmacodynamic study of intravenously administered Ad5CMV-p53, an adenoviral vector containing the wild-type

p53 gene, in patients with advanced cancer. *Journal of Clinical Oncology* 24, 2052–2058 (2006).

3. Prabha, S. and Labhasetwar, V. Nanoparticle-mediated wild-type p53 gene delivery results in sustained antiproliferative activity in breast cancer cells. *Mol. Pharm.* 1, 211–219 (2004).

4. Gurunathan, S., Klinman, D.M., Seder, R.A. DNA Vaccines: Immunology, application, and optimization. *Annual Review of Immunology* 18, 927–974 (2000).

5. Rice, J., Ottensmeier, C.H., Stevenson, F.K. DNA vaccines: Precision tools for activating effective immunity against cancer. *Nat Rev Cancer* 8, 108–120 (2008).

6. King, G.D., Muhammad, A.K.M.G., Larocque, D., Kelson, K.R., Xiong, W., Liu, C., Sanderson, N.S.R., Kroeger, K.M., Castro, M.G., Lowenstein, P.R. Combined Flt3L/TK gene therapy induces immunological surveillance which mediates an immune response against a surrogate brain tumor neoantigen. *Mol. Ther.* 19, 1793–1801 (2011).

7. Okada, H., Villa, L., Attanucci, J., Erff, M., Fellows, W.K., Lotze, M.T., Pollack, I.F., Chambers, W.H. Cytokine gene therapy of gliomas: Effective induction of therapeutic immunity to intracranial tumors by peripheral immunization with interleukin-4 transduced glioma cells. *Gene Ther.* 8, 1157–66 (2001).

8. Zarogoulidis, P., Darwiche, K., Sakkas, A., Yarmus, L., Huang, H., Li, Q., Freitag, L., Zarogoulidis, K., Malecki, M. Suicide gene therapy for cancer — current strategies. *J Genet Syndr Gene Ther* 9, 16849 (2013).

9. Tzeng, S.Y., Hung, B.P., Grayson, W.L., Green, J.J. Cystamine-terminated poly(beta-amino ester)s for siRNA delivery to human mesenchymal stem cells and enhancement of osteogenic differentiation. *Biomaterials* 33, 8142–51 (2012).

10. Bhise, N.S., Wahlin, K., Zack, D., Green, J.J. Evaluating the potential of poly(beta-amino ester) nanoparticles for reprogramming human fibroblasts to become induced pluripotent stem cells. *International Journal of Nanomedicine* 8, 4641–4658 (2013).

11. Wu, W., Sun, M., Zou, G.M., Chen, J. MicroRNA and cancer: Current status and prospective. *Int. J. Cancer* 120, 953–60 (2007).

12. Yadav, S., van Vlerken, L.E., Little, S.R., Amiji, M.M. Evaluations of combination MDR-1 gene silencing and paclitaxel administration in biodegradable polymeric nanoparticle formulations to overcome multidrug resistance in cancer cells. *Cancer Chemother. Pharmacol.* 63, 711–22 (2009).

13. Kuwabara, P.E., Coulson, A. RNAi–prospects for a general technique for determining gene function. *Parasitol. Today* 16, 347–9 (2000).

14. Thomas, C.E., Ehrhardt, A., Kay, M.A. Progress and problems with the use of viral vectors for gene therapy. *Nat. Rev. Genet.* 4, 346–358 (2003).

15. Goodman, J.C., Trask, T.W., Chen, S.H., Woo, S.L., Grossman, R.G., Carey, K.D., Hubbard, G.B., Carrier, D.A., Rajagopalan, S., Aguilar-Cordova, E., Shine, H.D. Adenoviral-mediated thymidine kinase gene transfer into the primate brain followed by systemic ganciclovir: Pathologic, radiologic, and molecular studies. *Hum Gene Ther* 7, 1241–50 (1996).

16. Pack, D.W., Hoffman, A.S., Pun, S., Stayton, P.S. Design and development of polymers for gene delivery. *Nat. Rev. Drug Discov.* 4, 581–593 (2005).

17. Alving, C.R., Steck, E.A., Chapman, W.L., Jr., Waits, V.B., Hendricks, L.D., Swartz, G.M., Jr., Hanson, W.L. Therapy of leishmaniasis: Superior efficacies of liposome-encapsulated drugs. *Proc Natl Acad Sci USA.* 75, 2959–63 (1978).
18. Scherphof, G.L., Dijkstra, J., Spanjer, H.H., Derksen, J.T., Roerdink, F.H. Uptake and intracellular processing of targeted and nontargeted liposomes by rat Kupffer cells in vivo and in vitro. *Ann N Y Acad Sci.* 446, 368–84 (1985).
19. Vadiei, K., Lopez-Berestein, G., Perez-Soler, R., Luke, D.R. *In vitro* evaluation of liposomal cyclosporine. *Int. J. Pharm* 57, 133–138 (1989).
20. Ma, Z., Li, J., He, F.T., Wilson, A., Pitt, B., Li, S. Cationic lipids enhance siRNA-mediated interferon response in mice. *Biochem. Biophys. Res. Commun.* 330, 755–759 (2005).
21. Dalby, B., Cates, S., Harris, A., Ohki, E.C., Tilkins, M.L., Price, P. J., Ciccarone, V.C. Advanced transfection with Lipofectamine 2000 reagent: Primary neurons, siRNA, and high-throughput applications. *Methods* 33, 95–103 (2004).
22. Palliser, D., Chowdhury, D., Wang, Q.Y., Lee, S.J., Bronson, R.T., Knipe, D.M., Lieberman, J. An siRNA-based microbicide protects mice from lethal herpes simplex virus 2 infection. *Nature* 439, 89–94 (2006).
23. Chono, S., Li, S.D., Conwell, C.C., Huang, L. An efficient and low immunostimulatory nanoparticle formulation for systemic siRNA delivery to the tumor. *J. Control. Release* 131, 64–69 (2008).
24. Judge, A.D., Bola, G., Lee, A.C. H., MacLachlan, I. Design of noninflammatory synthetic siRNA mediating potent gene silencing *in vivo*. *Mol. Ther.* 13, 494–505 (2006).
25. Chen, C., Okayama, H. High-efficiency transformation of mammalian cells by plasmid DNA. *Mol. Cell. Biol.* 7, 2745–2752 (1987).
26. Jordan, M., Schallhorn, A., Wurm, F.M. Transfecting mammalian cells: Optimization of critical parameters affecting calcium-phosphate precipitate formation. *Nucleic Acids Res.* 24, 596–601 (1996).
27. Tolou, H. Administration of oligonucleotides to cultured cells by calcium phosphate precipitation method. *Anal. Biochem.* 215, 156–158 (1993).
28. Ghosh, P.S., Kim, C.K., Han, G., Forbes, N.S., Rotello, V.M. Efficient gene delivery vectors by tuning the surface charge density of amino acid-functionalized gold nanoparticles. *ACS Nano.* 25, 2213–2218 (2008).
29. Love, K.T., Mahon, K.P., Levins, C.G., Whitehead, K.A., Querbes, W., Dorkin, J.R., Qin, J., Cantley, W., Qin, L.L., Racie, T., Frank-Kamenetsky, M., Yip, K.N., Alvarez, R., Sah, D.W. Y., de Fougerolles, A., Fitzgerald, K., Koteliansky, V., Akinc, A., Langer, R., Anderson, D.G. Lipid-like materials for low-dose, *in vivo* gene silencing. *Proc. Natl. Acad. Sci.* 107, 1864–1869 (2010).
30. Mirkin, C.A., Letsinger, R.L., Mucic, R.C., Storhoff, J.J. A DNA-based method for rationally assembling nanoparticles into macroscopic materials. (1996).
31. Daniel, M.C., Astruc, D. Gold nanoparticles: Assembly, supramolecular chemistry, quantum-size-related properties, and applications toward biology, catalysis, and nanotechnology. *Chemical Reviews-Columbus* 104, 293 (2004).
32. Connor, E.E., Mwamuka, J., Gole, A., Murphy, C.J., Wyatt, M.D. Gold nanoparticles are taken up by human cells but do not cause acute cytotoxicity. *Small* 1, 325–327 (2005).

33. Derfus, A.M., Chan, W.C. W., Bhatia, S.N. Intracellular delivery of quantum dots for live cell labeling and organelle tracking. *Adv. Mater.* 16, 961–966 (2004).

34. Gao, X., Cui, Y., Levenson, R.M., Chung, L.W. K., Nie, S. *In vivo* cancer targeting and imaging with semiconductor quantum dots. *Nat. Biotechnol.* 22, 969–976 (2004).

35. Kakizawa, Y., Kataoka, K. Block copolymer self-assembly into monodispersive nanoparticles with hybrid core of antisense DNA and calcium phosphate. *Langmuir* 18, 4539–4543 (2002).

36. Derfus, A.M., Chen, A.A., Min, D.H., Ruoslahti, E., Bhatia, S.N. Targeted quantum dot conjugates for siRNA delivery. *Bioconjugate Chem.* 18, 1391–1396 (2007).

37. Elbakry, A., Zaky, A., Liebl, R., Rachel, R., Goepferich, A., Breunig, M. Layer-by-layer assembled gold nanoparticles for siRNA delivery. *Nano Lett.* 9, 2059–2064 (2009).

38. Lee, J.S., Green, J.J., Love, K.T., Sunshine, J., Langer, R., Anderson, D.G. Gold, poly (beta-amino ester) nanoparticles for small interfering RNA delivery. *Nano Lett.* 9, 2402–2406 (2009).

39. Kaneda, Y. Gene therapy: A battle against biological barriers. *Curr. Mol. Med.* 1, 493–9 (2001).

40. Yin, H., Kanasty, R.L., Eltoukhy, A.A., Vegas, A.J., Dorkin, J.R., Anderson, D.G. Non-viral vectors for gene-based therapy. *Nature Reviews Genetics* 15, 541–55 (2014).

41. Kawabata, K., Takakura, Y., Hashida, M. The fate of plasmid DNA after intravenous injection in mice: Involvement of scavenger receptors in its hepatic uptake. *Pharmaceutical Research* 12, 825–30 (1995).

42. Chang, C., Weiskopf, M., Li, H.J. Conformational studies of nucleoprotein. Circular dichroism of deoxyribonucleic acid base pairs bound by polylysine. *Biochemistry* 12, 3028–32 (1973).

43. Li, H.J., Chang, C., Weiskopf, M. Helix-coil transition in nucleoprotein-chromatin structure. *Biochemistry* 12, 1763–72 (1973).

44. Laemmli, U.K. Characterization of DNA condensates induced by poly(ethylene oxide) and polylysine. *Proc. Natl. Acad. Sci. USA* 72, 4288–92 (1975).

45. Wagner, E., Cotten, M., Foisner, R., Birnstiel, M.L. Transferrin-polycation-DNA complexes: The effect of polycations on the structure of the complex and DNA delivery to cells. *Proc. Natl. Acad. Sci. USA* 88, 4255–9 (1991).

46. Wu, G.Y., Wu, C.H. Receptor-mediated in vitro gene transformation by a soluble DNA carrier system. *The Journal of Biological Chemistry* 262, 4429–32 (1987).

47. Midoux, P., Monsigny, M. Efficient gene transfer by histidylated polylysine/pDNA complexes *Bioconjugate Chemistry* 10, 406–11 (1999).

48. Okuda, T., Sugiyama, A., Niidome, T., Aoyagi, H. Characters of dendritic poly(L-lysine) analogues with the terminal lysines replaced with arginines and histidines as gene carriers in vitro. *Biomaterials* 25, 537–44 (2004).

49. Wagner, E., Plank, C., Zatloukal, K., Cotten, M., Birnstiel, M.L. Influenza virus hemagglutinin HA-2 N-terminal fusogenic peptides augment gene transfer by transferrin-polylysine-DNA complexes: Toward a synthetic virus-like gene-transfer vehicle. *Proc. Natl. Acad. Sci. USA* 89, 7934–8 (1992).

50. Meyer, M., Dohmen, C., Philipp, A., Kiener, D., Maiwald, G., Scheu, C., Ogris, M., Wagner, E. Synthesis and biological evaluation of a bioresponsive and endosomolytic siRNA-polymer conjugate. *Molecular Pharmaceutics* 6, 752–762 (2009).

51. Boussif, O., Lezoualc'h, F., Zanta, M.A., Mergny, M.D., Scherman, D., Demeneix, B., Behr, J.-P. A versatile vector for gene and oligonucleotide transfer into cells in culture and *in vivo*: Polyethylenimine. *Proc. Natl. Acad. Sci. USA* 92, 7297–7301 (1995).

52. Boeckle, S., Katharina von, G., Silke van der, P., Culmsee, C., Wagner, E., Ogris, M. Purification of polyethylenimine polyplexes highlights the role of free polycations in gene transfer. *Journal of Gene Medicine* 6, 1102–11 (2004).

53. Kunath, K., von Harpe, A., Fischer, D., Petersen, H., Bickel, U., Voigt, K., Kissel, T. Low-molecular-weight polyethylenimine as a non-viral vector for DNA delivery: Comparison of physicochemical properties, transfection efficiency and *in vivo* distribution with high-molecular-weight polyethylenimine. *Journal of Controlled Release* 89, 113–125 (2003).

54. Felgner, P.L., Gadek, T.R., Holm, M., Roman, R., Chan, H.W., Wenz, M., Northrop, J.P., Ringold, G.M., Danielsen, M. Lipofection: A highly efficient, lipid-mediated DNA-transfection procedure. *Proc. Natl. Acad. Sci. USA* 84, 7413–7 (1987).

55. Cohen, H., Levy, R.J., Gao, J., Fishbein, I., Kousaev, V., Sosnowski, S., Slomkowski, S., Golomb, G. Sustained delivery and expression of DNA encapsulated in polymeric nanoparticles. *Gene Therapy* 7, 1896–905 (2000).

56. Chen, Z.Y., He, C.Y., Ehrhardt, A., Kay, M.A. Minicircle DNA vectors devoid of bacterial DNA result in persistent and high-level transgene expression *in vivo*. *Mol. Ther.* 8, 495–500 (2003).

57. Keeney, M., Ong, S.G., Padilla, A., Yao, Z., Goodman, S., Wu, J.C., Yang, F. Development of poly(beta-amino ester)-based biodegradable nanoparticles for nonviral delivery of minicircle DNA. *ACS Nano.* 7, 7241–50 (2013).

58. Zhang, C., Gao, S., Jiang, W., Lin, S., Du, F., Li, Z., Huang, W. Targeted minicircle DNA delivery using folate-poly(ethylene glycol)-polyethylenimine as non-viral carrier. *Biomaterials* 31, 6075–86 (2010).

59. Fire, A., Xu, S., Montgomery, M.K., Kostas, S.A., Driver, S.E., Mello, C.C. Potent and specific genetic interference by double-stranded RNA in Caenorhabditis elegans. *Nature* 391, 806–11 (1998).

60. Hagerman, P.J. Flexibility of RNA. *Annu. Rev. Biophys. Biomol. Struct.* 26, 139–56 (1997).

61. Kebbekus, P., Draper, D.E., Hagerman, P. Persistence length of RNA. *Biochemistry* 34, 4354–7 (1995).

62. Bolcato-Bellemin, A.L., Bonnet, M.E., Creusat, G., Erbacher, P., Behr, J.P. Sticky overhangs enhance siRNA-mediated gene silencing. *Proc. Natl. Acad. Sci. USA* 104, 16050–5 (2007).

63. Mok, H., Lee, S.H., Park, J.W., Park, T.G. Multimeric small interfering ribonucleic acid for highly efficient sequence-specific gene silencing. *Nat. Mater.* 9, 272–8 (2010).

64. Lee, J.B., Hong, J., Bonner, D.K., Poon, Z., Hammond, P.T. Self-assembled RNA interference microsponges for efficient siRNA delivery. *Nat. Mater.* 11, 316–22 (2012).

65. Kawasaki, H., Taira, K. Short hairpin type of dsRNAs that are controlled by tRNA(Val) promoter significantly induce RNAi-mediated gene silencing in the cytoplasm of human cells. *Nucleic Acids Research* 31, 700–7 (2003).

66. Alemany, R., Suzuki, K., Curiel, D.T. Blood clearance rates of adenovirus type 5 in mice. *The Journal of General Virology* 81, 2605–9 (2000).

67. Mahato, R.I., Anwer, K., Tagliaferri, F., Meaney, C., Leonard, P., Wadhwa, M.S., Logan, M., French, M., Rolland, A. Biodistribution and gene expression of lipid/plasmid complexes after systemic administration. *Human Gene Therapy* 9, 2083–99 (1998).

68. Miyata, K., Nishiyama, N., Kataoka, K. Rational design of smart supramolecular assemblies for gene delivery: Chemical challenges in the creation of artificial viruses. *Chemical Society Reviews* 41, 2562–74 (2012).

69. Ogris, M., Steinlein, P., Kursa, M., Mechtler, K., Kircheis, R., Wagner, E. The size of DNA/transferrin-PEI complexes is an important factor for gene expression in cultured cells. *Gene Therapy* 5, 1425–33 (1998).

70. Dash, P.R., Read, M.L., Barrett, L.B., Wolfert, M.A., Seymour, L. W. Factors affecting blood clearance and in vivo distribution of polyelectrolyte complexes for gene delivery. *Gene Ther* 6, 643–50 (1999).

71. Davis, F.F. The origin of pegnology. *Advanced Drug Delivery Reviews* 54, 457–8 (2002).

72. Veronese, F.M., Harris, J.M. Introduction and overview of peptide and protein pegylation. *Advanced Drug Delivery Reviews* 54, 453–6 (2002).

73. Ogris, M., Brunner, S., Schuller, S., Kircheis, R., Wagner, E. PEGylated DNA/transferrin-PEI complexes: Reduced interaction with blood components, extended circulation in blood and potential for systemic gene delivery. *Gene Therapy* 6, 595–605 (1999).

74. Itaka, K., Yamauchi, K., Harada, A., Nakamura, K., Kawaguchi, H., Kataoka, K. Polyion complex micelles from plasmid DNA and poly(ethylene glycol)-poly(L-lysine) block copolymer as serum-tolerable polyplex system: Physicochemical properties of micelles relevant to gene transfection efficiency. *Biomaterials* 24, 4495–506 (2003).

75. MacKay, J.A., Deen, D.F., Szoka, F.C., Jr. Distribution in brain of liposomes after convection enhanced delivery; modulation by particle charge, particle diameter, and presence of steric coating. *Brain Res* 1035, 139–53 (2005).

76. Tseng, Y.C., Mozumdar, S., Huang, L. Lipid-based systemic delivery of siRNA. *Advanced Drug Delivery Reviews* 61, 721–31 (2009).

77. Hosseinkhani, H., Azzam, T., Tabata, Y., Domb, A.J. Dextran-spermine polycation: An efficient nonviral vector for in vitro and in vivo gene transfection. *Gene Therapy* 11, 194–203 (2004).

78. Trubetskoy, V.S., Wong, S.C., Subbotin, V., Budker, V.G., Loomis, A., Hagstrom, J.E., Wolff, J.A. Recharging cationic DNA complexes with highly charged polyanions for in vitro and in vivo gene delivery. *Gene Therapy* 10, 261–71 (2003).

79. Geng, Y., Dalhaimer, P., Cai, S., Tsai, R., Tewari, M., Minko, T., Discher, D.E. Shape effects of filaments versus spherical particles in flow and drug delivery. *Nature Nanotechnology* 2, 249–55 (2007).

80. Jiang, X., Qu, W., Pan, D., Ren, Y., Williford, J.M., Cui, H., Luijten, E., Mao, H.Q. Plasmid templated shape control of condensed DNA-block copolymer nanoparticles. *Adv Mater* 25, 227–32 (2013).

81. Chauhan, N.B. Trafficking of intracerebroventricularly injected antisense oligonucleotides in the mouse brain. *Antisense & Nucleic Acid Drug Development* 12, 353–7 (2002).

82. Dai, H., Jiang, X., Tan, G.C., Chen, Y., Torbenson, M., Leong, K.W., Mao, H.Q. Chitosan DNA nanoparticles delivered by intrabiliary infusion enhance liver-targeted gene delivery. *Int. J. Nanomedicine* 1, 507–22 (2006).

83. Coll, J.L., Chollet, P., Brambilla, E., Desplanques, D., Behr, J.P., Favrot, M. In vivo delivery to tumors of DNA complexed with linear polyethylenimine. *Human Gene Therapy* 10, 1659–66 (1999).

84. Zhang, Y., Satterlee, A., Huang, L. In vivo gene delivery by nonviral vectors: Overcoming hurdles? *Mol. Ther.* 20, 1298–304 (2012).

85. Suh, W., Han, S.O., Yu, L., Kim, S.W. An angiogenic, endothelial-cell-targeted polymeric gene carrier. *Molecular Therapy: The Journal of the American Society of Gene Therapy* 6, 664–72 (2002).

86. Reynolds, A.R., Moein Moghimi, S., Hodivala-Dilke, K. Nanoparticle-mediated gene delivery to tumour neovasculature. *Trends in Molecular Medicine* 9, 2–4 (2003).

87. Ogris, M., Walker, G., Blessing, T., Kircheis, R., Wolschek, M., Wagner, E. Tumor-targeted gene therapy: Strategies for the preparation of ligand-polyethylene glycol-polyethylenimine/DNA complexes. *Journal of Controlled Release: Official Journal of the Controlled Release Society* 91, 173–81 (2003).

88. Green, J.J., Chiu, E., Leshchiner, E.S., Shi, J., Langer, R., Anderson, D.G. Electrostatic ligand coatings of nanoparticles enable ligand-specific gene delivery to human primary cells. *Nano Lett.* 7, 874–9 (2007).

89. Shmueli, R.B., Anderson, D.G., Green, J.J. Electrostatic surface modifications to improve gene delivery. *Expert Opin Drug Deliv* 7, 535–50 (2010).

90. Farokhzad, O.C., Karp, J.M., Langer, R. Nanoparticle-aptamer bioconjugates for cancer targeting. *Expert Opin Drug Deliv* 3, 311–24 (2006).

91. Kunath, K., von Harpe, A., Fischer, D., Kissel, T. Galactose-PEI-DNA complexes for targeted gene delivery: Degree of substitution affects complex size and transfection efficiency. *Journal of Controlled Release: Official Journal of the Controlled Release Society* 88, 159–72 (2003).

92. Zhang, Y., Jeong Lee, H., Boado, R.J., Pardridge, W.M. Receptor-mediated delivery of an antisense gene to human brain cancer cells. *The Journal of Gene Medicine* 4, 183–94 (2002).

93. Lee, L.K., Siapati, E.K., Jenkins, R.G., McAnulty, R.J., Hart, S. L., Shamlou, P.A. Biophysical characterization of an integrin-targeted non-viral vector. *Medical Science Monitor: International Medical Journal of Experimental and Clinical Research* 9, BR54–61 (2003).

94. Hakkarainen, T., Hemminki, A., Pereboev, A.V., Barker, S.D., Asiedu, C.K., Strong, T.V., Kanerva, A., Wahlfors, J., Curiel, D.T. CD40 is expressed on ovarian cancer cells and can be utilized for targeting adenoviruses. *Clinical Cancer Research: An Official Journal of the American Association for Cancer Research* 9, 619–24 (2003).

95. Jaafari, M.R., Foldvari, M. Targeting of liposomes to melanoma cells with high levels of ICAM-1 expression through adhesive peptides from immunoglobulin domains. *Journal of Pharmaceutical Sciences* 91, 396–404 (2002).

96. Guerrero-Cázares, H., Tzeng, S.Y., Young, N.P., Abutaleb, A.O., Quiñones-Hinojosa, A., Green, J.J. Biodegradable polymeric nanoparticles show high efficacy and specificity at DNA delivery to human glioblastoma in vitro and in vivo. *ACS Nano* 8, 5141–53 (2014).

97. Tzeng, S.Y., Higgins, L.J., Pomper, M.G., Green, J.J. Student award winner in the Ph.D. category for the 2013 society for biomaterials annual meeting and exposition, april 10–13, 2013, Boston, Massachusetts: Biomaterial-mediated cancer-specific DNA delivery to liver cell cultures using synthetic poly(beta-amino ester)s. *J. Biomed. Mater. Res. A* 101, 1837–1845 (2013).

98. Shmueli, R.B., Sunshine, J.C., Xu, Z., Duh, E.J., Green, J.J. Gene delivery nanoparticles specific for human microvasculature and macrovasculature. *Nanomedicine: Nanotechnology, Biology, and Medicine* 2012, doi:10.1016/j.nano.2012.01.006.

99. Green, J.J. 2011 Rita Schaffer Lecture: Nanoparticles for Intracellular Nucleic Acid Delivery. *Ann Biomed Eng* 40, 1408–18 (2012).

100. Haase, R., Magnusson, T., Su, B., Kopp, F., Wagner, E., Lipps, H., Baiker, A., Ogris, M. Generation of a tumor- and tissue-specific episomal non-viral vector system. *BMC Biotechnology* 13, 49 (2013).

101. Amyere, M., Mettlen, M., Van Der Smissen, P., Platek, A., Payrastre, B., Veithen, A., Courtoy, P.J. Origin, originality, functions, subversions and molecular signalling of macropinocytosis. *International Journal of Medical Microbiology: IJMM* 291, 487–94 (2002).

102. Goncalves, C., Mennesson, E., Fuchs, R., Gorvel, J.P., Midoux, P., Pichon, C. Macropinocytosis of polyplexes and recycling of plasmid via the clathrin-dependent pathway impair the transfection efficiency of human hepatocarcinoma cells. *Molecular Therapy: The Journal of the American Society of Gene Therapy* 10, 373–85 (2004).

103. Verma, A., Stellacci, F. Effect of surface properties on nanoparticle-cell interactions. *Small* 6, 12–21 (2010).

104. Meyer, M., Wagner, E. pH-responsive shielding of non-viral gene vectors. *Expert Opin Drug Deliv* 3, 563–71 (2006).

105. Kursa, M., Walker, G.F., Roessler, V., Ogris, M., Roedl, W., Kircheis, R., Wagner, E. Novel shielded transferrin-polyethylene glycol-polyethylenimine/DNA complexes for systemic tumor-targeted gene transfer. *Bioconjugate Chemistry* 14, 222–31 (2003).

106. Marsh, M., McMahon, H.T. The structural era of endocytosis. *Science* 285, 215–20 (1999).

107. Rejman, J., Conese, M., Hoekstra, D. Gene transfer by means of lipo- and polyplexes: Role of clathrin and caveolae-mediated endocytosis. *Journal of Liposome Research* 16, 237–47 (2006).

108. Durymanov, M.O., Beletkaia, E.A., Ulasov, A.V., Khramtsov, Y.V., Trusov, G.A., Rodichenko, N.S., Slastnikova, T.A., Vinogradova, T.V., Uspenskaya, N.Y., Kopantsev, E.P., Rosenkranz, A.A., Sverdlov, E.D., Sobolev, A.S. Subcellular trafficking and transfection efficacy of polyethylenimine-polyethylene glycol polyplex nanoparticles with a ligand to melanocortin receptor-1. *J. Control. Release* 163, 211–9 (2012).

109. Gabrielson, N.P., Pack, D.W. Efficient polyethylenimine-mediated gene delivery proceeds via a caveolar pathway in HeLa cells. *J. Control. Release* 136, 54–61 (2009).

110. Pelkmans, L., Helenius, A. Endocytosis via caveolae. *Traffic* 3, 311–20 (2002).

111. van der Aa, M.A., Huth, U.S., Hafele, S.Y., Schubert, R., Oosting, R.S., Mastrobattista, E., Hennink, W.E., Peschka-Suss, R., Koning, G.A., Crommelin, D.J. Cellular uptake of cationic polymer-DNA complexes via caveolae plays a pivotal role in gene transfection in COS-7 cells. *Pharmaceutical Research* 24, 1590–8 (2007).

112. Conner, S.D., Schmid, S.L. Regulated portals of entry into the cell. *Nature* 422, 37–44 (2003).
113. Kim, J., Sunshine, J.C., Green, J.J. Differential polymer structure tunes mechanism of cellular uptake and transfection routes of poly(beta-amino ester) polyplexes in human breast cancer cells. *Bioconjug Chem* 25, 43–51 (2014).
114. Sonawane, N.D., Szoka, F.C., Jr., Verkman, A.S. Chloride accumulation and swelling in endosomes enhances DNA transfer by polyamine-DNA polyplexes. *The Journal of Biological Chemistry* 278, 44826–31 (2003).
115. Midoux, P., Mendes, C., Legrand, A., Raimond, J., Mayer, R., Monsigny, M., Roche, A.C. Specific gene transfer mediated by lactosylated poly-L-lysine into hepatoma cells. *Nucleic Acids Research* 21, 871–8 (1993).
116. Sunshine, J.C., Peng, D.Y., Green, J.J. Uptake and transfection with polymeric nanoparticles are dependent on polymer end-group structure, but largely independent of nanoparticle physical and chemical properties. *Mol Pharm* 9, 3375–83 (2012).
117. Benjaminsen, R.V., Mattebjerg, M.A., Henriksen, J.R., Moghimi, S.M., Andresen, T.L. The possible "proton sponge" effect of polyethylenimine (PEI) does not include change in lysosomal pH. *Molecular Therapy: The Journal of the American Society of Gene Therapy* 21, 149–57 (2013).
118. Wyman, T.B., Nicol, F., Zelphati, O., Scaria, P.V., Plank, C., Szoka, F.C. Design, synthesis, and characterization of a cationic peptide that binds to nucleic acids and permeabilizes bilayers. *Biochemistry* 36, 3008–3017 (1997).
119. Alhakamy, N.A., Nigatu, A.S., Berkland, C.J., Ramsey, J.D. Noncovalently associated cell-penetrating peptides for gene delivery applications. *Therapeutic Delivery* 4, 741–57 (2013).
120. Maitani, Y., Igarashi, S., Sato, M., Hattori, Y. Cationic liposome (DC-Chol/DOPE=1:2) and a modified ethanol injection method to prepare liposomes, increased gene expression. *International Journal of Pharmaceutics* 342, 33–9 (2007).
121. Farhood, H., Serbina, N., Huang, L. The role of dioleoyl phosphatidylethanolamine in cationic liposome mediated gene transfer. *Biochimica et Biophysica Acta* 1235, 289–95 (1995).
122. Gary, D.J., Puri, N., Won, Y.Y. Polymer-based siRNA delivery: Perspectives on the fundamental and phenomenological distinctions from polymer-based DNA delivery. *J. Control. Release* 121, 64–73 (2007).
123. Luo, D., Saltzman, W.M. Synthetic DNA delivery systems. *Nat. Biotechnol.* 18, 33–7 (2000).
124. Bishop, C.J., Ketola, T.M., Tzeng, S.Y., Sunshine, J.C., Urtti, A., Lemmetyinen, H., Vuorimaa-Laukkanen, E., Yliperttula, M., Green, J.J. The effect and role of carbon atoms in poly(beta-amino ester)s for DNA binding and gene delivery. *Journal of the American Chemical Society* 135, 6951–7 (2013).
125. Hill, I.R., Garnett, M.C., Bignotti, F., Davis, S.S. In vitro cytotoxicity of poly(amidoamine)s: Relevance to DNA delivery. *Biochim. Biophys. Acta* 1427, 161–74 (1999).
126. Grayson, A.C.R., Doody, A.M., Putnam, D. Biophysical and structural characterization of polyethylenimine-mediated siRNA delivery in vitro. *Pharmaceut. Res.* 23, 1868–1876 (2006).

127. Sutton, D., Kim, S., Shuai, X., Leskov, K., Marques, J.T., Williams, B. R., Boothman, D.A., Gao, J. Efficient suppression of secretory clusterin levels by polymer-siRNA nanocomplexes enhances ionizing radiation lethality in human MCF-7 breast cancer cells in vitro. *Int. J. Nanomedicine* 1, 155–62 (2006).

128. Lim, Y.B., Han, S.O., Kong, H.U., Lee, Y., Park, J.S., Jeong, B., Kim, S.W. Biodegradable polyester, poly[alpha-(4 aminobutyl)-L-glycolic acid], as a non-toxic gene carrier. *Pharmaceut. Res.* 17, 811–816 (2000).

129. Forrest, M.L., Koerber, J.T., Pack, D.W. A degradable polyethylenimine derivative with low toxicity for highly efficient gene delivery. *Bioconjugate Chem.* 14, 934–940 (2003).

130. Lynn, D.M., Langer, R. Degradable poly (beta-amino esters): Synthesis, characterization, and self-assembly with plasmid DNA. *J. Am. Chem. Soc.* 122, 10761–10768 (2000).

131. Green, J.J., Langer, R., Anderson, D.G. A combinatorial polymer library approach yields insight into nonviral gene delivery. *Accounts of Chemical Research* 41, 749–759 (2008).

132. Wu, D., Liu, Y., Jiang, X., Chen, L., He, C., Goh, S.H., Leong, K.W. Evaluation of hyperbranched poly(amino ester)s of amine constitutions similar to polyethylenimine for DNA delivery. *Biomacromolecules* 6, 3166–3173 (2005).

133. Woodrow, K.A., Cu, Y., Booth, C.J., Saucier-Sawyer, J.K., Wood, M.J., Mark Saltzman, W. Intravaginal gene silencing using biodegradable polymer nanoparticles densely loaded with small-interfering RNA. *Nat Mater* 8, 526–533 (2009).

134. Nguyen, J., Steele, T.W. J., Merkel, O., Reul, R., Kissel, T. Fast degrading polyesters as siRNA nano-carriers for pulmonary gene therapy. *J. Control. Release* 132, 243–251 (2008).

135. Oster, C., Wittmar, M., Unger, F., Barbu-Tudoran, L., Schaper, A., Kissel, T. Design of amine-modified graft polyesters for effective gene delivery using DNA-loaded nanoparticles. *Pharmaceut. Res.* 21, 927–931 (2004).

136. Yang, T.-f., Chin, W.-k., Cherng, J.-y., Shau, M.-d. Synthesis of novel biodegradable cationic polymer: N,N-diethylethylenediamine polyurethane as a gene carrier. *Biomacromolecules* 5, 1926–1932 (2004).

137. Kim, Y.H., Park, J.H., Lee, M., Kim, Y.-H., Park, T.G., Kim, S.W. Polyethylenimine with acid-labile linkages as a biodegradable gene carrier. *J. Control. Release* 103, 209–219 (2005).

138. Wang, C., Ge, Q., Ting, D., Nguyen, D., Shen, H.-R., Chen, J., Eisen, H. N., Heller, J., Langer, R., Putnam, D. Molecularly engineered poly(ortho ester) microspheres for enhanced delivery of DNA vaccines. *Nat Mater* 3, 190–196 (2004).

139. Heller, J., Barr, J., Ng, S.Y., Abdellauoi, K.S., Gurny, R. Poly(ortho esters): Synthesis, characterization, properties and uses. *Adv. Drug Deliv. Rev.* 54, 1015–1039 (2002).

140. Griffith, O.W. Biologic and pharmacologic regulation of mammalian glutathione synthesis. *Free Radical Bio. Med.* 27, 922–935 (1999).

141. Tzeng, S.Y., Yang, P.H., Grayson, W.L., Green, J.J. Synthetic poly(ester amine) and poly(amido amine) nanoparticles for efficient DNA and siRNA delivery to human endothelial cells. *Int. J. Nanomedicine* 6, 3309–22 (2012).

142. Trubetskoy, V.S., Loomis, A., Slattum, P.M., Hagstrom, J.E., Budker, V.G., Wolff, J.A. Caged DNA does not aggregate in high ionic strength solutions. *Bioconjugate Chem.* 10, 624–8 (1999).

143. McKenzie, D.L., Smiley, E., Kwok, K.Y., Rice, K.G. Low molecular weight disulfide cross-linking peptides as nonviral gene delivery carriers. *Bioconjugate Chem.* 11, 901–909 (2000).

144. Matsumoto, S., Christie, R.J., Nishiyama, N., Miyata, K., Ishii, A., Oba, M., Koyama, H., Yamasaki, Y., Kataoka, K. Environment-responsive block copolymer micelles with a disulfide cross-linked core for enhanced siRNA delivery. *Biomacromolecules* 10, 119–127 (2009).

145. Miyata, K., Kakizawa, Y., Nishiyama, N., Harada, A., Yamasaki, Y., Koyama, H., Kataoka, K. Block catiomer polyplexes with regulated densities of charge and disulfide cross-linking directed to enhance gene expression. *J. Am. Chem. Soc.* 126, 2355–2361 (2004).

146. Breunig, M., Hozsa, C., Lungwitz, U., Watanabe, K., Umeda, I., Kato, H., Goepferich, A. Mechanistic investigation of poly (ethylene imine)-based siRNA delivery: disulfide bonds boost intracellular release of the cargo. *J. Control. Release* 130, 57–63 (2008).

147. Mok, H., Park, T.G. Self-crosslinked and reducible fusogenic peptides for intracellular delivery of siRNA. *Biopolymers* 89, 881–888 (2008).

148. Jeong, J.H., Christensen, L.V., Yockman, J.W., Zhong, Z.Y., Engbersen, J.F.J., Kim, W.J., Feijen, J., Kim, S.W. Reducible poly(amido ethylenimine) directed to enhance RNA interference. *Biomaterials* 28, 1912–1917 (2007).

149. Vader, P., van der Aa, L.J., Engbersen, J.F.J., Storm, G., Schiffelers, R.M. Disulfide-based poly(amido amine)s for siRNA delivery: Effects of structure on siRNA complexation, cellular uptake, gene silencing and toxicity. *Pharmaceut. Res.* 28, 1013–1022 (2011).

150. van der Aa, L.J., Vader, P., Storm, G., Schiffelers, R.M., Engbersen, J.F.J. Optimization of poly(amido amine)s as vectors for siRNA delivery. *Journal of Controlled Release* 150, 177–186 (2011).

151. Emilitri, E., Ranucci, E., Ferruti, P. New poly(amidoamine)s containing disulfide linkages in their main chain. *Journal of Polymer Science Part a-Polymer Chemistry* 43, 1404–1416 (2005).

152. Tzeng, S.Y., Green, J.J. Subtle changes to polymer structure and degradation mechanism enable highly effective nanoparticles for siRNA and DNA delivery to human brain cancer. *Adv. Healthcare Mater.* 2, 467 (2013).

153. Kozielski, K.L., Tzeng, S.Y., Green, J.J. A bioreducible linear poly(beta-amino ester) for siRNA delivery. *Chem. Commun.* 49, 5319–5321 (2013).

154. Kozielski, K.L., Tzeng, S.Y., Mendoza, B.A.H.d., Green, J.J. Bioreducible cationic polymer-based nanoparticles for efficient and environmentally triggered cytoplasmic siRNA delivery to primary human brain cancer cells. *ACS Nano* 8, 3232–3241 (2014).

155. Kim, H.S., Yoo, H.S. MMPs-responsive release of DNA from electrospun nanofibrous matrix for local gene therapy: In vitro and in vivo evaluation. *J. Control. Release* 145, 264–271 (2010).

156. Li, H.-J., Wang, H.-X., Sun, C.-Y., Du, J.-Z., Wang, J. Shell-detachable nanoparticles based on a light-responsive amphiphile for enhanced siRNA delivery. *Royal Society of Chemistry Advances* 4, 1961–1964 (2014).

157. Zanta, M.A., Belguise-Valladier, P., Behr, J.P. Gene delivery: a single nuclear localization signal peptide is sufficient to carry DNA to the cell nucleus. *Proc Natl Acad Sci USA* 96, 91–6 (1999).

158. Collas, P., Husebye, H., Alestrom, P. The nuclear localization sequence of the SV40 T antigen promotes transgene uptake and expression in zebrafish embryo nuclei. *Transgenic Research* 5, 451–8 (1996).

159. Remy, J.S., Kichler, A., Mordvinov, V., Schuber, F., Behr, J.P. Targeted gene transfer into hepatoma cells with lipopolyamine-condensed DNA particles presenting galactose ligands: a stage toward artificial viruses. *Proc. Natl. Acad. Sci. USA* 92, 1744–8 (1995).

160. Moffatt, S., Wiehle, S., Cristiano, R.J. A multifunctional PEI-based cationic polyplex for enhanced systemic p53-mediated gene therapy. *Gene Ther.* 13, 1512–23 (2006).

161. Guo, J., Cheng, W.P., Gu, J., Ding, C., Qu, X., Yang, Z., O'Driscoll, C. Systemic delivery of therapeutic small interfering RNA using a pH-triggered amphiphilic polyllysine nanocarrier to suppress prostate cancer growth in mice. *European Journal of Pharmaceutical Sciences* 45, 521–532 (2012).

162. Watanabe, K., Harada-Shiba, M., Suzuki, A., Gokuden, R., Kurihara, R., Sugao, Y., Mori, T., Katayama, Y., Niidome, T. In vivo siRNA delivery with dendritic poly (L-lysine) for the treatment of hypercholesterolemia. *Molecular BioSystems* 5, 1306–1310 (2009).

163. Subbarao, N.K., Parente, R.A., Szoka Jr, F.C., Nadasdi, L., Pongracz, K. The pH-dependent bilayer destabilization by an amphipathic peptide. *Biochemistry* 26, 2964–2972 (1987).

164. EL Andaloussi, S., Lehto, T., Mäger, I., Rosenthal-Aizman, K., Oprea, I.I., Simonson, O.E., Sork, H., Ezzat, K., Copolovici, D.M., Kurrikoff, K., Viola, J.R., Zaghloul, E.M., Sillard, R., Johansson, H.J., Said Hassane, F., Guterstam, P., Suhorutšenko, J., Moreno, P.M.D., Oskolkov, N., Hälldin, J., Tedebark, U., Metspalu, A., Lebleu, B., Lehtiö, J., Smith, C.I.E., Langel, Ü. Design of a peptide-based vector, PepFect6, for efficient delivery of siRNA in cell culture and systemically in vivo. *Nucleic Acids Research* 39, 3972–3987 (2011).

165. Schellinger, J.G., Pahang, J.a., Shi, J., Pun, S.H. Block copolymers containing a hydrophobic domain of membrane-lytic peptides form micellar structures and are effective gene delivery agents. *ACS Macro Letters* 2, 725–730 (2013).

166. Dohmen, C., Edinger, D., Fröhlich, T., Schreiner, L., Lächelt, U., Troiber, C., Rädler, J., Hadwiger, P., Vornlocher, H.-P., Wagner, E. Nanosized multifunctional polyplexes for receptor-mediated siRNA delivery. *ACS Nano* 6, 5198–208 (2012).

167. Chu, D.S.H., Johnson, R.N., Pun, S.H. Cathepsin B-sensitive polymers for compartment-specific degradation and nucleic acid release. *Journal of Controlled Release: Official Journal of the Controlled Release Society* 157, 445–54 (2012).

168. Breunig, M., Lungwitz, U., Liebl, R., Goepferich, A. Breaking up the correlation between efficacy and toxicity for nonviral gene delivery. *Proc. Natl. Acad. Sci. USA* 104, 14454–14459 (2007).

169. Abdallah, B., Hassan, A., Benoist, C., Goula, D., Behr, J.P., Demeneix, B.A. A powerful nonviral vector for in vivo gene transfer into the adult mammalian brain: Polyethylenimine. *Human Gene Therapy* 7, 1947–1954 (1996).

170. Fischer, D., Bieber, T., Li, Y., Elsasser, H.-P., Kissel, T. A novel non-viral vector for DNA delivery based on low molecular weight, branched polyethylenimine: Effect of molecular weight on transfection efficiency and cytotoxicity. *Pharmaceutical Research* 16, 1273–9 (1999).

171. Gosselin, M.A., Guo, W., Lee, R.J. Efficient gene transfer using reversibly cross-linked low molecular weight polyethylenimine. *Bioconjugate Chemistry* 12, 989–994 (2001).

172. Kloeckner, J., Wagner, E., Ogris, M. Degradable gene carriers based on oligomerized polyamines. *European Journal of Pharmaceutical Sciences* 29, 414–425 (2006).

173. Russ, V., Elfberg, H., Thoma, C., Kloeckner, J., Ogris, M., Wagner, E. Novel degradable oligoethylenimine acrylate ester-based pseudodendrimers for in vitro and in vivo gene transfer. *Gene Therapy* 15, 18–29 (2007).

174. Liu, J., Jiang, X., Xu, L., Wang, X., Hennink, W.E., Zhuo, R. Novel reduction-responsive cross-linked polyethylenimine derivatives by click chemistry for nonviral gene delivery. *Bioconjugate Chemistry* 21, 1827–1835 (2010).

175. Anwer, K., Barnes, M.N., Fewell, J., Lewis, D.H., Alvarez, R.D. Phase-I clinical trial of IL-12 plasmid/lipopolymer complexes for the treatment of recurrent ovarian cancer. *Gene Therapy* 17, 360–369 (2010).

176. Anderson, D.G., Lynn, D.M., Langer, R. Semi-automated synthesis and screening of a large library of degradable cationic polymers for gene delivery. *Angewandte Chemie International Edition* 42, 3153–3158 (2003).

177. Greenland, J.R., Liu, H., Berry, D., Anderson, D.G., Kim, W.-K., Irvine, D.J., Langer, R., Letvin, N.L. Beta-amino ester polymers facilitate *in vivo* DNA transfection and adjuvant plasmid DNA immunization. *Molecular Therapy: The Journal of the American Society of Gene Therapy* 12, 164–170 (2005).

178. Jere, D., Xu, C.-X., Arote, R., Yun, C.-H., Cho, M.-H., Cho, C.-S. Poly(β-amino ester) as a carrier for si/shRNA delivery in lung cancer cells. *Biomaterials* 29, 2535–2547 (2008).

179. Tzeng, S.Y., Guerrero-Cázares, H., Martinez, E.E., Sunshine, J.C., Quiñones-Hinojosa, A., Green, J.J. Non-viral gene delivery nanoparticles based on poly(β-amino esters) for treatment of glioblastoma. *Biomaterials* 32, 5402–10 (2011).

180. Kim, T.-I., Seo, H.J., Choi, J.S., Yoon, J.K., Baek, J.-U., Kim, K., Park, J.-S. Synthesis of biodegradable cross-linked poly(β-amino ester) for gene delivery and its modification, inducing enhanced transfection efficiency and stepwise degradation. *Bioconjugate Chemistry* 16, 1140–1148 (2005).

181. Huang, Y.-H., Zugates, G.T., Peng, W., Holtz, D., Dunton, C., Green, J.J., Hossain, N., Chernick, M.R., Padera, R.F., Langer, R., Anderson, D.G., Sawicki, J.A. Nanoparticle-delivered suicide gene therapy effectively reduces ovarian tumor burden in mice. *Cancer Research* 69, 6184–91 (2009).

182. Cohen, J.A., Beaudette, T.T., Cohen, J.L., Broaders, K.E., Bachelder, E.M., Fréchet, J.M.J. Acetal-modified dextran microparticles with controlled degradation kinetics and surface functionality for gene delivery in phagocytic and non-phagocytic cells. *Advanced Materials* 22, 3593–3597 (2010).

183. Little, S.R., Lynn, D.M., Ge, Q., Anderson, D.G., Puram, S.V., Chen, J., Eisen, H.N., Langer, R. Poly-β amino ester-containing microparticles enhance the activity of non-viral genetic vaccines. *Proc. Natl. Acad. Sci. USA* 101, 9534–9539 (2004).

184. Zhang, J., Chua, L.S., Lynn, D.M. Multilayered thin films that sustain the release of functional DNA under physiological conditions. *Langmuir* 20, 8015–8021 (2004).

185. Jewell, C.M., Zhang, J., Fredin, N.J., Lynn, D.M. Multilayered polyelectrolyte films promote the direct and localized delivery of DNA to cells. *Journal of Controlled Release* 106, 214–223 (2005).

186. Flessner, R.M., Jewell, C.M., Anderson, D.G., Lynn, D.M. Degradable polyelectrolyte multilayers that promote the release of siRNA. *Langmuir* 27, 7868–7876 (2011).

187. DeMuth, P.C., Min, Y., Huang, B., Kramer, J.A., Miller, A.D., Barouch, D.H., Hammond, P.T., Irvine, D.J. Polymer multilayer tattooing for enhanced DNA vaccination. *Nat Mater* 12, 367–76 (2013).

188. Bayele, H.K., Sakthivel, T., O'Donell, M., Pasi, K.J., Wilderspin, A.F., Lee, C.A., Toth, I., Florence, A.T. Versatile peptide dendrimers for nucleic acid delivery. *Journal of Pharmaceutical Sciences* 94, 446–457 (2005).

189. Chen, J., Wu, C., Oupicky, D. Bioreducible hyperbranched poly(amido amine)s for gene delivery. *Biomacromolecules* 10, 2921–2927 (2009).

190. Barnard, A., Posocco, P., Pricl, S., Calderon, M., Haag, R., Hwang, M.E., Shum, V.W.T., Pack, D.W., Smith, D.K. Degradable self-assembling dendrons for gene delivery: Experimental and theoretical insights into the barriers to cellular uptake. *Journal of the American Chemical Society* 133, 20288–300 (2011).

191. Arote, R., Kim, T.-H., Kim, Y.-K., Hwang, S.-K., Jiang, H.-L., Song, H.-H., Nah, J.-W., Cho, M.-H., Cho, C.-S. A biodegradable poly(ester amine) based on polycaprolactone and polyethylenimine as a gene carrier. *Biomaterials* 28, 735–744 (2007).

192. Lai, W.-F., Lin, M.C.-M. Nucleic acid delivery with chitosan and its derivatives. *Journal of Controlled Release: Official Journal of the Controlled Release Society* 134, 158–68 (2009).

193. Yuan, X., Shah, B.A., Kotadia, N.K., Li, J., Gu, H., Wu, Z. The development and mechanism studies of cationic chitosan-modified biodegradable PLGA nanoparticles for efficient siRNA drug delivery. *Pharmaceutical Research* 27, 1285–95 (2010).

194. Mao, H.-Q., Roy, K., Troung-Le, V.L., Janes, K.A., Lin, K.Y., Wang, Y., August, J.T., Leong, K.W. Chitosan-DNA nanoparticles as gene carriers: Synthesis, characterization and transfection efficiency. *Journal of Controlled Release* 70, 399–421 (2001).

195. Roy, K., Mao, H.Q., Huang, S.K., Leong, K.W. Oral gene delivery with chitosan–DNA nanoparticles generates immunologic protection in a murine model of peanut allergy. *Nat Med* 5, 387–91 (1999).

196. Rojanarata, T., Opanasopit, P., Techaarpornkul, S., Ngawhirunpat, T., Ruktanonchai, U. Chitosan-thiamine pyrophosphate as a novel carrier for siRNA delivery. *Pharmaceutical Research* 25, 2807–2814 (2008).

197. Wardwell, P.R., Bader, R.A. Immunomodulation of cystic fibrosis epithelial cells via NF-κB decoy oligonucleotide-coated polysaccharide nanoparticles. *Journal of Biomedical Materials Research. Part A* 1–10, (2014).

198. Raviña, M., Cubillo, E., Olmeda, D., Novoa-Carballal, R., Fernandez-Megia, E., Riguera, R., Sánchez, A., Cano, A., Alonso, M.J. Hyaluronic acid/chitosan-g-

poly(ethylene glycol) nanoparticles for gene therapy: An application for pDNA and siRNA delivery. *Pharmaceutical Research* 27, 2544–55 (2010).

199. Ganesh, S., Iyer, A.K., Morrissey, D.V., Amiji, M.M. Hyaluronic acid based self-assembling nanosystems for CD44 target mediated siRNA delivery to solid tumors. *Biomaterials* 34, 3489–3502 (2013).

200. Park, K., Lee, M.-Y., Kim, K.S., Hahn, S.K. Target specific tumor treatment by VEGF siRNA complexed with reducible polyethyleneimine–hyaluronic acid conjugate. *Biomaterials* 31, 5258–5265 (2010).

201. Tokatlian, T., Cam, C., Segura, T. Non-viral DNA delivery from porous hyaluronic acid hydrogels in mice. *Biomaterials* 35, 825–835 (2014).

202. Bartlett, D.W., Davis, M.E. Physicochemical and biological characterization of targeted, nucleic acid-containing nanoparticles. *Bioconjugate Chemistry* 18, 456–68 (2007).

203. Davis, M.E. The first targeted delivery of siRNA in humans via a nanoparticle: From concept to clinic. 6, 659–668 (2009).

204. Zuckerman, J.E., Gritli, I., Tolcher, a., Heidel, J.D., Lim, D., Morgan, R., Chmielowski, B., Ribas, a., Davis, M.E., Yen, Y. Correlating animal and human phase Ia/Ib clinical data with CALAA-01, a targeted, polymer-based nanoparticle containing siRNA. *Proc. Natl. Acad. Sci. USA* 111, (2014).

205. Davis, M.E., Zuckerman, J.E., Choi, C.H. J., Seligson, D., Tolcher, A., Alabi, C.a., Yen, Y., Heidel, J.D., Ribas, A. Evidence of RNAi in humans from systemically administered siRNA via targeted nanoparticles. *Nature* 464, 1067–70 (2010).

206. Bachelder, E.M., Beaudette, T.T., Broaders, K.E., Dashe, J., Fréchet, J.M.J. Acetal-derivatized dextran: An acid-responsive biodegradable material for therapeutic applications. *Journal of the American Chemical Society* 130, 10494–5 (2008).

207. Cohen, J.L., Schubert, S., Wich, P.R., Cui, L., Cohen, J.a., Mynar, J.L., Fréchet, J.M.J. Acid-degradable cationic dextran particles for the delivery of siRNA therapeutics. *Bioconjugate Chemistry* 22, 1056–65 (2011).

208. Azzam, T., Eliyahu, H., Shapira, L., Linial, M., Barenholz, Y., Domb, A. J. Polysaccharide Oligoamine Based Conjugates for Gene Delivery. *Journal of Medicinal Chemistry* 45, 1817–1824 (2002).

209. Lee, H., Lytton-Jean, A.K., Chen, Y., Love, K.T., Park, A.I., Karagiannis, E.D., Sehgal, A., Querbes, W., Zurenko, C.S., Jayaraman, M., Peng, C.G., Charisse, K., Borodovsky, A., Manoharan, M., Donahoe, J.S., Truelove, J., Nahrendorf, M., Langer, R., Anderson, D.G. Molecularly self-assembled nucleic acid nanoparticles for targeted in vivo siRNA delivery. *Nature Nanotechnology* 7, 389–93 (2012).

210. Jensen, S.A., Day, E.S., Ko, C.H., Hurley, L.A., Luciano, J.P., Kouri, F.M., Merkel, T.J., Luthi, A.J., Patel, P.C., Cutler, J.I., Daniel, W.L., Scott, A.W., Rotz, M.W., Meade, T.J., Giljohann, D.A., Mirkin, C.A., Stegh, A.H. Spherical nucleic acid nanoparticle conjugates as an RNAi-based therapy for glioblastoma. *Science Translational Medicine* 5, 209ra152 (2013).

Hydrogels for Drug Delivery 7

Yuyan Weng*, Yue Lu† and Zhen Gu†

**Center for Soft Condensed Matter Physics and Interdisciplinary Research,
Soochow University, Suzhou 215006, China
†Joint Department of Biomedical Engineering, University of North Carolina
at Chapel Hill and North Carolina State University, Raleigh, NC 27695, USA
Pharmaceutics Division, Eshelman School of Pharmacy, University of North Carolina
at Chapel Hill, Chapel Hill, NC 27599, USA*

Abstract

Hydrogels, consisting of three-dimensional, cross-linked polymeric networks, are highly absorbent of aqueous solvents within their structures. Due to their biocompatibility, unique physical properties and tunable chemical composites, hydrogels are of particular interest for biomedical applications, including tissue engineering, drug delivery and diagnostics. In this chapter, recent advances in the development of hydrogel-based drug delivery systems (DDSs) involving micro- or nanotechnologies are surveyed. The fundamentals of hydrogels, their drug release mechanisms, and their stimuli-responsive behaviors are summarized. Advantages and limitations of hydrogel-based approaches, as well as future opportunities and challenges, are also discussed.

7.1 Introduction

In recent years, considerable efforts have been made to develop drug delivery systems (DDSs) that can release therapeutics in a site-specific and controlled manner.[1–7] Various approaches have been pursued to achieve the following goals: (1) maintaining the drug concentration within the therapeutic window during the delivery period; (2) protecting the active pharmaceutical ingredients from premature degradation; (3) enhancing drug efficiency and reducing drug dose; and (4) avoiding frequent administration. To date, a number of versatile delivery tools have been investigated from inorganic particles to polymeric matrices, and from natural substances to synthetic materials. Among these tools, hydrogels, consisting of three-dimensional, cross-linked polymeric networks, have attracted extensive attention due to their unique properties.[8–17]

Hydrogels can be formulated in a great many physical forms, including slabs,[15–17] nanoparticles,[18,19] coatings[20–22] and films.[23,24] They are used in both clinical practice and experimental medicine with a broad spectrum of applications, including biosensors,[25–27] tissue engineering[28,29] and regenerative medicine,[30,31] diagnostics,[32,33] cellular immobilization,[34–36] separation of biomolecules or cells[37,38] and barrier materials to regulate biological adhesions.[39–41] Among these, hydrogel-based delivery systems[8–14] have garnered increasing interest. The high water content and porous structure of hydrogels make them biocompatible materials with high loading efficiency of therapeutics. Compared to other delivery systems, where preparation conditions are sometimes destructive to these active agents, hydrogel preparation procedures are superior in maintaining drug activity and stability, as mild physiological conditions are typically adopted. The unique structure of hydrogels limits the mobility of therapeutic agents, which would be favorable for their preservation. In addition, hydrogels can protect active pharmaceutical ingredients from harsh environments during administration, such as enzymes in the serum and low pH in the stomach. More importantly, by choosing the appropriate cross-linking method and corresponding structure, drug release from hydrogels[1,8,42–45] can be "intelligently" triggered by different stimuli, such as pH, temperature, ionic strength, magnetic field, light, electric field, mechanical force, ultrasound, enzymes and specific molecular recognition.

Polymers utilized to form hydrogels can be divided into two broad categories: natural polymers[45,46] and synthetic polymers.[45,47,48] Natural polymers, such as polysaccharides [chitosan (CTS), hyaluronic acid (HA), alginate (ALG)], and proteins [gelatin (GEL), heparin (HEP), chondroitin sulfate (CS)], have many advantages such as biocompatibility and biodegradability. Synthetic polymers, such as polypeptides, polyesters and polyphosphazenes, usually present controllable degradation rates, mechanical strength and microstructures.[42,49,50] Polymers that are often used in hydrogel formation are listed in Table 7.1.

In this chapter, recent advances in the development of hydrogel-based DDSs, with an emphasis on their biomedical applications, are surveyed. Their drug release mechanisms, hydrogel formation methods and stimuli responsive behaviors are summarized. Advantages and limitations of hydrogel-based approaches, as well as future opportunities and challenges, are also discussed.

Table 7.1:   Typical polymers used in the formation of hydrogels for drug delivery.

Type	Hydrogel	Chemical Structure
Natural materials	Chitosan	
	Alginate	
	Dextran	
	β-Cyclodextrin (β-CD)	
	Hyaluronic acid (HA)	
	Chondroitin sulfate	

(Continued)

Table 7.1: *(Continued)*

Type	Hydrogel	Chemical Structure
Synthetic materials	Poly(hydroxyethyl methacrylate) (pHEMA)	
	Poly(vinyl alcohol) (PVA)	
	Poly(vinyl pyrrolidone) (PVP)	
	Polyurethane (PU)	

7.2   Drug Release Mechanism

Based on experimental and theoretical fundamentals,[51,52] many release mechanisms for hydrogel-based DDSs have been established. They are typically classified into three main release mechanisms: diffusion-controlled mechanism, swelling-controlled mechanism and chemically-controlled mechanism. The release mechanisms depend on the characteristics of both the polymeric network and the drug. When the hydrogel pores are bigger than the hydrodynamic radius of the drug, diffusion is the driving force for release. In contrast, when the hydrogel pores are smaller than the drug radius, swelling or degradation of the bulk or surface is essential for release. Such behaviors can be triggered by specific stimuli. Besides these two mechanisms, chemically-controlled release is also often present in hydrogel-based DDSs.

7.2.1   *Diffusion-controlled Release*

Diffusion is the most common mechanism governing drug release from hydrogels, as shown in Figure 7.1. It is specifically useful for analyzing the reservoir system and 3D-based matrix system. In this case, drug release from each type of system occurs by diffusion through the macromolecular network or water-filled pores.

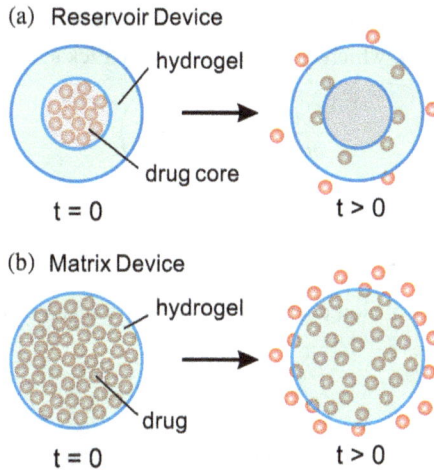

Figure 7.1: Schematic representation of two hydrogel-based diffusional controlled models: (a) reservoir device and (b) matrix device.

In the reservoir system, the drug depot is encapsulated within the core of the polymer gel, which can be described by Fick's first law.[53]

$$J_A = -D\frac{dC_A}{dx} \tag{7.1}$$

J_A is the molar flux of the drug (mol/cm²s), D is the drug diffusion coefficient in the hydrogel, C_A is the drug concentration. The diffusion coefficient is generally considered to be a constant.

In a steady-state diffusion process, by integrating the upper equation, the following expression will be given as:

$$J_A = K\frac{Dt,.C_A}{\delta} \tag{7.2}$$

where δ is the thickness of the reservoir layer in the hydrogel, and K is equal to the ratio of drug concentration in the hydrogel to that in the solution, called the partition coefficient. From this equation, the drug flux from the reservoir cannot remain constant, unless the concentration difference $t,.C_A$ remains constant. To achieve this goal, a device with excess solid drug core could be satisfied, whose internal solution will always be kept saturated. Its drug release rate will be time-independent and zero-order release kinetics can be achieved.

In the matrix drug delivery system, the drug molecules uniformly disperse throughout the hydrogel, which can be described by Fick's second law.[53,54]

$$\frac{\partial C_A}{\partial t} = \frac{\partial}{\partial x}\left(D\frac{\partial C_A}{\partial x}\right) \tag{7.3}$$

This equation is suitable during one-dimensional transport with non-moving boundaries, and can be deduced into two cases depending on whether the diffusion coefficient D is constant or not.

For the constant diffusion coefficient, the equation can be deduced to:

$$\frac{dC_A}{dt} = D\frac{d^2 C_A}{dx^2} \tag{7.4}$$

However, in most cases, the drug diffusion coefficient is dependent on concentration. Based on free volume theories[55] that can account for void space in the gel structure, researchers proposed the diffusion coefficient on the gel property. Following Higuchi's kinetics,[56] the gel matrices exhibited diffusion-controlled release, where drug release is proportional to the square root of time. Ritger and Peppas demonstrated the drug release mechanism by the time dependent power law function. By fitting the release data to the Ritger–Peppas equation,[57] the mechanism responsible for release can be determined:

$$\frac{M_t}{M_\infty} = k \cdot t^n \tag{7.5}$$

where M_t is the amount of drug released at time t, and M_∞ is the total amount of released drug, so their ratio is the drug fractional release at time t. The constant k is a kinetic constant based on the structural and geometric characteristic of the device. The release index n of this drug release system corresponds to different mechanisms of drug release.

7.2.2 Swelling-controlled Release

Once the hydrogels swell, a transformation will take place from glassy state to rubbery state, corresponding to a transformation from drug encapsulation state to fast diffusing state.

In polymeric matrices, both drug diffusion and polymer chain relaxation determine drug release. In swelling polymeric matrices, the time scale for polymer relaxation (λ) is the rate-limiting step; while in diffusion-controlled release systems, the time-scale of drug diffusion t is the rate-limiting step. The Deborah number (D_e) is used to compare these two time scales:

$$D_e = \frac{\lambda}{t} = \frac{\lambda D}{\delta(t)^2} \tag{7.6}$$

where $t = \delta(t)^2/D$ and $\delta(t)$ is the time-dependent thickness of the swollen phase. In diffusion-controlled systems, when $D_e \leq 1$, the molecule release process is dominated by Fick's diffusion.

In swelling-controlled systems, when $D_e \geq 1$, the molecule release rate is controlled by the swelling rate of the polymer networks.

Taking drug diffusion and chain relaxation into consideration, Peppas and Sahlin deformed the Ritger–Peppas equation to describe the swelling-controlled releasing system by the empirical power law model.[58]

$$\frac{M_t}{M_\infty} = k_1 t^m + k_2 t^{2m} \tag{7.7}$$

k_1, k_2, m are constants, and the two parts on the right side of the equation correspond to the diffusion contribution and the polymer relaxation, respectively. However, the circumstance with the moving-boundary is much more complicated, and this formula is not appropriate in cases where the gel swells heterogeneously. A dimensionless swelling interface number S_w that correlates the moving boundary phenomena to hydrogel swelling was introduced by Korsmeyer and Peppas:

$$S_w = \frac{V\delta(t)}{D} \tag{7.8}$$

where V is the hydrogel's front-moving velocity and D is the drug diffusion coefficient in the swollen phase. Take the slab system, where $S_w \ll 1$, for example, drug diffuses much faster than the interface movement of the glassy-rubbery state and thus a zero-order release profile is expected.

Then, Siepmann and Peppas proposed a sequential layer model to describe a system, building on several modeling iterations and considering drug diffusion, polymer chain relaxation and dissolution. For example, in cylindrical geometry with concentration-dependent diffusion coefficients, the releasing equation can be deduced to:[59]

$$\frac{\partial C_k}{\partial t} = \frac{\partial}{\partial r}\left(D_k \frac{\partial C_k}{\partial r}\right) + \frac{D_k}{r}\frac{\partial C_k}{\partial r} + \frac{\partial}{\partial z}\left(D_k \frac{\partial C_k}{\partial z}\right) \tag{7.9}$$

where C_k and D_k are the concentration and diffusivity of the drug, respectively.

As the dependence of the diffusion coefficients and the concentration are hard to express in the analysis formula, the above equations can only be solved numerically and tested experimentally for agreement.

In ionizable networks such as HPMC matrix tablets distributed with KCl,[60] the moving boundary and the chemical potential should be taken into account. Moreover, a mathematical model by Wu *et al.* was developed to explain swelling-controlled release.[61] The moving boundary condition was derived from the volume balance in this model. Assuming a homogeneous mixture of drug polymer in the tablet at $t = 0$ with perfect sink conditions, the model was verified in poly (ethylene oxide) (PEO) hydrogels with different molecular weights. The results of water uptake, swelling and dissolution of PEO matrices as well as drug release are consistent with the mathematical model.

7.2.3 *Chemically-controlled Delivery*

Chemically-controlled release systems can be further divided into two types: erodible drug delivery systems and pendant chain systems. In erodible systems, the drug is released with hydrogel degradation or dissolution. In pendant chain systems, where drug molecules are affixed to the polymer backbone, the drug is released as these linkages degrade.

Erodible drug release systems

In erodible drug release systems (either matrix or reservoir), drug release is not only controlled by drug diffusion through the gel, but also determined by material erosion. Supposing matrix erosion is the sole reason for drug release, a model developed by Hopfenberg can be expressed as:[62]

$$\frac{M_t}{M_\infty} = 1 - \left(1 - \frac{k_a t}{C_0 a_0}\right)^n \tag{7.10}$$

where a_0 stands for the geometrical dimension. For spherical or cylindrical geometry, a_0 is the radius; while for slab geometry, a_0 is the half-thickness. C_0 is the drug concentration in the surface erosion device. n is the index, equal to 1, 2, 3 for slab, cylinder, or sphere, respectively. Drug release presents a zero-order profile when $n = 1$ in a slab system. However, the erosion process is heterogeneous in most cases, and so a general mathematical model was further developed by Katzhendler *et al.*[63] For a flat tablet, different radial and vertical erosion were taken into account:

$$\frac{M_t}{M_\infty} = 1 - \left(1 - \frac{k_a t}{C_0 a_0}\right)^2 1 - \frac{2K_b t}{C_0 b_0} \tag{7.11}$$

where, k_a and k_b are degradation constants, radial and axial. a_0 is the initial tablet radius and b_0 is the tablet thickness. Later, a generalized statistical *co*-kinetic model was developed by Martens *et al.* to predict the degradation behaviors of acrylated poly(vinyl alcohol) (PVA) hydrogels.[64] In this model, a statistical approach predicted the configuration of the cross-linking molecules and kinetic chains as well as the probability of an intact degradable linkage.

Pendant chain systems

In a pendent chain system, the drug molecule is chemically linked to the backbone of polymers to form hydrogels. Drug release may be accompanied by chemical or enzymatic hydrolysis. In a biodegradable system, gel polymers gradually decompose and controlled drug release takes place. The release rate of covalently attached drugs is often determined by the degradation rate of polymer-drug linkages. In hydrolysis degradation systems, the release rate can be illustrated by a simple first-order kinetic relationship. However, in the case of enzymatic degradation, release kinetics are more complicated. A statistical-*co*-kinetic model has been developed to describe the effects of hydrolytic or enzymatic degradation on the macroscopic properties of hydrogels.[65]

7.3   Formation of Hydrogels

The methods used to form hydrogels are numerous and there are various classification standards for these methods. Here, we will summarize these methods in two basic classes: non-*in situ* formation and *in situ* formation, based on the location where the gelling process happens.

7.3.1   *Non-*in situ *Hydrogel Formation*

Traditionally, hydrogels for drug delivery are formed by non-*in situ* (*in vitro*) processes and encapsulated with target drugs before being administered into the body. Hydrogels can be formed non-*in situ* by a wide range of cross-linking approaches, including UV photopolymerization and various chemical cross-linking techniques,[66] embuing them with design flexibility. However, this strategy has several limitations for clinical applications. A major concern is the difficulty in extruding formed hydrogels through a needle due to the defined dimensionality and typically high elasticity of hydrogels. One way to address this issue is to convert the formed hydrogels into injectable micro- or nano-scale formations. Several examples will be given in Section 5.4. The inevitable leaching of drugs out of the formed hydrogels during the removal of toxic reagents before hydrogel administration also restricts the efficiency of non-*in situ* gelled hydrogels for drug delivery applications. An alternative approach is to perform the gelling process *in situ*, which will be discussed here.

7.3.2 In situ *Hydrogel Formation*

In situ hydrogel formation approaches have several advantages over conventional implantable approaches due to the fact that they can be administered via a mini-mally invasive route to achieve sustained release of therapeutic drugs.[67] Different *in situ* formation methods of hydrogels can be divided into two main categories: chemical cross-linking and physical cross-linking. The covalent bonds produced during chemical cross-linking may improve the mechanical properties of hydrogels. A variety of reactions have been introduced to obtain cross-linked polymers. Besides conventional coupling reactions, recently developed reactions such as click chem-istry and native chemical ligation have become increasingly popular due to their ease of use and high conversion rates. On the other hand, non-covalent interactions are key in physical cross-linking, and interactions such as hydrophobic interactions, ionic interactions, hydrogen bonding and host-guest interactions may play indepen-dent roles or show combined effects. In this section, we will briefly introduce some representative cross-linking methods (Figure 7.2).

7.3.2.1 *Chemically cross-linked hydrogels*

7.3.2.1.1 Photopolymerization

Hydrogels can be formed *in situ* by photopolymerization with UV/visible radi-ation and with the aid of photoinitiators. Radiation can be carried out *in vivo* in a minimally-invasive manner by laparoscopic devices, catheters or transder-mal illumination. For these hydrogels, the photocuring process can be completed fast enough with precise timing and spatial control. Since Hubbell *et al.*[69] pio-neered this crosslinking method, photopolymerization of gel crosslinking *in situ* has been widely used. They designed a polymer with a central PEG chain and lat-eral oligomeric blocks of hydrolysable α-hydroxy acid. Upon UV radiation, rapid gelation was observed. To date, a number of hydrogels polymerized from different monomers or polymers have been designed, such as thermo-sensitive methacry-lated poly(*N*-(2-hydroxylpropyl)methacrylamide lactate)−PEG−poly(*N*-(2-hydr-oxylpropyl) methacrylamide lactate) [*p*(HPMAm-lac)−PEG−*p*(HPMAm-lac)], acrylated four-arm PEG, methacrylated dextran, methacrylated dextran−HEMA−

(a) **Photo-crosslinked hydrogel**

Macromer

Vinyl group

hv
initiator

Sol   **Gel**

(b) **Thermosensitive hydrogel**

Hydrophilic
block

Hydrophobic
block

ΔT

Sol   **Gel**

(c) **Steteocomplexed hydrogel**

Dextran

Oligo (D-lactiide)

Oligo (L-lactiide)

+   Stereo-
complex

Sol   **Gel**

(d) **pH-sensitive hydrogel**

Hydrophobic   Hydrophilic
block   block

Ionized pH-sensitive block

Non-ionized
pH-sensitive block

ΔT/pH

Sol   **Gel**

(e) **Peptide-based hydrogel**

Hydrophobic amino acid

Positively   Negatively
charged   charged
amino acid   amino acid

ΔT/pH/salt

Sol   **Gel**

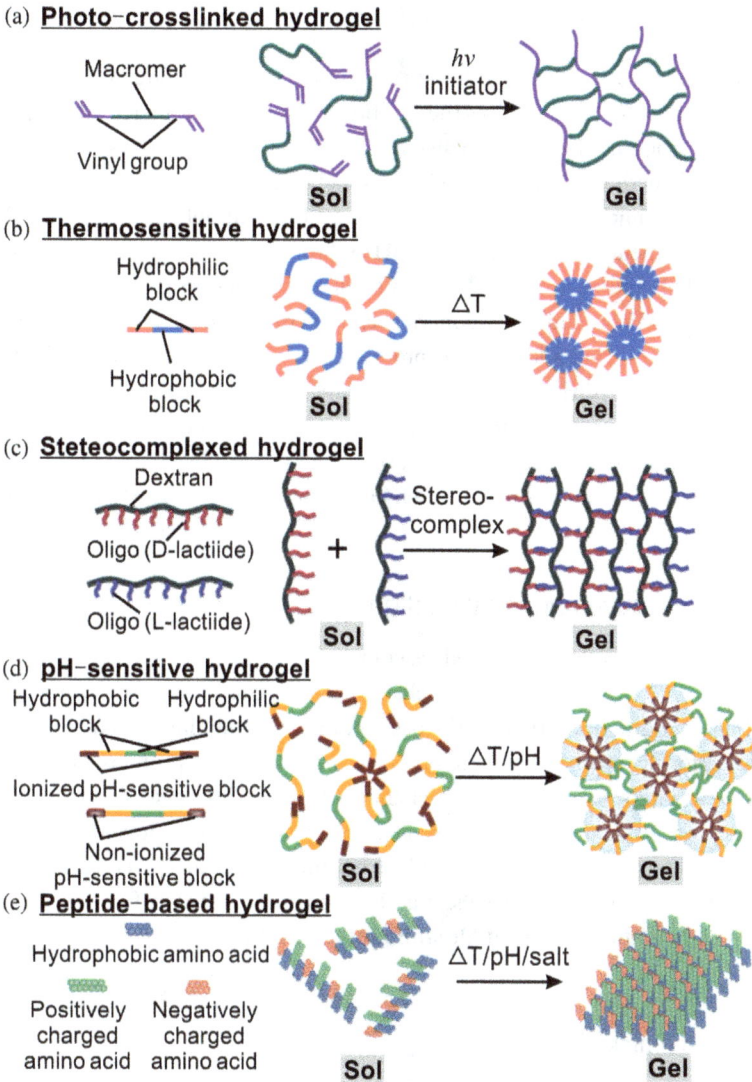

Figure 7.2: Representative methods for the formation of *in situ* hydrogels.[68]

dimethylaminoethyl (dex−HEMA−DMAE), and methacrylated eight-arm PEG−
poly(lactic acid) (PEG−PLA) star block copolymers.[70–73] However, a concern
regarding photopolymerization is that during UV treatment, the gel matrix
structure and drug activity may be affected.

7.3.2.1.2   Michael addition

Michael addition is a unique addition reaction that can be conducted under mild conditions. This reaction involves the addition of a nucleophile or activated olefin to a carbon–carbon double bond on alkenes. Its high selectivity makes this chemical cross-linking method favorable for DDSs. Hubbell and co-workers used this reaction to prepare injectable matrices by reacting PEG–dithiol with PEG–acrylates in an aqueous medium at physiological pH and room temperature. In a subsequent study, Hubbell and co-workers[74] used a similar approach to cross-link hydrogels by combining Michael addition donors such as pentaerythritol tetrakis 3'-mercaptopropionate (QT) and addition acceptors such as poly(ethylene glycol) diacrylate (PEGDA), pentaerythritol triacrylate (TA) and poly(propylene glycol) diacrylate (PPODA), which gave rise to hydrogels that showed much higher mechanical strength. Hahn *et al.* developed a novel sustained release formulation for erythropoietin (EPO) by using hyaluronic acid hydrogels cross-linked by Michael addition.

7.3.2.1.3   Native chemical ligation

The native chemical ligation-based reaction, which was first reported by Dawson *et al.*[75] in 1994, has a much higher chemoselective coupling ability compared to the Michael addition reaction. In the past decade, this coupling method has mainly been used for the synthesis of large peptides, moderate sized proteins and peptide-based dendrimers. Recently, native chemical ligation has been investigated as a cross-linking mechanism for hydrogels. Messersmith *et al.*[75] mixed four-armed PEG macromers, either functionalized with a thioester or an *N*-terminal cysteine peptide, to yield strong hydrogels. The high chemoselectivity of native chemical ligation makes it highly compatible with proteins or drugs.

7.3.2.1.4   Click chemistry

Click chemistry has gained traction in the past decade because of an increased emphasis on precise cross-linking position. Compared to radical or photochemical cross-linking approaches, a more controlled distribution of cross-links over the network and better control over mesh size distribution can be obtained using click chemistry. The process is fast and can be completed under physiological conditions. In the early stages of development of this method, the Cu(I) catalyst was a limitation[77] due to its cytotoxicity. The more recent Cu(I)-free click chemistry may be relatively slow and inefficient, but its biocompatibility makes it desirable.[78,79]

7.3.2.2 *Physically cross-linked hydrogels*

Compared with chemical cross-linking approaches, cross-linking methods that utilize different types of physical interactions are superior as there is no need for chemical modification or addition of cross-linking entities *in vivo* during the gelation process. Like chemically cross-linked hydrogels, physically cross-linked hydrogels have their own limitations, such as poor design flexibility and short tissue dwell time. We will briefly introduce some common physical cross-linking methods.

7.3.2.2.1 Hydrophobic interactions

Polymers with hydrophobic domains can be cross-linked by reverse thermal gelation for hydrogel formation in aqueous environments.[66] To design a hydrophobicity-driven gelling system, the hydrophobic segment is usually coupled to a hydrophilic polymer segment using graft polymerization or copolymerization. Typically, in response to increased temperature, hydrophobic domains aggregate to minimize the hydrophobic surface, resulting in a decrease in the amount of structured water surrounding the hydrophobic domains as well as a decrease in maximum solvent entropy.[66] The gelation temperature can be adjusted by tuning the copolymer concentration, the proportion of the hydrophobic block, and the chemical structure of the polymer.

Hydrogels based on poly(N-isopropylacrylamide) (PNIPAAm), a thermosensitive polymer, exhibit reversible sol-gel transition behaviors via increased inter- and intra-molecular hydrophobic interactions when the environmental temperature rises above the lower critical solution temperature (LCST). Similarly, other thermogelling hydrophobic blocks can be used to design hydrogel systems capable of *in situ* gelling through hydrophobic interactions. Chemical structures of common thermogelling hydrophobic blocks are shown in Figure 7.3. The most commonly used thermoreversible gels are those prepared from poly (ethyleneoxide)-*b*-poly(propyleneoxide)-*b*-poly(ethyleneoxide) (PEO-PPO-PEO, Pluronics®, Tetronics®, poloxamer). PEO-PPO-PEO triblock copolymers display sol-gel transition behaviors as a result of increased intermolecular hydrophobic interactions between PPO segments in response to raising the temperature.[66] The polymer solution is a free-flowing liquid at ambient temperature and gels at body temperature, which makes it easy to administer into any desired body cavity. A major concern of this copolymer-based hydrogel system is the rapid dissolution or dissipation of such copolymers *in vivo*. One approach to address this issue is to chemically cross-link the network. Alternatively, PEG-PLGA-PEG or PLGA-PEG-PLGA copolymers have been used to form hydrogels *in situ* via hydrophobic interactions due to the increased hydrophobicity of PLGA.[80] These PLGA/PEG tri-block copolymers

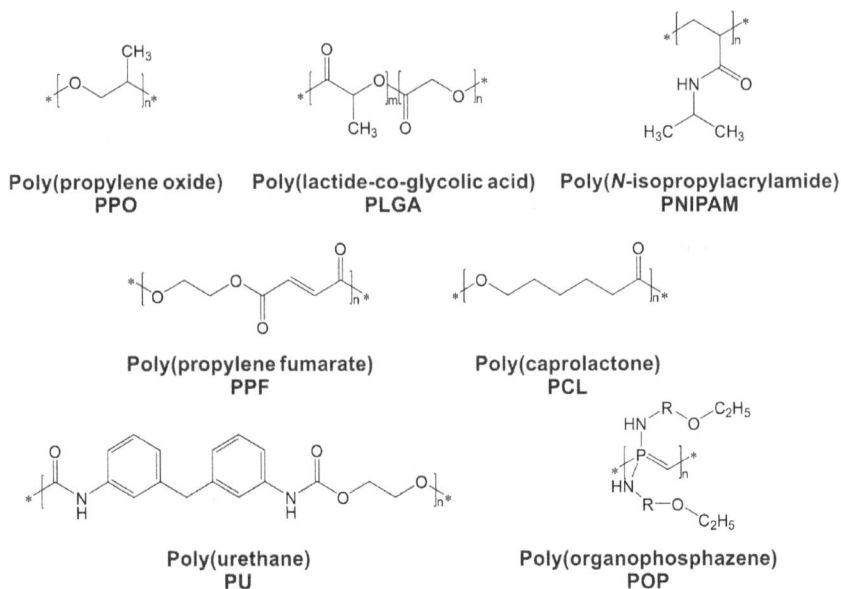

Figure 7.3: Chemical structures of typical thermogelling hydrophobic blocks.[66]

can achieve sustained release of various proteins,[81] and thus show great potential for the design and development of DDSs.

In the last decade, the Stupp group has carried out systematic studies focusing on peptide amphiphile (PA)-based hydrogels. PA contains a short peptide sequence covalently attached to a saturated alkyl chain and is thus capable of self-assembly into fiber-like nanostructures.[82] Recently, they reported a PA-based nanofiber gel and demonstrated its efficacy for controlled release of the anti-inflammatory drug dexamethasone (Dex).[83] In addition to polymeric gelators, small molecule gelators have also been used successfully to form hydrogels utilizing hydrophobic interactions. Xing et al. developed hydrogels based on vancomycin, an important class of antibiotics, by carefully balancing hydrophobic interactions and hydrogen bonds in water.[84]

7.3.2.2.2 Ionic interactions

Ionic interactions between a polymer and a small molecule, or between polymers, have also been widely used in the *in situ* formation of hydrogels. Ionic interaction-induced gelation can be achieved by introducing ionic groups to the end or side chains of the polymer structure. The gelation process can thus be reversibly controlled by environment pH, due to a change between ionized and non-ionized states.

A representative example of this approach is the use of oppositely-charged ionic ABA triblock copolyelectrolytes, derived from a common triblock copolymer precursor.[85] The polymers were mixed equimolarly, leading to coacervate cross-linking and network formation. Here, the properties of the resulting hydrogel can be adjusted by fine-tuning the ionic group and polymer composition, making this strategy modular.

7.3.2.2.3 Hydrogen bonding

Hydrogen bonding, via the blending two or more natural polymers, can be utilized to achieve *in situ* gelation. Hydrogen bonding interactions between polymer chains, facilitated by the compatible geometries of the mixed polymers, can lead to the increased gel-like viscoelastic properties of polymer blends compared to the individual polymers.[86] Blends of natural polymers including HA/emethylcellulose and gelatin/agar are representative examples of injectable hydrogels formed via hydrogen bonding interactions.[87,88] It is worth noting that dilution and dispersion caused by water influx limit the application of this method to short term drug delivery.

7.3.2.2.4 Stereocomplexation

Stereocomplexation is the synergistic interaction between small molecules or polymer chains that share the same chemical composition but have different stereochemistry.[66] Stereocomplexation of poly(L-lactide) (PLLA) and poly(D-lactide) (PDLA) as stereocomplex crystals was first reported during the preparation of *in situ* forming hydrogels.[89] PLLA and PDLA can be coupled to dextran to form hydrogels, and this gelation mechanism was first described by De Jong *et al.*[90] Their applications for the delivery of lysozyme, immunoglobulin G (IgG) and recombinant human IL-2 (rhIL-2) have been widely reported. Hiemstra and colleagues formed biodegradable hydrogels *in situ* from the stereocomplexation of PEG-PLLA and PEG-PDLA star block copolymers.[91] The protein loading capability and drug release profile of the hydrogels were investigated using two model proteins, lysozyme and IgG.

7.3.2.2.5 Supramolecular chemistry

Supramolecular chemistry is a novel approach to hydrogel formation *in situ*. A classic cross-linking interaction in this category is the formation of β-CD inclusion complexes. The inclusion complexes were cross-linked into gels *in situ*.[92] Star-shaped PEG polymers end-functionalized by β-CD and cholesterol were prepared for protein delivery.[93,94] The gel-sol transition was reversible and controllable, based on polymer concentration and PEG content of graft copolymers. The hydrogels were observed to be both pH- and thermo-responsive with the addition of a pH-responsive poly(D-lysine) (PL) moiety.[95] Polypeptides are biodegradable and biocompatible, and possess these diverse structures: α-helix, β-sheet and random coil, which create opportunities for complex architectures to be built. Schneider et al.[96] developed a metal ion-responsive peptide-based hydrogel that self-assembled into a β-sheet-rich fibrillar hydrogel upon the addition of zinc ions. Besides well-known peptide—peptide interactions such as coiled-coil and β-sheets, specific peptide—polysaccharide interactions can be used to build supramolecular hydrogels. Kiick et al.[97] showed an elegant example where he utilized the interaction between low molecular weight heparin and heparin binding peptide sequences to couple them into star PEGs. These gels were able to bind to and release heparin-binding growth factors in a controlled fashion.

7.4 Stimuli-responsive Hydrogels

7.4.1 Temperature-responsive Hydrogels

In the transition between room temperature and body temperature, polymers that exhibit lower critical solution temperature (LCST) behavior will ideally form gels suitable for administration. In contrast, polymers that exhibit upper critical solution temperature (UCST) will undergo phase separation below a specific temperature. Most thermo-sensitive polymers used in biomedical applications show LCST behavior. Temperature-dependent hydrogels are made from moderately hydrophobic groups or a mixture of hydrophilic and hydrophobic segments. At lower temperatures, hydrogen bonding dominates and leads to enhanced dissolution in water. As the temperature increases, hydrophobic interactions among hydrophobic segments strengthen, while hydrogen bonding becomes weaker. The net result is that the hydrogels shrink due to inter-polymer chain association through hydrophobic interactions.

Negatively thermo-sensitive hydrogels behave similarly, shrinking at high temperatures and swelling at low temperatures. By changing the targeting positions, different temperature response behaviors can be observed in these hydrogels. The on-off drug release profile is the most common one in response to stepwise temperature change. Thermo-sensitive monolithic hydrogels used in these studies include crosslinked P(NIPAAm–co–BMA) hydrogels, as well as interpenetrating polymer networks (IPNs) of P(NIPAAm) and poly(tetramethylene ether glycol) (PTMEG).[98,99] By introducing BMA, a hydrophobic co-monomer, the mechanical strength of NIPAAm gels can be increased. Release of indomethacin from these matrices was achieved at low temperature (on) and stopped at high temperature (off), and can be explained by the skin-type barrier. Temperature-sensitive hydrogels can also be placed inside the holes or apertures of a rigid capsule.[100,101] Changes in volume define the on–off states of release. Temperature-sensitive hydrogels can also be placed inside a rigid matrix or be grafted onto the surface of rigid membranes. The release of the model drug 4-acetamidophen was dependent on temperature. A similar approach was used to develop a reservoir-type microcapsule by encapsulating ethyl-cellulose containing nano-sized PNIPAAm hydrogel particles within the drug core.[102] To make stable thermally-controlled on–off devices, PNIPAAm hydrogels can be grafted onto the entire surface of rigid porous polymer membranes.[103] Yang and colleagues studied the interactions between proteins and PNIPAAm hydrogels cross-linked using N, N'-methylenebisacrylamide, and found that they displayed low LCST. Insulin and BSA were chosen as model proteins to examine the ability of PNIPAAm hydrogels to be protein delivery carriers. Protein release was incomplete as a result of strong interactions between the polymer and the cargo protein.[104] Xu *et al.* achieved sufficient concentration of surface-coupled ATRP initiators [porous polycaprolactone (PPCL)-Br surfaces] via a reaction between hydroxyl groups on PPCL films (prepared by using PEG as the pore-forming agent) and 2-bromoisobutyryl bromide, for surface-initiated ATRP of NIPAAm to be subsequently achieved.[105,106] The resultant PNIPAAm-grafted PPCL films possessed an interconnected porous structure and exhibited low LCST of about 32°C. The potential for this grafted polymer film to be a controlled protein delivery system was then demonstrated using BSA as a model protein.

Positive thermo-sensitivity, which refers to swelling at high temperature and shrinking at low temperature, can be observed in hydrogels formed by IPNs. IPNs of poly(acrylic acid) and polyacrylamide (PAAm) or P(AAm-co-BMA) show positive temperature-dependent swelling.[107] By increasing the BMA content, the transition temperature was shifted to a higher temperature. The transition of these hydrogels was reversible and responded to stepwise temperature changes. In a device for ketoprofen release, a reversible release rate was achieved.

7.4.2 pH-responsive Hydrogels

pH-sensitive polymers contain pendant acidic or basic groups that can undergo ionization during changes in pH. Since the swelling of polyelectrolyte hydrogels is mainly due to electrostatic repulsion among the charges present on the polymer chain, the extent of swelling is influenced by changes in electrostatic repulsion, such as pH, ionic strength and the type of counter ion. The swelling and pH-responsiveness of polyelectrolyte hydrogels can be adjusted by incorporating neutral co-monomers, such as 2-hydroxyethyl methacrylate, methyl methacrylate and maleic anhydride. Different co-monomers provide different hydrophobicity to the polymer chain, leading to different pH-sensitive behaviors. Grafted with poly(ethylene glycol), poly(methacrylic acid) is a typical example of a pH-sensitive hydrogel,[108] which can shrink at low pH and swell at high pH. This principle is not limited to this system and can be applied to other IPN systems, as the hydrogen binding of both polymer chains is tuned by pH.

A typical application of pH-sensitive hydrogels is in oral drug delivery. As the pH in the stomach is lower than in other parts of the body, pH-sensitive behavior can be observed. For poly-cationic hydrogels,[109] the degree of swelling reached a minimum at neutral pH and drug release was inhibited. This property has proved to be useful in preventing foul-tasting drugs from leaking into the mouth. When caffeine was loaded into hydrogels of methyl methacrylate and N,N'-dimethylaminoethyl methacrylate (DMAEM) copolymers, it was not released at neutral pH, but in a zero-order profile over pH 3–5 at which DMAEM was ionized. Semi-IPN of cross-linked chitosan and PEO showed significant swelling in the stomach,[110] which was ideal for the localized release of antibiotics such as amoxicillin and metronidazole for the treatment of *Helicobacter pylori*. In contrast, hydrogels of poly(acrylic acid) (PAA) or poly(methacrylic acid) (PMA) can be used to release drugs in a neutral pH environment.[111] Polyanion hydrogels like PAA were developed for site-specific drug delivery in the colon.[112] Contrary to the phenomenon described above, swelling of the hydrogels reached a minimum in the stomach and drug release was inhibited. Once the hydrogels passed through the stomach, a rise in pH led to ionization of the carboxylic groups. Only in the colon was the azoaromatic cross-links degraded by azoreductase, an enzyme produced by the microbial flora of the colon.

Poly(vinylacetaldiethylaminoacetate) (PVD) has pH-dependent aqueous solubility, forming a gel when the pH was changed from 4 to 7.4. The release of chlorpheniramine maleate was fast once the PVD solution was introduced into a buffer solution at pH = 7.4, but became very slow after a PVD hydrogel was formed. If the sol-to-gel transition time is shortened, the synthesized PVD will be mucoadhesive, making it an ideal system for nasal delivery.[113,114]

Synthetic polypeptides have also been used in biodegradable pH-sensitive hydrogels because their structures are more regular with fewer amino acid residues compared with natural peptides. Examples of such synthetic polypeptide hydrogels include poly(hydroxyl-L-glutamate), poly(L-ornithine), poly(aspartic acid), poly(L-lysine) and poly(L-glutamic acid). In addition to regular electrostatic effects associated with most pH-sensitive synthetic polymer hydrogels, secondary structures of the polypeptide backbone may also contribute to the pH-sensitive swelling behavior observed. The overall extent of pH-responsive swelling could be engineered by modifying the polypeptide's hydrophobicity and degree of ionization.[115]

Combining pH-sensitivity and biodegradability is one of the most important strategies when designing hydrogels for controlled drug delivery. For example, Gao *et al.* developed a biodegradable and pH-sensitive hydrogel, comprised of four types of pH-sensitive polyacrylic acid derivatives (PAAD) and a biodegradable poly(L-glutamic acid) (PGA) crosslinker, for oral delivery of proteins and peptides.[106] Insulin was loaded into the hydrogels as a model protein. The PAAD-based hydrogels with higher hydrophobicity exhibited lower swelling ratios and reduced biodegradation in conditions mimicking the stomach environment, and higher swelling ratios and greater biodegradation in conditions mimicking the intestinal environment. When the insulin-loaded hydrogels were orally administered to diabetic rats, a significant hypoglycemic effect was observed within 7 h.

7.4.3 *Light-responsive Hydrogels*

Light-based stimulus offers hydrogels the ability to spatiotemporally control drug release. Light-responsive hydrogels can be separated into ultraviolet (UV)-responsive, visible light-responsive and infrared (IR) light-responsive hydrogels.

A leuco-derivative molecule, bis(4-dimethylamino) phenylmethyl leucocyanide, was synthesized for UV sensitivity. At a fixed temperature, the hydrogels swelled under UV irradiation and shrank without radiation. The UV light-induced swelling was due to an increase in osmotic pressure within the hydrogel caused by UV-generated cyanide ions.[116]

Visible light-responsive hydrogels were prepared by introducing a light-sensitive chromophore such as trisodium salt of copper chlorophyllin to poly(N-isopropylacrylamide) (PNIPAAm) hydrogels.[117] When light (488 nm) was applied to the hydrogel, the chromophore absorbed light and increased the local temperature, which activated the swelling behavior of thermo-sensitive PNIPAAm hydrogels. In addition, IR light can also be used to trigger the volume changing response without chromophores. When PNIPAAm hydrogels without any chromophores were irradiated by a CO_2 laser infrared beam, a volume phase transition was observed together with gel bending towards the laser beam. Lo and colleagues reported a light-responsive PNIPAAm hydrogel nanocomposite incorporated with glycidyl methacrylate functionalized graphene oxide (GO–GMA).[118] In their work, the GO–GMA incorporated hydrogels, which were prepared via photopolymerization of NIPAAm in GO-GMA dispersed dimethyl sulfoxide (DMSO) solution, displayed significant volumetric change upon exposure to IR light as a result of efficient GO-GMA photothermal conversion.

Zhao *et al.* investigated the reversible light-responsive gel-to-sol and sol-to-gel phase transitions of a deoxycholate-β-CD and azobenzene-branched poly(acrylic acid)copolymer-based hydrogel, induced by the trans-cis photoisomerization of its azobenzene units.[119] When triggered by UV light, the hydrogel is converted into the sol phase as a result of the photochemical conversion of trans-azobenzene units into cis configurations, and the resulting dissociation of cis-azobenzene units. Upon photoirradiation with visible light at 450 nm, the hydrogel recovered from the sol phase. This light-responsive hydrogel system has a promising role in the development of DDS.

7.4.4 *Enzyme-responsive Hydrogels*

Enzymes play a central role in biochemical processes, and are therefore important targets for drug development.[120] When enzymatic activity is associated with a particular tissue, or when an enzyme is found at a higher concentration at the target site, sol-to-gel and gel-to-sol transition systems can be designed to deliver drugs via specific enzymatic conversion. Compared with other trigger-responsive hydrogel-based DDSs, the enzyme-based approach provides both high sensitivity and selectivity for programmed drug delivery to enzyme-overexpression target sites, which has led to a growing interest in developing enzyme-responsive hydrogels as smart DDSs.[121]

A biomimetic DDS designed by Biswas *et al.* for protein delivery consisted of a nano capsule crosslinked by bisacrylated peptides with specific sensitivity towards furin,[122] a ubiquitous intracellular protease that specifically localizes to the cell membrane, early endosome and Golgi complex. The researchers delivered both cytosolic and nuclear proteins in their active forms to different cell lines using enhanced green fluorescent protein (EGFP), CP3, BSA and the transcription factor KLF4 as model proteins. Aimetti *et al.* presented a method of fabricating enzyme-responsive PEG hydrogels for controlled protein delivery via thiol-ene photopolymerization.[123] To treat local inflammation, they synthesized a PEG hydrogel via thiol-ene photopolymerization of human neutrophil elastase (HNE)-sensitive peptide crosslinkers, allowing the gel to degrade at sites of inflammation. Protein therapeutics physically entrapped within the network would be selectively released upon exposure to HNE. Using this PEG hydrogel system, the controlled delivery of a model protein, BSA, was demonstrated. Thornton and colleagues conjugated amino-functionalized poly (ethylene glycol acrylamide) (PEGA) hydrogel particles with peptide actuators that cause charge-induced swelling and cargo release upon exposure to enzyme stimuli.[124] The peptide actuators here were designed based on both the specificity of the target enzyme and the charge properties of the to-be released protein cargo, thereby allowing for tunable release. Fluorescently-labeled albumin and avidin, proteins of similar size but opposite charge, were released at a rate that was regulated by the peptide actuator linked to the polymer carrier, thus exhibiting a highly controlled release mechanism.

7.4.5 *Electro-responsive Hydrogels*

Hydrogels sensitive to electric current are usually made of polyelectrolytes. Electro-responsive hydrogels undergo shrinking or swelling in the presence of an applied electric field. A change in hydrogel shape is isotropic upon its local electric field. Partially hydrolyzed polyacrylamide hydrogels touching both electrodes will undergo a volume response from a tiny change in electric potential. When a potential is applied, hydrated hydrogen ions migrate toward the cathode, resulting in a loss of water at the anode. At the same time, the electrostatic attraction of negatively-charged acrylic acid groups towards the anode surface creates uniaxial stress along the gel axis, mostly at the anode. These two simultaneous events lead to hydrogel shrinking at the anode.[125,126] In another case, hydrogels are placed in an electric field but without touching both electrodes. When a hydrogel made of sodium acrylic acid–acrylamide copolymers is placed in an aqueous solution, hydrogel deformation depends on the concentration of electrolytes. In the absence of electrolytes or in the presence of a very low concentration of electrolytes, application of an electric field causes the hydrogel to shrink.[127] In the presence of a high concentration of electrolytes in solution, swelling is more prominent at the hydrogel side facing the anode. If a cationic surfactant such as *n*-dodecylpyridinium chloride is added to the aqueous solution, swelling occurs at the side facing the cathode.[128]

By separating the solid polymer complex into two water-soluble polymers, Kwon *et al.*[129] designed a polymeric system that shows a sensitive and rapid response to electric currents. The system, based on poly(ethyloxazoline) that formed gels with poly(acrylic acid) or poly(methacrylic acid), was demonstrated to modulate insulin release.

7.4.6 *Magnetic-responsive Hydrogels*

The concept of utilizing magnetic fields to achieve pulsatile drug release from polymer composites was first proposed by Langer and coworkers. They demonstrated controlled insulin release from polymeric matrices embedded with magnetic beads, in response to a low frequency oscillating magnetic field.[130] A similar approach can be taken for the development of magnetic hydrogel-based DDSs using magnetic force as a drug release trigger. Magnetic hydrogels can also be targeted to specific sites such as tumors under the guidance of a magnetic field. Thus, designing vehicles based on magnetic-responsive hydrogels is a rather appealing approach in DDS development.

Typically, magnetic hydrogels are prepared by incorporating magnetic particles into hydrogels. Satarkar *et al.* incorporated superparamagnetic iron oxide (Fe_3O_4) nanoparticles into a thermo-sensitive N-isopropylacrylamide (NIPAAm)-based matrix.[131] When a high frequency alternating magnetic field (AMF) is applied, Fe_3O_4 nanoparticles will generate heat, raising the temperature in the polymer matrix and leading to the collapse of the hydrogel and subsequent drug release. Vitamin B12 and methylene blue were chosen as model drugs, and remote-controlled pulsatile drug release was demonstrated. Zhang and colleagues developed a magnetic hydrogel, consisting of chitosan (CS), β-glycerophosphate (GP) and Fe_3O_4 magnetic nanoparticles, for the controlled delivery of bacillus Calmette–Guérin (BCG).[132] Recently, researchers developed a new type of magnetic hydrogel using magnetic particles as cross-linking reagents. In this approach, the magnetic nanoparticles themselves are used as cross-linkers, instead of adding additional cross-linkers. To prepare these ferrohydrogels, Messing *et al.* added magnetic $CoFe_2O_4$ nanoparticles into polyacylamide hydrogels as cross-linkers during the polymerization process.[133]

7.4.7 *Glucose-responsive Hydrogels*

Glucose-responsive insulin delivery systems, or closed-loop smart insulin delivery systems, have attracted much attention recently. They can mimic normal pancreatic β-cell function and "secrete" insulin in response to blood glucose levels. Currently, most of the synthetic closed-loop insulin delivery systems have been developed from hydrogels, the matrix of which contains glucose-sensitive elements, such as glucose oxidase (GOx), phenylboronic acid (PBA) or glucose binding proteins (GBP). The rate of insulin release can be tuned by changes to the matrix structure, polymer degradation or glucose binding competition (Figure 7.4).

GOx is a glucose-specific enzyme that converts glucose into gluconic acid,[135] whose generation rate is directly correlated to the external glucose concentration.[136–138] If GOx is integrated into hydrogels, a glucose-triggered volume change or degradation in the pH-sensitive polymeric matrices can result in insulin release. In addition, another enzyme catalase (CAT) is often co-encapsulated to scavenge

(a) **Glucose-specific enzyme-based**

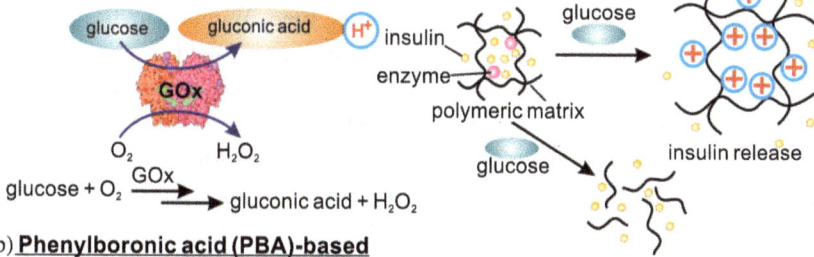

$$glucose + O_2 \xrightarrow{\text{GOx}} gluconic\ acid + H_2O_2$$

(b) **Phenylboronic acid (PBA)-based**

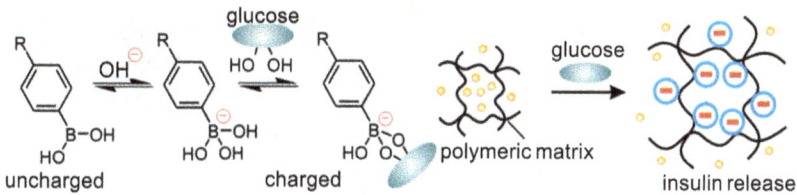

(c) **Glucose binding protein (GBP)-based**

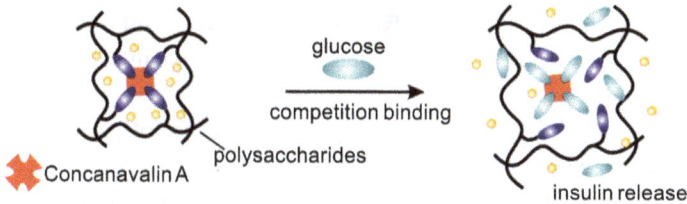

Figure 7.4: Schematic illustration of the insulin release mechanisms used in glucose-responsive (a) enzyme-, (b) PBA-, and (c) GBP-based synthetic closed-loop systems.[134]

for hydrogen peroxide (H_2O_2), which is generated during the catalysis of GOx. For example, N, N-Dimethylaminoethyl methacrylate (DMAEMA) is often introduced into copolymeric hydrogels, rendering them pH-sensitive. When glucose diffuses into the hydrogel, it is converted into gluconic acid, leading to the protonation of tertiary amine groups in DMAEMA. Hydrogels swell due to increased electrostatic chain repulsion, resulting in larger pores in the gel and the release of encapsulated insulin. Recently, Gu *et al.* developed chitosan-based sponge-like microgels for glucose-responsive delivery of insulin.[139] Monodispersed microgels ($256 \pm 18\,\mu m$), consisting of chitosan, enzyme-based nanocapsules and insulin, were fabricated by a one-step electrospraying method. Of note, both enzymes are covalently encapsulated into the nanogel-based capsules to enhance enzymatic stability, avoid denaturation, shield from immunogenicity, and minimize enzyme diffusion from the polymeric matrix. Microgels incubated under hyperglycemic conditions (400 mg dL^{-1}) swelled steadily over time. Within 3 h, the particles exhibited a 1.7-fold change in diameter, corresponding to a 5-fold change in volume. Importantly, this system was reversible and under normoglycemic conditions, the microgels shrank and insulin release was inhibited.

The reversible interactions between PBA and sugars (with diol) was first discovered by Lorand and Edwards.[140] For glucose-responsive insulin delivery, PBA is often incorporated into the crosslinked matrix; interaction with glucose gives it a negative charge, which usually causes swelling.[141–143] For example, Kitano et al.[144] utilized poly(N-vinyl-2-pyrrolidone-co-phenylboronic acid) [p(NVP−PBA)] for self-regulated insulin delivery. A remaining challenge for PBA-based systems is to optimize the chemical structure of the gel so that it can exert glucose sensitivity under physiological pH, since the pK_a of PBA is usually around 8–9. To address this, Matsumoto et al. have recently developed a new PBA derivative, 4-(2-acrylamidoethylcarbamoyl)-3-fluorophenylboronic acid (AmECFPBA) with paracarbamoyl and meta-fluoro substituents, which has strong electron-withdrawing substituents and exhibits a pK of 7.2. Hydrogels composed of AmECFPBA showed a volume phase transition in response to glucose concentration under physiological conditions.

Brownlee et al.[145] and Kim et al.[146] pioneered the use of glucose-sensitive hydrogels containing lecitin, a glucose-binding protein. The hydrogel, composed of cross-linked concanavalin A (a lectin having four binding sites) complexed to a glycosylated insulin, was loaded into a Durapore membrane pouch. In the presence of glucose, it competitively bound to concanavalin and triggered the release of insulin. For example, Yin and coworkers developed glucose-responsive concanavalin A-based hydrogels for self-regulated insulin delivery.[147–149] When the glucose concentration rose, the hydrogels swelled or hydrolyzed and therefore promoted insulin release.

7.4.8 Mechanical-responsive Hydrogels

Similar to the hydrogels mentioned above, hydrogels with mechanic force sensitivity have also been investigated and utilized in the development of DDSs. Lee and colleagues reported an alginate hydrogel-based system capable of pulsatile release of vascular endothelial growth factor (VEGF) in response to compressive forces with varying strain amplitudes.[150] Upon compression, the encapsulated cargo protein was released. Once the strain was removed, the hydrogel resumed its initial volume. In vivo, VEGF released from this system led to a significant increase in the formation of new blood vessels in non-obese diabetic mice.

7.5   Summary and Outlook

Hydrogels exhibit tremendous potential for controlled drug delivery, taken from both a research and clinical perspective. It is encouraging to see that substantial advances have been made towards developing versatile hydrogel-based scaffolds for the delivery of a variety of therapeutics, ranging from small drugs to biomacro-molecules. From the angle of material and formulation design, numerous synthesis strategies and drug loading techniques have been developed. Considerable insights have been gleaned into the loading efficiency, release mechanisms, and delivery pathways of hydrogels.

Despite these efforts, further development is required for the field to achieve its full clinical potential. Ideally, hydrogel formulations should systematically integrate analyses of drug characteristics, material properties and the environment of admin-istration sites. Figure 7.5 summarizes the typical administration sites and relevant material properties for engineering hydrogel-based DDSs. For example, ease of administration, by avoiding the risk of premature gelation inside the needle during injection, is extremely important for subcutaneous drug delivery. On the one hand, developing new approaches to trigger the crosslinking of the gel and drugs inside the body is essential; on the other hand, exploring novel hydrogel-based administration

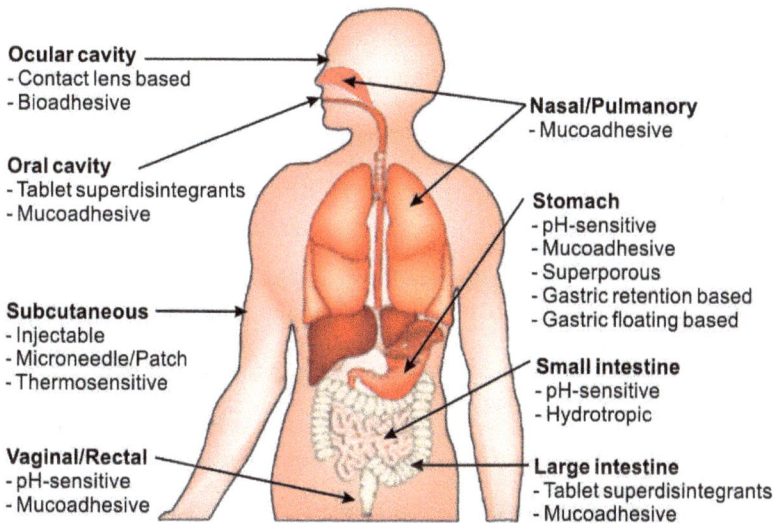

Figure 7.5:   Schematic illustration of typical administration routes for hydrogel-based DDSs and general design requirements associated with hydrogel properties.[153,154]

methods, such as a painless microneedle array[151] made out of degradable materials, hold great promise for clinical treatment.

Tailoring the kinetics of drug release by tuning material structure remains an important pursuit in this field. Take for example glucose-responsive hydrogels used in insulin delivery; an ideal design should involve both long-term release and pulsatile release mediated by blood glucose levels. Therefore, the swelling property or degradability of hydrogels should be carefully adjusted to achieve both long-term release as well as quick release in response to glucose levels.

Lastly, programmable delivery systems consisting of dual or multiple stimuli-responsive elements for enhanced efficacy remains the most valuable of delivery goals and are particularly useful for anticancer treatment. For example, Gu and coworkers have recently developed a core-shell-based nanocapsule (Gelipo) consisting of a liposomal core and a crosslinked-gel shell for the sequential and site-specific delivery of dual anticancer therapeutics. Gelipo is programmed to release proteins and small molecule drugs successively and to transport them to their distinct targets, the plasma membrane and the nucleus, respectively, for synergetic anticancer activity.[152]

References

1. He, C., et al. In situ gelling stimuli-sensitive block copolymer hydrogels for drug delivery. Journal of Controlled Release 127, 3189–207 (2008).
2. Hillebrenner, H., et al. Corking nano test tubes by chemical self-assembly. Journal of the American Chemical Society 128, 13, 4236–4237 (2006).
3. Qiu, Y. and Park, K. Environment-sensitive hydrogels for drug delivery. Advanced Drug Delivery Reviews 53, 3, 321–339 (2001).
4. Slowing, I.I., et al. Mesoporous silica nanoparticles as controlled release drug delivery and gene transfection carriers. Advanced Drug Delivery Reviews 60(11), 1278–1288 (2008).
5. Vivero-Escoto, J.L., et al. Mesoporous silica nanoparticles for intracellular controlled drug delivery. Small 6(18), 1952–1967 (2010).
6. Wilson, D.S., et al. Orally delivered thioketal nanoparticles loaded with TNF-alpha-siRNA target inflammation and inhibit gene expression in the intestines. Nature Materials 9(11), 923–928 (2010).
7. Zhang, S. and Uludag, H. Nanoparticulate systems for growth factor delivery. Pharmaceutical Research 26(7), 1561–1580 (2009).
8. Lin, C.-C. and Metters, A.T. Hydrogels in controlled release formulations: Network design and mathematical modeling. Advanced Drug Delivery Reviews 58(12–13), 1379–1408 (2006).
9. Ballios, B.G., et al. A hydrogel-based stem cell delivery system to treat retinal degenerative diseases. Biomaterials 31(9), 2555–2564 (2010).
10. Chen, M.-C., et al. A nanoscale drug-entrapment strategy for hydrogel-based systems for the delivery of poorly soluble drugs. Biomaterials 30(11), 2102–2111 (2009).

11. Wu, J., *et al*. A thermosensitive hydrogel based on quaternized chitosan and poly(ethylene glycol) for nasal drug delivery system. *Biomaterials* 28(13), 2220–2232 (2007).

12. Brandl, F., *et al*. Hydrogel-based drug delivery systems: Comparison of drug diffusivity and release kinetics. *Journal of Controlled Release* 142(2), 221–228 (2010).

13. He, H.Y., Cao, X. and Lee, L.J. Design of a novel hydrogel-based intelligent system for controlled drug release. *Journal of Controlled Release* 95(3), 391–402 (2004).

14. Dong, L. and Jiang, H. Autonomous microfluidics with stimuli-responsive hydrogels. *Soft Matter* 3(10), 1223–1230 (2007).

15. Birgersson, E., Lib, H. and Wua, S. Transient analysis of temperature-sensitive neutral hydrogels. *Journal of the Mechanics and Physics of Solids* 56(2), 444–466 (2008).

16. Boztas, A.O. and A. Guiseppi-Elie, Immobilization and release of the redoxmediator ferrocene monocarboxylic acid from within cross-linked p(HEMA-co-PEGMA-co-HMMA) hydrogels. *Biomacromolecules* 10(8), 2135–2143 (2009).

17. Wu, X.Y. and Zhou, Y. Finite element analysis of diffusional drug release from complex matrix systems. II. Factors influencing release kinetics. *Journal of Controlled Release* 51(1), 57–71 (1998).

18. Agnihotri, S.A., Mallikarjuna, N.N. and Aminabhavi, T.M. Recent advances on chitosan-based micro- and nanoparticles in drug delivery. *Journal of Controlled Release* 100(1), 5–28 (2004).

19. Merkel, T.J., *et al*. Using mechanobiological mimicry of red blood cells to extend circulation times of hydrogel microparticles. *Proceedings of the National Academy of Sciences of the United States of America* 108(2), 586–591 (2011).

20. Patil, S.D., Papadmitrakopoulos, F. and Burgess, D.J. Concurrent delivery of dexamethasone and VEGF for localized inflammation control and angiogenesis. *Journal of Controlled Release* 117(1), 68–79 (2007).

21. Picart, C., *et al*. Controlled degradability of polysaccharide multilayer films in vitro and in vivo. *Advanced Functional Materials* 15(11), 1771–1780 (2005).

22. Kim, D.H. and Martin, D.C. Sustained release of dexamethasone from hydrophilic matrices using PLGA nanoparticles for neural drug delivery. *Biomaterials* 27(15), 3031–3037 (2006).

23. Luo, Y., Kirker, K.R. and Prestwich, G.D. Cross-linked hyaluronic acid hydrogel films: new biomaterials for drug delivery. *Journal of Controlled Release* 69(1), 169–184 (2000).

24. Rolland, J.P., *et al*. Direct fabrication and harvesting of monodisperse, shape-specific nanobiomaterials. *Journal of the American Chemical Society* 127(28), 10096–10100 (2005).

25. Zhang, Z., Chen, S. and Jiang, S. Dual-functional biomimetic materials: Nonfouling poly(carboxybetaine) with active functional groups for protein immobilization. *Biomacromolecules* 7(12), 3311–3315 (2006).

26. Sharma, A.C., *et al*. A general photonic crystal sensing motif: Creatinine in bodily fluids. *Journal of the American Chemical Society* 126(9), 2971–2977 (2004).

27. Shen, C.Y. and Kostic, N.M. Kinetics of photoinduced electron-transfer reactions within sol-gel silica glass doped with zinc cytochrome c. Study of electrostatic effects in confined liquids. *Journal of the American Chemical Society* 119(6), 1304–1312 (1997).

28. Suh, J.K.F. and Matthew, H.W.T. Application of chitosan-based polysaccharide biomaterials in cartilage tissue engineering: a review. *Biomaterials* 21(24), 2589–2598 (2000).

29. Nowak, A.P., *et al.* Rapidly recovering hydrogel scaffolds from self-assembling diblock copolypeptide amphiphiles. *Nature* 417(6887), 424–428 (2002).

30. Slaughter, B.V., *et al.* Hydrogels in regenerative medicine. *Advanced Materials* 21(32–33), 3307–3329 (2009).

31. Lin, C.-C. and Anseth, K.S. PEG hydrogels for the controlled release of biomolecules in regenerative medicine. *Pharmaceutical Research* 26(3), 631–643 (2009).

32. Dong, L., *et al.* Adaptive liquid microlenses activated by stimuli-responsive hydrogels. *Nature* 442(7102), 551–554 (2006).

33. Ulijn, R.V., *et al.* Bioresponsive hydrogels. *Materials Today* 10(4), 40–48 (2007).

34. Luo, Y. and Shoichet, M.S. Light-activated immobilization of biomolecules to agarose hydrogels for controlled cellular response. *Biomacromolecules* 5(6), 2315–2323 (2004).

35. Gupta, N., *et al.* A versatile approach to high-throughput microarrays using thiol-ene chemistry. *Nature Chemistry* 2(2), 138–145 (2010).

36. Miller, J.S., *et al.* Bioactive hydrogels made from step-growth derived PEG-peptide macromers. *Biomaterials* 31(13), 3736–3743 (2010).

37. Fini, M., *et al.* The healing of confined critical size cancellous defects in the presence of silk fibroin hydrogel. *Biomaterials* 26(17), 3527–3536 (2005).

38. Sangeetha, N.M. and Maitra, U. Supramolecular gels: Functions and uses. *Chemical Society Reviews* 34(10), 821–836 (2005).

39. Elbert, D.L., Herbert, C.B. and Hubbell, J.A. Thin polymer layers formed by poly-electrolyte multilayer techniques on biological surfaces. *Langmuir* 15(16), 5355–5362 (1999).

40. Li, H., *et al.* Synthesis and biological evaluation of a cross-linked hyaluronan-mitomycin C hydrogel. *Biomacromolecules* 5(3), 895–902 (2004).

41. Cencetti, C., *et al.* Preparation and characterization of a new gellan gum and sulphated hyaluronic acid hydrogel designed for epidural scar prevention. *Journal of Materials Science-Materials in Medicine* 22(2), 263–271 (2011).

42. Kiser, P.F., Wilson, G. and Needham, D. A synthetic mimic of the secretory granule for drug delivery. *Nature* 394(6692), 459–462 (1998).

43. Traitel, T., Cohen, Y. and Kost, J. Characterization of glucose-sensitive insulin release systems in simulated in vivo conditions. *Biomaterials* 21(16), 1679–1687 (2000).

44. Xiao, X.C., *et al.* Positively thermo-sensitive monodisperse core-shell microspheres. *Advanced Functional Materials* 13(11), 847–852 (2003).

45. Peppas, N.A., *et al.* Hydrogels in biology and medicine: From molecular principles to bionanotechnology. *Advanced Materials* 18(11), 1345–1360 (2006).

46. Jeong, B., Kim, S.W. and Bae, Y.H. Thermosensitive sol-gel reversible hydrogels. *Advanced Drug Delivery Reviews* 54(1), 37–51 (2002).

47. Oh, J.K., *et al.* The development of microgels/nanogels for drug delivery applications. *Progress in Polymer Science* 33(4), 448–477 (2008).

48. Hamidi, M., Azadi, A. and Rafiei, P. Hydrogel nanoparticles in drug delivery. *Advanced Drug Delivery Reviews* 60(15), 1638–1649 (2008).

49. Hawker, C.J. and Wooley, K.L. The convergence of synthetic organic and polymer chemistries. *Science* 309(5738), 1200–1205 (2005).

50. Moon, J.J., *et al.* Interbilayer-crosslinked multilamellar vesicles as synthetic vaccines for potent humoral and cellular immune responses. *Nature Materials* 10(3), 243–251 (2011).

51. Baroli, B., Hydrogels for tissue engineering and delivery of tissue-inducing substances. *J. Pharm. Sci.* 96, 2197–2223 (2007).

52. Hoffman, A.S., Hydrogels for biomedical applications. *Advanced Drug Delivery Reviews* 64, 18–23 (2012).

53. Lin, C.C. and Matters, A.T., Hydrogels in controlled release formulations: Network design and mathematical modeling. *Advanced Drug Delivery Reviews* 58, 1379–1408 (2006).

54. Costa, P. and Sousa Lobo, J.M., Modeling and comparison of dissolution profiles. *Eur. J. Pharm. Sci.* 13, 123–133 (2001).

55. Lustig, S.R. and Peppas, N.A., Solute diffusion in swollen membranes. 9. Scaling laws for solute diffusion in gels. *J. Appl. Polym. Sci.* 36, 735–747 (1988).

56. Grassi, M. and Grassi, G., Mathematical modelling and controlled drug delivery: Matrix systems. *Curr. Drug Deliv.* 2, 97–116 (2005).

57. Peppas, N.A. Leobandung, P.B.W., Ichikawa, H. Hydrogels in pharmaceutical formulations. *Eur. J. Pharm. Biopharm.* 50, 27–46 (2000).

58. Peppas, N.A. and Sahlin, J.J., A simple equation for the description of solute release 3 Coupling of diffusion and relaxation. *Int. J. Pharm.* 57, 169–172 (1989).

59. Siepmann, J., P.N.A., Hydrophilic matrices for controlled drug delivery: an improved mathematical model to predict the resulting drug release kinetics (the "sequential layer" model). *Pharm. Res.* 69, 455–468 (2000).

60. Siepmann, J., Bodmeier, H.K.R. and Peppas, N.A. HPMC matrices for controlled drug delivery: a new model combining diffusion, swelling, and dissolution mechanisms and predicting the release kinetics. *Pharm. Res.* 16, 1748–1756, (1999).

61. Wu, N., Wang, L.S., Tan, D.C.W., Moochhala, S.M. and Yang, Y.Y. Mathematical modeling and in vitro study of controlled drug release via a highly swellable and dissoluble polymer matrix: polyethylene oxide with high molecular weights. *J. Control. Release* 102, 569–581 (2005).

62. Li H., Luo, R. and Lam, K.Y. Modeling of environmentally sensitive hydrogels for drug delivery: An overview and recent developments. *Front. Drug Des. Discov.* 2, 295–331 (2006).

63. Katzhendler, I., Hoffman, A., Goldberger, A. and Friedman, M., Modeling of drug release from erodible tablets. *J. Pharm. Sci.* 86, 110–115 (1997).

64. Martens, P.J. C.N.B. and Anseth, K.S. Degradable networks formed from multi-functional poly(vinyl alcohol) macromers: comparison of results from a generalized bulkdegradation model for polymer networks and experimental data. *Polymer* 45, 3377–3387 (2004).

65. DuBose, J.W., Cutshall, C. and Metters, A.T. Controlled release of tethered molecules via engineered hydrogel degradation: Model development and validation. *Journal of Biomedical Materials Research Part A* 74A(1), 104–116 (2005).

66. Hoare, T.R. and Kohane, D.S. Hydrogels in drug delivery: Progress and challenges. *Polymer* 49(8), 1993–2007 (2008).

67. Peng, K., *et al.* Cyclodextrin-dextran based in situ hydrogel formation: a carrier for hydrophobic drugs. *Soft Matter* 61, 85–87.

68. Chung, H.J. and Park, T.G. Self-assembled and nanostructured hydrogels for drug delivery and tissue engineering. *Nano Today* 4(5), 429–437 (2009).

69. Sawhney, A.S., Pathak, C.P., and Hubbell, J.A., Bioerodible hydrogels based on photopolymerized poly(ethylene glycol)-co-poly(.alpha.-hydroxy acid) diacrylate macromers. *Macromolecules* 26(4), 581–587 (1993).

70. Hiemstra, C., *et al.* Rapidly *in situ* forming biodegradable robust hydrogels by combining stereocomplexation and photopolymerization. *Journal of the American Chemical Society* 129(32), 9918–9926 (2007).

71. Kamoun, E.A. and Menzel, H. Crosslinking behavior of dextran modified with hydroxyethyl methacrylate upon irradiation with visible light — Effect of concentration, coinitiator type, and solvent. *Journal of Applied Polymer Science* 117(6), 3128–3138 (2010).

72. Van Tomme, S.R., Storm, G. and Hennink, W.E. In situ gelling hydrogels for pharmaceutical and biomedical applications. *International Journal of Pharmaceutics* 355(1–2), 1–18 (2008).

73. Vermonden, T., Censi, R. and Hennink, W.E. Hydrogels for protein delivery. *Chemical Reviews* 112(5), 2853–2888 (2012).

74. Elbert, D.L., Pratt, A.B., Lutolf, M.P., Halstenberg, S. and Hubbell, J.A. Protein delivery from materials formed by self-selective conjugate addition reactions. *J. Controlled Release* 76(1–2), 11–25 (2001).

75. Dawson, P.E., Muir, T.W., Clark-Lewis, I., and Kent, S.B.H., Synthesis of proteins by native chemical ligation. *Science* 266, 776–779 (1994).

76. Su, J., Hu, B.-H., Lowe, W.L. Jr, Kaufman, D.B., and Messersmith, P.B., Anti-inflammatory peptide-functionalized hydrogels for insulin-secreting cell encapsulation. *Biomaterials* 31(2), 308–314 (2010).

77. Malkoch, M., V.R., Gupta, N., Mespouille, L., Dubois, P., Mason, A.F., Hedrick, J.L., Liao, Q., Frank, C. W., Kingsbury, K., and Hawker, C.J., Synthesis of well-defined hydrogel networks using Click chemistry. *Chem. Commun.* 2774, (2006).

78. Clark, M., K.P., In situ crosslinked hydrogels formed using Cu(I)-free Huisgen cycloaddition reaction. *Polym. Int.* 58(10), 1190–1195 (2009).

79. Lallana, E., Fernandez-Megia, E., and Riguera, R., Surpassing the use of copper in the click functionalization of polymeric nanostructures: A strain-promoted approach. *J. Am. Chem. Soc.* 131(16), 5748–5750 (2009).

80. Jeong, B., Bae, Y.H. and Kim, S.W. Thermoreversible gelation of PEG-PLGA-PEG triblock copolymer aqueous solutions. *Macromolecules* 32(21), 7064–7069 (1999).

81. Qiao, M., *et al.* Injectable biodegradable temperature-responsive PLGA–PEG–PLGA copolymers: Synthesis and effect of copolymer composition on the drug release from the copolymer-based hydrogels. *International Journal of Pharmaceutics* 294(1–2), 103–112 (2005).

82. Hartgerink, J.D., Beniash, E. and Stupp, S.I. Self-assembly and mineralization of peptide-amphiphile nanofibers. *Science* 294(5547), 1684–1688 (2001).

83. Webber, M.J., *et al.* Controlled release of dexamethasone from peptide nanofiber gels to modulate inflammatory response. *Biomaterials* 33(28), 6823–6832 (2012).

84. Xing, B., *et al.* Hydrophobic interaction and hydrogen bonding cooperatively confer a vancomycin hydrogel: A potential candidate for biomaterials. *Journal of the American Chemical Society* 124(50), 14846–14847 (2002).

85. Hunt, J.N., *et al.* Tunable, high modulus hydrogels driven by ionic coacervation. *Advanced Materials* 23(20), 2327–2331 (2011).

86. Ricciardi, R., *et al.* Investigation of the relationships between the chain organization and rheological properties of atactic poly(vinyl alcohol) hydrogels. *Polymer* 44(11), 3375–3380 (2003).

87. Gupta, D., Tator, C.H. and Shoichet, M.S. Fast-gelling injectable blend of hyaluronan and methylcellulose for intrathecal, localized delivery to the injured spinal cord. *Biomaterials* 27(11), 2370–2379 (2006).

88. Liu, J., *et al.* Release of theophylline from polymer blend hydrogels. *International Journal of Pharmaceutics* 298(1), 117–125 (2005).

89. Brizzolara, D., C.H.-J., Diederichs, K., Keller, E., and Domb, A.J., Mechanism of the stereocomplex formation between enantiomeric poly(lactide)s. *Macromolecules* 29(1), 191–197 (1996).

90. De Jong, S.J., D.S.S.C., Wahls, M.W.C., Demeester, J., Kettenes-van Den Bosch, J.J., and Hennink, W.E., Novel self-assembled hydrogels by stereocomplex formation in aqueous solution of enantiomeric lactic acid oligomers grafted to dextran. *Macromolecules* 33(10), 3680–3686 (2000).

91. Hiemstra, C., *et al.* Protein release from injectable stereocomplexed hydrogels based on PEG–PDLA and PEG–PLLA star block copolymers. *Journal of Controlled Release* 116(2), e19–e21, (2006).

92. Hashidzume, A., Tomatsu, I. and Harada, A., Interaction of cyclodextrins with side chains of water soluble polymers: A simple model for biological molecular recognition and its utilization for stimuli-responsive systems. *Polymer* 47(17), 6011–6027 (2006).

93. Van de Manakker, F., van der Pot, M., Vermonden T., Van Nostrum, C.F., and Hennink, W.E., Self-assembling hydrogels based on β-cyclodextrin/cholesterol inclusion complexes. *Macromolecules* 41(5), 1766–1773 (2008).

94. van de Manakker, F., Vermonden, T., el Morabit, N., van Nostrum, C.F., and Hennink, W.E., Rheological behavior of self-assembling PEG-β-cyclodextrin/PEG-cholesterol hydrogels. *Langmuir* 24(21), 12559–12567 (2008).

95. Choi, H.S., Yamamoto, K., Ooya, T., and Yui, N., Synthesis of poly(ε-lysine)-grafted dextrans and their pH- and thermosensitive hydrogelation with cyclodextrins. *Chem. Phys. Chem.* 6(6), 1081–1086 (2005).

96. Micklitsch, C.M., Knerr, P.J., Branco, M.C., Nagarkar, R., Pochan, D.J., and Schneider, J.P., Zinc-triggered hydrogelation of a self-assembling β-hairpin peptide. *Angew. Chem.* 123(7), 1615–1617 (2011).

97. Kiik, K.L. Peptide- and protein-mediated assembly of heparinized hydrogels. *Soft Matter* 4(1), 29–37 (2008).

98. Okano, T. Y.H.B., Jacobs, H., and Kim, S.W. Thermally on–off switching polymers for drug permeation and release. *J. Controlled Release* 11, 255–265, (1990).

99. Gutowska, A.Y.H.B., Feijen, J., and Kim, S.W. Heparin release from thermosensitive hydrogels. *J. Controlled Release* 22, 95–104 (1992).

100. Dinarvand, R.D.A.E., Use of thermoresponsive hydrogels for on–off release of molecules. *J. Controlled Release* 36, 221–227 (1995).

101. Gutowska, A.J.S.B., Kwon, I.C. Bae, Y.H., and Kim, S.W. Squeezing hydrogels for controlled oral drug delivery. *J. Controlled Release* 48, 141–148 (1997).

102. Ichikawa, H.Y.F., Novel positively thermosensitive controlled-release microcapsule with membrane of nano-sized poly(N-isopropylacrylamide) gel dispersed in ethylcellulose matrix. *J. Controlled Release* 63, 107–119 (2000).

103. Spohr, R.N.R., Wolf, A., Alder, G.M., Ang, V., Bashford, C.L., Pasternak, C.A., Omichi, H., and Yoshida, M. Thermal control of drug release by a responsive ion track membrane observed by radio tracer flow dialysis. *J. Control. Release* 50, 1–11 (1998).

104. Wu, J.-Y., *et al.* Evaluating proteins release from, and their interactions with, thermosensitive poly (N-isopropylacrylamide) hydrogels. *Journal of Controlled Release* 102(2), 361–372 (2005).

105. Hu, Y., *et al.* Temperature-responsive porous polycaprolactone-based films via surface-initiated ATRP for protein delivery. *Journal of Materials Chemistry* 22(39), 21257–21264 (2012).

106. Gao, X., *et al.* Biodegradable pH-responsive polyacrylic acid derivative hydrogels with tunable swelling behavior for oral delivery of insulin. *Polymer* 54(7), 1786–1793 (2013).

107. Katono, H.A.M., Sanui, K., Okano, T., and Sakurai, Y. Thermo-responsive swelling and drug release switching of interpenetrating polymer networks composed of poly(acrylamide–co–butyl methacrylate) and poly(acrylic acid). *J. Control. Release* 16, 215–227 (1991).

108. Peppas, N.A.J.K., Controlled release by using poly(methacrylic acid–g–ethylene glycol) hydrogels. *J. Control. Release* 16, 203–214 (1991).

109. Patel, V.R. and Amiji, M.M., Preparation and characterization of freeze-dried chitosan–poly(ethylene oxide) hydrogels for site-specific antibiotic delivery in the stomach. *Pharm. Res.* 13, 588–593 (1996).

110. Carelli, V.S.C., Di Colo, G., Nannipieri, E., and Serafini, M.F. Silicone microspheres for pH-controlled gastrointestinal drug delivery. *Int. J. Pharm.* 179, 73–83 (1999).

111. Brannon-Peppas, L. and Peppas, N.A., Dynamic and equilibrium swelling behaviour of pH-sensitive hydrogels containing 2-hydroxyethyl methacrylate. *Biomaterials* 11, 635–644 (1990).

112. Akala, E.O., Kopecková, P. and Kopecek, J. Novel pH-sensitive hydrogels with adjustable swelling kinetics. *Biomaterials* 19, 1037–1047 (1998).

113. Aikawa, K., *et al.* Hydrogel formation of the pH response polymer polyvinylacetal diethylaminoacetate (AEA). *International Journal of Pharmaceutics* 167(1–2), 97–104, (1998).

114. Aikawa, K., *et al.* Drug release from pH-response polyvinylacetal diethylaminoacetate hydrogel, and application to nasal delivery. *International Journal of Pharmaceutics* 168(2), 181–188 (1998).

115. Markland, P., *et al.* A pH- and ionic strength-responsive polypeptide hydrogel: Synthesis, characterization, and preliminary protein release studies. *Journal of Biomedical Materials Research* 47(4), 595–602 (1999).

116. Mamada, A., *et al.* Photoinduced phase transition of gels. *Macromolecules* 23(5), 1517–1519 (1990).

117. Suzuki, A. and Tanaka, T. Phase transition in polymer gels induced by visible light. *Nature* 346(6282), 345–347 (1990).

118. Lo, C.-W., Zhu, D. and Jiang, H. An infrared-light responsive graphene-oxide incorporated poly(N-isopropylacrylamide) hydrogel nanocomposite. *Soft Matter* 7(12), 5604–5609 (2011).

119. Zhao, Y.-L. and Stoddart, J.F. Azobenzene-based light-responsive hydrogel system. *Langmuir* 25(15), 8442–8446 (2009).

120. de la Rica, R., Aili, D. and Stevens, M.M. Enzyme-responsive nanoparticles for drug release and diagnostics. *Advanced Drug Delivery Reviews* 64(11), 967–978 (2012).

121. Guo, D.-S., *et al.* Cholinesterase-responsive supramolecular vesicle. *Journal of the American Chemical Society* 134(24), 10244–10250 (2012).

122. Biswas, A., *et al.* Endoprotease-mediated intracellular protein delivery using nanocapsules. *ACS Nano* 5(2), 1385–1394 (2011).

123. Aimetti, A.A., Machen, A.J. and Anseth, K.S. Poly(ethylene glycol) hydrogels formed by thiol-ene photopolymerization for enzyme-responsive protein delivery. *Biomaterials* 30(30), 6048–6054 (2009).

124. Thornton, P.D., *et al.* Enzyme-responsive hydrogel particles for the controlled release of proteins: designing peptide actuators to match payload. *Soft Matter* 4(4), 821–827 (2008).

125. Gong, J.P., Nitta, T. and Osada, Y. Electrokinetic modeling of the contractile phenomena of polyelectrolyte gels. One-dimensional capillary model. *The Journal of Physical Chemistry* 98(38), 9583–9587 (1994).

126. Tanaka, T., *et al.* Collapse of gels in an electric field. *Science* 218(4571), 467–469 (1982).

127. Shiga, T., *et al.* Electric field-associated deformation of polyelectrolyte gel near a phase transition point. *Journal of Applied Polymer Science* 46(4), 635–640 (1992).

128. Osada, Y., Okuzaki, H. and Hori, H. A polymer gel with electrically driven motility. *Nature* 355(6357), 242–244 (1992).

129. Kwon, I.C., B.Y.H., Kim, S.W, Electrically credible polymer gel for controlled release of drugs. *Nature* 354, 291 (1991).

130. Kost, J., Wolfrum, J. and Langer, R. Magnetically enhanced insulin release in diabetic rats. *Journal of Biomedical Materials Research* 21(12), 1367–1373 (1987).

131. Satarkar, N.S. and Hilt, J.Z. Magnetic hydrogel nanocomposites for remote controlled pulsatile drug release. *Journal of Controlled Release* 130(3), 246–251 (2008).

132. Zhang, D., *et al.* A magnetic chitosan hydrogel for sustained and prolonged delivery of Bacillus Calmette–Guérin in the treatment of bladder cancer. *Biomaterials* 34(38), 10258–10266 (2013).

133. Messing, R., *et al.* Cobalt ferrite nanoparticles as multifunctional cross-linkers in PAAm ferrohydrogels. *Macromolecules* 44(8), 2990–2999 (2011).

134. Mo, R., Jiang, T., Di, J. Tai, W. and Gu, Z. Emerging micro- and nanotechnology based synthetic approaches for insulin delivery. *Chem. Soc. Rev.* 43, 3595–3629 (2014).

135. Bankar, S.B., *et al.* Glucose oxidase — An overview. *Biotechnology Advances* 27(4), 489–501 (2009).

136. Ravaine, V., Ancla, C. and Catargi, B. Chemically controlled closed-loop insulin delivery. *Journal of Controlled Release* 132(1), 2–11 (2008).

137. Steiner, M.-S., Duerkop, A. and Wolfbeis, O.S. Optical methods for sensing glucose. *Chemical Society Reviews* 40(9), 4805–4839 (2011).

138. Wu, Q., *et al.* Organization of glucose-responsive systems and their properties. *Chemical Reviews* 111(12), 7855–7875 (2011).

139. Gu, Z., *et al.* Glucose-responsive microgels integrated with enzyme nanocapsules for closed-loop insulin delivery. *ACS Nano* 7(8), 6758–6766 (2013).

140. Lorand, J.P. and Edwards, J.O. Polyol complexes and structure of the benzeneboronate ion. *The Journal of Organic Chemistry* 24(6), 769–774 (1959).
141. Miyake, K., Tanaka, T. and McNeil, P.L. Lectin-based food poisoning: A new mechanism of protein toxicity. *PLoS ONE* 2(8), e687 (2007).
142. Vaz, A.F.M., *et al.* High doses of gamma radiation suppress allergic effect induced by food lectin. *Radiation Physics and Chemistry* 85(0), 218–226 (2013).
143. Vilarem, M.J., *et al.* Differential effects of lectins on the *in vitro* growth of normal mouse lung cells and low- and high-cancer-derived cell lines. *Cancer Research* 38(11 Part 1), 3960–3965 (1978).
144. Shiino, D., M.Y., Kataoka, K., Koyama, Y., Yokoyama, M., Okano, T., Sakurai, Y. Preparation and characterization of a glucose-responsive insulin-releasing polymer device. *Biomaterials* 15, 121 (1994).
145. Brownlee, M., C.A., A glucose-controlled insulin-delivery system: semisynthetic insulin bound to lectin. *Science* 206, 1190 (1979).
146. Kim, S.W., P.C.M., Makino, K., Seminoff, L.A., Holmberg, D.L., Gleeson, J.M., Wilson, D.E., Mack, E.J., Self-regulated glycosylated insulin delivery. *J. Control. Release* 11, 193 (1990).
147. Yin, R., *et al.* Glucose-responsive microhydrogels based on methacrylate modified dextran/concanavalin A for insulin delivery. *J Control Release* 152(Suppl 1), e163–5 (2011).
148. Yin, R., *et al.* Glucose and pH dual-responsive concanavalin A based microhydrogels for insulin delivery. *International Journal of Biological Macromolecules* 49(5), 1137–1142 (2011).
149. Yin, R., *et al.* Glucose-responsive insulin delivery microhydrogels from methacrylated dextran/concanavalin A: Preparation and in vitro release study. *Carbohydrate Polymers* 89(1), 117–123 (2012).
150. Lee, K.W., *et al.* Sustained release of vascular endothelial growth factor from calcium-induced alginate hydrogels reinforced by heparin and chitosan. *Transplantation Proceedings* 36(8), 2464–2465 (2004).
151. Prausnitz, M.R. and Langer, R. Transdermal drug delivery. *Nat Biotech* 26(11), 1261–1268 (2008).
152. Jiang, T., *et al.* Gel–liposome-mediated co-delivery of anticancer membrane-associated proteins and small-molecule drugs for enhanced therapeutic efficacy. *Advanced Functional Materials* 2014.
153. Owens, D.R., New horizons alternative routes for insulin therapy. *Nat Rev Drug Discov* 1(7), 529–540 (2002).
154. Simões, S., Ana, F. and Francisco, V. Modular hydrogels for drug delivery. *Journal of Biomaterials and Nanobiotechnology* 3, 185–199 (2012).

Cell Ghosts: Cellular Membranes for Drug Delivery

8

Beth Schoen and Marcelle Machluf

The Lab for Cancer Drug Delivery & Cell Based Technologies,
Faculty of Biotechnology & Food Engineering,
Technion — Israel Institute of Technology
Haifa, 3200003, Israel

Abstract

Despite many achievements in the development of effective drug delivery vehicles (DDVs), their clinical application is compromised by considerable targeting, delivery and safety concerns. DDVs must target a specific subset of cells or organs, requiring intervention and transport and uptake, while having versatile loading capacities for a variety of therapeutics with different compositions, stabilities and solubilities. Synthetic particulate delivery systems (e.g. lipoplexes, lipidoids, cationic liposomes, polymeric particles, micelles, etc.) can be carefully formulated to overcome some of these limitations. Nonetheless, and despite the flexibility in their production, most of them are still restricted from clinical use due to other shortcomings, such as unknown tissue interactions, low loading efficiencies, rapid clearance, non-biodegradability, adsorption of serum proteins and, most importantly, lack of specific and active targeting to the tissue of interest.

This chapter reviews a new class of targeted delivery systems called nanoerythrosomes and nanoghosts that are based on, or produced from, the cell membrane of different mammalian cell types (e.g. erythrosomes and mesenchymal stem cells), or bud from various cells through spontaneous or inducible biological processes (exosomes and other extracellular shed vesicles). The type of cells used to produce such vesicles is of great importance, as these cells target specific tissues or are involved in crucial processes such as inflammation and tumorgenesis. Therefore, the specific cells selected as the platform material for the production or isolation of these nanocarriers play a critical role in determining the targeting capabilities of these nanoparticles as well as their biodistribution and biocompatibility. This chapter will provide an overview of the choice of mammalian cells used as DDVs themselves, and also describe the new class of nano delivery carriers designed or isolated from some of these cells.

8.1 Whole-living Cells as Drug Carriers

A variety of carrier cells (CCs) have been engineered for use as DDVs, based on their tropism towards a location of interest and their ability to release a desired therapeutic reagent.[1] CCs can be autologous or from donor sources, and propagated and engineered by viral or non-viral methods *in vitro*. They are capable of delivering therapeutic genes and proteins,[1] suicide genes,[2] oncolytic viruses,[3] antibodies,[4] microRNA,[5] immunotoxins[6] and engulfing nanoparticles (NPs).[7] The whole cell drug carriers described in this section are developed *ex vivo*, created for the treatment of human diseases, and containing whole-living cells that have not been devoid of their organelles. The following subsections have been divided accordingly: (1) mesenchymal stem cells (MSCs), (2) neural stem cells (NSCs), (3) macrophages, (4) lymphocytes, (5) endothelial cells (ECs), and (6) other cell types.

8.1.1 *Mesenchymal Stem Cells*

MSCs are adult, non-hematopoietic cells originally derived from the bone marrow (BM).[8] Other sources include adipose tissue, umbilical cord blood, peripheral blood, placenta, brain, lung, kidney, liver, periodontal ligament, hair follicles and amniotic fluid. They are capable of differentiating into cells of connective tissues such as cartilage, bone, adipose and marrow stroma,[1] and provide microenvironmental support for hematopoietic stem cells (HSCs).[9] MSCs are known for their natural tropism to sites of inflammation and tissue injury, which are prominent in the tumor microenvironment and various other diseases. They are plastic-adherent and capable of self-renewal, making them easy to obtain and expand *in vitro*. They are also immune evasive, and display immunosuppressive properties upon translation.[8]

Ample studies have shown that allogeneically-transplanted MSCs can selectively target pathological tissues and have produced positive clinical trial outcomes with different diseases, including solid and hematological tumors, degenerative diseases and immune disorders.[10,11] Regardless of any specific mechanisms by which MSCs may affect these conditions, to do so they must first reach the vicinity of their targets.[12,13] This targeting ability is probably attributed to the fact that MSC-targetable pathologies involve angiogenesis, inflammation and/or remodeling of the extracellular matrix (ECM), all of which require MSC support. The mechanisms responsible for this targeting were suggested to involve chemotaxis in response to soluble factors, as well as direct contact between MSCs and their target cells and ECM. Their immune evasiveness and immuno-modulatory capabilities — also attributed to both secreted and surface-bound factors, such as low

MHC-I levels and lack of MHC-II, CD80 or CD86 expression — are important features that have driven their utilization as allogeneic products for transplantation.[14–17] Although most research has focused on the effect of soluble factors, recent studies suggest that direct surface interactions are as important, if not more important, in governing both MSC targeting capabilities[18–23] and immunological properties.[13,23–25]

Engineering MSCs to deliver therapeutic agents has greatly expanded their therapeutic potential. In cancer therapy, it is ideal to deliver treatment directly to the tumor microenvironment while producing the lowest collateral toxicity, which can be easily accomplished by exploiting the natural ability of MSCs to home to inflammation sites. MSCs can be easily transduced (TD) or transfected (TF), and they can encapsulate NPs through cellular uptake (CU).[8] Consequently, MSCs have been used in a large variety of cancers and other diseases as DDVs (Table 8.1). For example, glioblastoma multiforme (GBM), the most common and aggressive form of glioma tumors, has attracted significant research with MSCs as the chosen DDV.[7] Decades of research and development have provided no cure, and median survival with current treatment is 14.6 months. Prior treatment methods have been unsuccessful due to their inability to cross the blood brain barrier (BBB) and effectively target tumor cells, as the tumor cells disseminate into the normal parenchyma of the brain at distant sites from the tumor mass. However, when administered intravascularly, the ability of MSCs to home to tumor sites, actively seek metastases, and migrate within glioma tissue has allowed MSCs to be affective DDVs for GBM.[7,26]

Although there has been significant progress in the use of MSCs as CCs, their use remains controversial. Some studies have demonstrated that MSCs promote tumor vascularization, cancer progression and metastasis.[8,27–30] In peritoneal cancers, MSCs located around the tumor cells and in the ascitic fluid were no longer able to differentiate.[31] These MSCs show characteristic markers of carcinoma–associated fibroblasts and are called carcinoma-associated MSCs. Other studies showed that MSCs can be involved in the development of chemoresistance through the release of specific factors in the vicinity of tumors.[32] However, additional findings have suggested that the source of MSCs plays a role in cancer progression or suppression. In one study, umbilical cord blood-derived MSCs inhibited GBM proliferation, whereas adipose tissue-derived MSCs supported it.[28] It is also possible that transformed MSCs can give rise to tissue-specific cancers.[33] Some researchers have proposed the use of a cellular suicide gene that will ultimately destroy the CC after delivery, or afford kinetic control over drug release within the NPs.[34] Slowing the rate of drug release from the NPs within the MSC will initially release little to no drug; later, as the rate of drug release accelerates, the drug will kill both the MSCs

Table 8.1: MSCs as CCs.

Type	Active Ingredient	Disease Treated*	Rationale/ Effect	Preparation
Therapeutic Gene/ Protein	TRAIL	Glioma, breast, cervical,[35] pancreatic, myeloma,[36] colon, squamous, lung,[37] esophogeal,[38] MFH,[39] liver,[40] MB[41]	• Induces apoptosis in tumor cells only • Binds 5 TNF receptors	TD-ADV, LV, RV, TF-CL, CP, Nuc
	IFN-α	Plasmacytoma[42]	• Apoptosis of tumor cells	TD-LV
	IFN-β	Breast, pancreatic, melanoma, prostate†,[36] glioma,[35] ovarian,[43] bladder[44]	• Inhibits growth of tumor cells	TD-ADV, AAV, LV
	IFN-γ	Lung,[45] prostate,[46] leukemia[47]	• Activates TRAIL pathway	TD-LV,ADV
	IL-2	Melanoma[36]	• Immune response through T and NK cells	TD-RV
	IL-4	Multiple sclerosis[48]	• Mediates remission of autoimmunity	TD-LV
	IL-10	MS,[49] distant sites of inflammation[50]	• Antiinflammatory cytokine	TD-LV, TF-CL
	IL-12	Glioma,[37] breast†, melanoma†, liver,[51] renal cell carcinoma[36]	• Activates T cells and NK cells	TD-ADV, RV
	IL-21	Ovarian[52]	• Induce antitumor immunity	TD-LV
	IL-24 (MDA-7)	Prostate[44]	• Induce antitumor immunity • Tumor suppressor gene	TD-LV
	PEDF	Prostate,[35] liver, lung, glioma[53]	• Fas/FasL death pathway–endothelial cell death • Balances pro/anti-angiogenic factors	TD-AAV, LV
	BMP4	Glioma[54]	• Tumor suppressive effects	TD-RV/LV
	CX3CL1	Lung[35]	• Inhibits metastasis	TD-ADV
	Endostatin	Ovarian[55]	• Most potent anti-angiogenesis agent	TD-ADV

iNOS	Fibrosarcoma[35]	• Anti-tumor effect	TD-LV
NK4	Pancreatic[56]	• Inhibits HGF-mediated cell proliferation and migration	TD-ADV
IGFIR	Liver[57]	• Prevention of hepatic metastasis	TD-RV
Erythopotein	AKI,[58] HR, ESRD anemia[59]	• Tissue protectiveness/organ reparative ability	TD-RV, TF-Nuc
IGF-I	Kidney-failure induced anemia[47]	• Regulates a number of cellular processes	TD-RV
TNF-α & CD40	Breast[60]	• Activation and maturation of DCs	TD-LV
TNF-α & tumstatin	Prostate[42]	• Induce apoptosis	TD-LV
Neurotrophin-3	Medullablastoma[61]	• Induce tumor cell apoptosis	TF-Nuc
ABOX	Colon[62]	• Modulator of tumor angiogenesis	TF-Nuc
LIGHT	Breast,[63] gastric[64]	• Inhibits tumor growth	TD-RV, LV
Pdx-1	Diabetes[65]	• Express insulin 1 and insulin 2	TD-LV

(Continued)

Table 8.1: (*Continued*)

Type	Active Ingredient	Disease/Disease Treated*	Rationale/Effect	Preparation
	α-melanocyte	Chronic inflammatory disease[66]	• Mediator of inflammation and immunity	TD-LV
	Trx-1	Acute radiation exposure[67]	• Scavenges reactive oxygen	TD-ADV
	E7, PE(ΔIII)–E7-KDEL3	Lung[37]	• Suppressed tumorigenesis and metastasis	TD-RV
	NIS	Breast,[42] liver[68]	• Noninvasive radionuclide imaging/therapy	TD-ADV, LV, TF-CL
	α1–antitrypsin	POC[42]	• Cytotoxic against HUVEC cells	TD-LV
Suicide gene/ Prodrug	CD & 5-FC	Glioma,[50] gastric[42]	• Induces proapoptotic genes in tumor cells	TD- LV, ADV, TF-LC
	CDy::UPRT & 5-FC	Glioma, prostate,[63] melanoma, colon[36]	• Kills dividing/nondividing cells	TD- RV
	CE & CPT-11	Glioma[36]	• Converts CPT-11 to cytotoxic drug SN-38	TF-Nuc
	HSV-tk & GC	Glioma[36,‡,69§,70], prostate,[42] breast,[71] ovarian,[37] renal cell carcinoma[+63]	• Inhibits target cellular DNA polymerases • dTRAIL–tk enhancing killing activity	TD- RV, LV, BV, ADV, TF-LC
	CMV-tk & GC	Melanoma[†63]	• Inhibit tumor growth through cation pullulan	TF-CP
	CD/HSV-tk & 5-FC/GC	Breast,[72] ovarian[73]	• Double suicide gene therapy	TD-LV

Oncolytic Virus	delta-24-RGD	Glioma,[50] breast, ovarian[37]	• Lyses tumor cells with inactivated Rb	TD-ADV
	CRAd	Glioma,[50] breast[74,†,37], colon[75]	• Selectively replicates and kills tumor cells	TD-ADV, IF
	Capsid-modified vMyxgfp (myxoma)	Breast, lung[37]	• Antitumor effect	TD-ADV
		Glioma[61]	• Relative selectivity for tumor cells	IF–vMxgfp
	Ad-hOC-E1 & D$_3$	Renal cell carcinoma[76]	• Tumor regression by OV induced by D$_3$	TD-ADV
	β-galactosidase	Melanoma[77]	• Inhibited tumor growth	TD-ADV
	Measles	Ovarian,[36] liver[78]	• Preferential killing of cancer cells	TD-LV
	ICOVIR	Neuroblastoma[†,50]	• Selectively replicates and kills tumor cells	TD-ADV
Synthetic Drug	Doxorubicin in silica NR	Glioma[36]	• Enhances tumor-cell apoptosis	CU
	Paclitaxel	Glioma, prostate, melanoma, ALL[75]	• Anti-cancer and antiogenic drug	CU

(Continued)

Table 8.1: (*Continued*)

Type	Active Ingredient	Disease Disease Treated*	Rationale/ Effect	Preparation
	Ferrociphenol in lipid NC	Glioma[50]	• Cytotoxic to glioma cells	CU
	Gd^{3+}, ^{64}Cu in silica NP	Glioma[79]	• Imaging agents for optical MR and PET	CU
	Purpurin-18 in silica NP	Breast[80]	• Generates cytotoxic reactive oxygen species	CU
	Ciprofloxacin	Chronic osteomyelitis[81]	• Penetrates poorly vascularized infection sites	CU
	Porphyrin in NP	Osteosarcoma,[78] breast[80]	• Cell death upon photoactivation	CU
AB	scFv EGFRvIII	Glioma[50]	• Delayed growth of tumor cells	TF-Nuc
	scFv, TRAIL, CD20	Non-Hodgkin's lymphoma[78]	• Mediates antigen specific apoptosis	TD-LV
mR	miR-124, miR-145[¶]	Glioma[5]	• Decrease GC/GSC migration/self-renewal	TF-Nuc
IM	EphrinA1-PE38	Glioma[6]	• Against EphA2 receptor	TD-ADV
	VEGF165-PE38	HUVEC-POC[¶][82]	• Capable of killing HUVECs	TF-LC

Notes: *Disease is cancer unless specified otherwise, †Metastasis, ‡Derived fom iPS/ESC, §Progenitor cells, ‖mimic mR, ¶mononuclear cells.

and the tumor cells. Besides prometastasis and tumor progression, MSCs also run the risk of off-target toxicity.[7] A patient with a recent surgery, injury or off-target inflammation is at risk of MSCs accumulating in this location. Off-target toxicity is also a possibility, as the NPs within the MSCs can induce a cytotoxic effect in off-target cells before the target is reached.

8.1.2 *Neural Stem Cells*

NSCs are multipotent, self-renewal cells found within regions of the brain, and can be derived from fetal, neonatal or postnatal tissue.[83] NSCs have the capacity to generate into three major CNS cell types: neurons, astrocytes and oligodendrocytes, and are physiologically responsible for neocortical neurogenesis to help replace damaged tissue. NSCs are known for their ability to seamlessly integrate into the host brain without disrupting normal function,[84] and for their unique ability to migrate towards areas of tumor metastasis and degeneration within the brain. They can be propagated *in vitro* as spherical floating clusters called neurospheres.[85]

NSCs can also be used as CCs by exploiting their natural capacity for tropism. NSCs transplanted into the brain of an animal model demonstrated the capacity to migrate towards metastatic tumor beds at distant locations from the original transplant sites.[86] Consequently, they have been used to deliver chemotherapeutic agents not only to glial neoplasms, but also to breast cancer[87] and melanoma metastasis,[88] intercerebral medulloblastomas[89] and disseminated neuroblastomas.[90] NSCs exhibited enhanced tropism towards intracranial glioma over MSCs,[91] and oncolytic adenovirus-modified NSCs showed a prolonged survival over MSCs.[92]

Although NSCs do not appear to have the potential to promote tumor growth, and may present a viable alternative to MSCs for this reason,[93] there are still significant challenges to their use as CCs. Primarily, it is very difficult to harvest NSCs due to ethical reasons and the technical difficulties associated with their isolation (neonatal tissue) and expansion, as well as problems associated with their immune response.[94,95] Alloimmunization has been observed in 4 out of 13 patients receiving transplanted NSCs for Huntington's disease.[96] One of the patients displayed a substantial rejection reaction, which was later resolved with immunosuppressive treatment. The strategy to isolate and expand autologous NSCs has been explored, but requires invasive procedures and can only be used to obtain a small quantity of cells.[97,98] Nonetheless, NSCs have been isolated and expanded from patients who have undergone unilateral olfactory bulb removal for benign conditions, but the technique requires further optimization.[99] Other researchers have attempted to generate NSCs from embryonic stem cells (ESCs) and induced pluripotent stem cells (iPSCs).[93] These studies have demonstrated that NSCs derived from these sources display similar tumor tropism and cell surface markers as their primary counterparts.[100,101]

8.1.3 *Monocytes/Macrophages*

Macrophages and their precursor cells, monocytes, are key players in the human immune system and are found in every tissue of the human body. They are responsible for defending against diverse pathogens, dead cells and foreign entities through endocytosis.[102] Circulating monocytes are short-lived and undergo daily spontaneous apoptosis, while macrophages have a longer lifespan. Macrophages play a crucial role in homeostatic, immunological and inflammatory processes.[103] They are rapidly recruited to diseased sites by signaling molecules such as cytokines, and are involved in various pathological conditions such as cancer, atherosclerosis and several inflammatory diseases. Macrophages are terminally differentiated, non-dividing cells, but can be easily obtained from circulating blood monocytes that are plastic-adherent, and differentiate into mature macrophages *in vitro*.[104]

Circulating monocytes are specifically recruited by hypoxic regions of the tumor microenvironment, which are virtually impossible to reach with chemotherapeutic drugs and radiation therapy. Once the circulating monocytes cross the endothelial basement membrane of the tumor they differentiate into macrophages.[105] Prolonged hypoxia causes the center of solid tumors to be largely necrotic, and drug delivery based on the EPR effect is not possible here. However, cancer cells can survive in this region, and they are thought to be the source of subsequent local recurrence and distant metastasis. Additionally, macrophages are active in HIV infections, since they express CD4 and HIV-1 co-receptors at the cell surface.[106] They are also found in the brain since monocytes exhibit the ability to cross the BBB.[107] Combining their natural tropism and ability to engulf drug-loaded NPs using their phagocytic capacity, macrophages are potential drug delivery agents to treat HIV/AIDS,[108] CNS diseases[109] and solid tumors.[105] Another method for drug delivery with macrophages is the use of phagocytosis-resistant backpacks (PRB).[103] This method avoids the phagosome, which can reduce the drug release rate, and prevents drug degradation, since PRBs are attached to the macrophage surface. PRBs are fabricated by a standard photolithography lift-off technique of layer-by-layer and spray-deposited film. The polyelectrolyte multilayers (PEM) that comprise the majority of the backpack have been used in many biomedical and drug delivery applications.

Yet, there are still significant challenges to the use of monocytes/macrophages as CCs. Macrophages within the tumor mass, or tumor-associated macrophages

(TAMs), can promote tumor progression. They have demonstrated the ability to promote invasion, proliferation, tumorigenesis, neoangiogenesis and metastasis.[105] Incredibly, 70% of the tumor mass in breast cancer may be comprised of macrophages,[110] demonstrating the need for methods to destroy TAMs. It is proposed that therapeutic TAMs can be destroyed by near infrared radiation (NIR), but that is dependent on the ability of the encapsulated NPs to absorb light in the near-infrared region.[105] Additionally, macrophages are not capable of taking up naked chemotherapeutics, and must therefore incorporate delivery agents within NPs.[111]

8.1.4 *Lymphocytes*

Lymphocytes are white blood cells of the immune system that are produced in the central lymphoid organs (thymus and bone marrow).[112] They can be found in circulation throughout the body or within specialized lymphoid organs, such as the spleen, tonsils, lymph nodes and gut-associated lymphoid tissues. They are active in both innate and adaptive immune functions, but are primarily involved in the adaptive immune response, defined as antigen specific, long-lasting immunity triggered by natural exposure to foreign molecules.[112] Innate lymphocytes, such as NK, $\gamma\delta$ T, invariant NK T, and B-1 cells, exhibit a quick response (1–3 days) at infection sites to a limited antigen pool.[113] Adaptive lymphocytes exhibit a slower response, but to a larger pool of antigens.[114] Lymphocytes are primarily isolated from human peripheral blood, separated from other components, and expanded in suspension *in vitro*.[115]

Lymphocytes are notorious for being used in adoptive cell transfer (ACT) therapy, or adoptive immunotherapy, in which autologous or allogeneic lymphocytes that carry specificity for tumor-associated antigens (TAAs) are expanded *in vitro* and subsequently administered to the patient.[115–117] Guided by TAAs, the lymphocytes home to the tumor. There, they induce an antitumor immune response and arm the immune system against the patient's own metastatic cancer.[116] However, there is a high potential for toxicity as a result of T cells localizing to healthy tissue, where TAAs are expressed at a low level.[117,118] To circumvent this problem, lymphocytes have been engineered to express specific T cell receptors (TCRs) that target TAAs. The expression of chimeric antigen receptors (CARs), which comprise an antigen-specific single-chain antibody variable fragment (scFv), is a major breakthrough allowing T cells to be targeted with high specificity to various surface TAAs.[119]

Although ACT therapy is linked to great success in cancer treatment, not every tumor cell exposed directly to these cytotoxic T cells is killed,[120] due to immune escape mechanisms of these tumor cells.[121] Therefore, using lymphocytes engineered for ACT therapy to also deliver therapeutic agents has proven to be a more successful strategy.[117] However, it is difficult to transduce fresh T cells with RV vectors,[122] and so a method known as viral hitch-hiking (VHH) has been developed.[117] VHH is the process of loading antigen-specific T cells with viral particles that adhere to the cell surface. Engineered lymphocytes have not demonstrated transformation or T cell malignancies, seen with other CCs, but impose their own set of challenges, such as off-site toxicity. Another challenge is that not all TAAs are expressed in all stages of the disease, which prohibits active targeting.[123] Additionally, T cells cannot be infected directly, RVs are inefficient for infection, and ADVs have to be used in large quantities.

8.1.5 *Endothelial Cells*

ECs form the inner lining of all blood vessels, or the endothelium, which provides an anticoagulant barrier between the vessel wall and blood.[124,125] ECs are responsible for regulating homeostasis, vasomotor tone, and immune and inflammatory responses through interactions with circulating physical and chemical stimuli. They are essential in angiogenesis and vasculogenesis, and can break through the basal lamina of existing blood vessels, migrate to an angiogenic signal, proliferate, and form tubular structures.[126] ECs are derived from plastic-adherent endothelial precursor cells (EPCs) isolated from peripheral blood, and cultured *in vitro*. They can be manipulated using viral vectors, and home to areas of ischemia in tumors.[127]

ECs are vital to neovascularization, which is crucial for malignant tumor growth.[126] During tumorigenesis, existing blood vessels form new vasculature through the process known as angiogenesis; during vasculogenesis, blood vessels are formed from circulating HSCs, or EPCs. Some researchers report that BM-derived EPCs contribute significantly to tumor vasculature.[128] ECs can be mobilized to tumor angiogenesis sites and access distant and undetectable microscopic metastases,[126] making them excellent candidates for use as CCs. Several ECs, such as EPCs, human umbilical vein endothelial cells (HUVECs)[129] and human adult blood late outgrowth endothelial cells (BOECs), have been studied for use as CCs.[130] Additionally, embryonic stem cell-derived EPCs (eEPCs) are actively recruited to tumor angiogenesis,[131] show an affinity for hypoxic lung metastases, and exhibit an immunoprivileged status. eEPCs are rapidly expanded *in vitro* and are easy to genetically manipulate.[132] However, the homing of ECs is not completely understood, and researchers question the exact role of EPCs in tumor vasculature growth.[133] Studies have shown that BOECs homed to most Lewis lung carcinoma (LLC) lung metastases, but less to liver or kidney metastases.[130] Other studies showed that only a fraction of administered BOECs (1 EC per 10^5 tumor cells) are incorporated into tumor vessels, while 60% were trapped in the lungs.[126] Researchers have proposed other areas for injection besides the tail vein, as well as larger doses, but bolus injections larger than 250,000 cells per injection have been associated with up to 10% of sudden deaths.[126] The source of the EC may also play a significant role in its success as a CC. BOECs expressing CD did not prolong the lifespan of mice bearing disseminated metastases of LLC, whereas eEPCs did.[130] ECs can also potentially form teratomas or other malignant tumors.[126]

8.1.6 *Other Carrier Cells*

Cancer cells can also be used as drug carriers. They are suitable for use as drug delivery devices to target the tumor and metastases *in vivo* because they already express the necessary receptors and effector molecules. Specific cancer cells that have been used for this purpose include the following: MDA-MB-231 (breast), PC3 (prostate), A549 (lung), L1210 (leukemia), CT26 (colon), MH392 (liver), MC38 (colon), DU145 (prostate), MM1 (myeloma) and PA-1 (teratocarcinoma). Additionally, some cancer cells, such as A549 or L1210, have also been used to deliver therapeutic agents to non-related cancers, due to their superior ability to carry the therapeutic agent.[134–137] Leukemic carriers can disseminate throughout the circulatory system, allowing them to deliver therapeutic agents to diverse anatomical locations.[137] However, since there is an inherent risk of developing new cancers from the CCs themselves, killing the CC is critical in this therapeutic strategy. Therefore, prodrug therapy and oncolytic virus therapy, which are known to kill both the tumor and the CCs, are used to kill the CCs; if necessary, irradiation therapy can be used to ensure complete killing.[138] Besides the concern of new malignancies, the source of the CCs also poses an obstacle: The CCs need to be autologous, but the patient is already ill.

8.2 Carrier Erythrocytes

Erythrocytes, or red blood cells (RBCs), are the most abundant cells in the human body, and are primarily responsible for the transport of O_2 from the lungs to the tissues, and the transport of CO_2 produced in the tissues back to the lungs.[139,140] They are biconcave discs that, unlike other cells, lack a nucleus, mitochondria and other organelles, which provides ample intracellular space for O_2 transport.[141] They can be isolated from whole blood and have a lifespan of 120 days. RBCs are the first and most commonly used cell for the delivery of therapeutic agents.[142,143] In 1973, both Ihler[144] and Zimmerman[145] independently published the first reports of RBCs for delivery purposes, and the term "carrier erythrocytes" was coined in 1979.[140] RBCs are in particular used to target the reticuloendothelial system (RES), or more specifically monocyte-derived macrophages, since aging and factors contributing to cellular membrane changes make them recognizable by phagocyting macrophages.[140] RBCs can be loaded with therapeutic agents using physical methods (i.e. electroporation), chemical methods (i.e. to induce membrane binding) or osmosis-based methods.[139] Osmosis-based methods are the most common, since they rely on the natural ability of RBCs to increase in volume when placed under conditions of reduced osmotic pressure, such as in a hypotonic solution.[146] Just before cell lysis, transient pores (200–500 Å in diameter) appear in the membrane through which drugs dissolved in the medium can enter. This swelling is reversible, and the membrane can be resealed after drug loading. There are a variety of osmosis-based loading methods, including: hypotonic dilution (HDn), hypotonic dialysis (HD), hypotonic preswelling (HP) and osmotic pulse (OP) (Table 8.2).[139] However, these methods are limited by the degree of water solubility of the drug, drug loading efficiency, and the extent of membrane destruction. Non-osmosis-based methods for loading RBCs include: electroporation (E),[141] cellular hitchhiking (CH),[147] endocytosis (End)[141] and membrane binding (MB)[142] (Table 8.3).

A variety of carrier erythrocytes have been developed to deliver therapeutic agents, provide enzyme replacement therapy, and improve O_2 delivery to tissues.[142] Many of these studies exploit the ability of RBCs to sufficiently target the RES, specifically in HIV/AIDs therapy.[148] Monocyte-derived macrophages are among the first cells to be infected by HIV and later act as a viral reservoir. For targeting organs outside of the RES system, different methods, such as co-encapsulation of the drugs with paramagnetic particles or photosensitive agents,[149] application of ultrasound waves,[150] and site-specific antibody attachment,[151] have been used. Besides targeted delivery, carrier erythrocytes can also be used as circulating bioreactors,[142] by loading the RBCs using a method that yields minimal membrane destruction. Thus, the drug can passively diffuse out of the loaded cell in circulation, diffuse into circulation after phagocytosis of the cells by the RES, accumulate in the RES upon lysis of the carrier and be slowly released into the system, or accumulate within the carrier in lymphatic nodes and be released upon hemolysis at the site.[142]

Although researchers have found great success using carrier erythrocytes, there are still complexities surrounding their use. Drugs escape from these cells rapidly, even during loading, especially when lipophilic drugs are used.[155] Drug escape can cause toxicological problems *in vivo*, prompting researchers to use drug-loaded

Table 8.2:   Osmosis based methods (OBMs) for RBC loading.

OBM	Description	% E	Advantages	Disadvantages
HDn	• RBC vol. diluted 2–20x's aq. drug soln. • Destructive to membrane, causing short life span- readily phagocytable by RES	20–40[152]	• Fastest • Simplest • RES targeting	• Low entrapment • Large hemoglobin loss • Very short life span
HD	• Isotonic drug/RBC soln. in dialysis tube • Placed in hypotonic buffer	30–50[140]	• Longer lifespan • Higher entrapment	• Long process time • Expensive equipment
HP	• RBC/hypotonic solution centrifuged • Supernatant discarded • Step add. 100–200 μ L aq. drug at lysis pt.	30–90[152]	• Minimal destruction • Highest entrapment • Normal lifespan	• Minimal cell destruction
OP	• Usually isotonic, but can be hypertonic[153] • Suspended in DMSO-rapidly diffused[140] • Isotonic soln.: causes transient gradient w. membrane, H_2O pushed to intracellular side • OP created: transient openings for loading	20–70[154]	• Better *in vivo* survival	• Membrane changes

Table 8.3: Physical and chemical methods for RBC loading.

Method	Description/Advantages	Disadvantages
E	• Erythrocyte membrane opened by dielectric breakdown • Pores can be resealed in isotonic media	• Permanent membrane damage
CH	• Non-covalent attachment of NP to RBCs • Longer *in vivo* circulation	• Method not fully understood
End	• 1 vol. RBCs to 9 vol. buffer • Pores resealed by using 154 mM NaCl and incubation at 37°C • Less membrane damage	• content of vesicle may release into cytoplasm of RBC
RCL	• Nondiffusible drugs • Based on 2 sequential hypotonic dilutions of washed erythrocytes • Followed by concentration with a hemofilter and isotonic resealing	• Similar to osmosis-based
MB	• Using avidin-biotin bridges (biotinylation) • Either covalent attachment to membrane amino groups through NHS-biotin • Or oxidation of induced aldehyde group by biotin hydrazide • High cell recovery and regular circulation time	• Drug not protected inside

NPs rather than free drug encapsulated within RBCs.[147,156,157] Others have used cross-linking agents such as glutaraldehyde to help prevent drug leakage.[158] Glutaraldehyde cross-linking has also been found to enhance the uptake of certain drugs such as daunorubicin *in vivo*.[158] The storage of loaded erythrocytes is also difficult, since they are viable cells and need to survive in circulation upon administration into the host.[142] The use of nucleosides, chelators, lyophilization and isotonic buffers containing essential nutrients have all been suggested as potential solutions to this problem. It is also possible that the blood can be contaminated due to its origin, which requires the implementation of rigorous controls for the safe handling and collection of RBCs. Most importantly, the RBCs need to be a direct match, or autologous, to the host. Thus, matching blood must be available at the time of treatment.

8.3 Exosomes

Exosomes and other extracellular vesicles that bud from various cells through spontaneous or inducible biological processes have recently been suggested as a natural, yet non-viral, alternative to synthetic vectors.[159] Due to their natural role in cell-to-cell communication, exosomes may possess some active targeting capabilities and the ability to traverse physiological and biological barriers.[160] Mammalian cells release relatively low quantities of exosomes; methods such as physical extrusion[161,162] and microfluidic chambers[163,164] have been used to create exosome-mimetic nanovesicles in larger quantities that demonstrate similar effects to natural exosomes. Already investigated in several clinical trials, exosomes derived from autologous or immune-evasive cells appear to be well tolerated, even after multiple administrations.[160] Nonetheless, the biological derivation of exosomes presents substantial challenges that currently limit their clinical use. To date, exosomes are mostly derived from immune cells that require an autologous source and thus cannot be used as an off-the-shelf pharmaceutical product. In addition, exosomes made from tumor cells or immune-evasive, immortalized primary cells may contain oncogenes and other hazardous impurities that originate from the cells from which they bud. Besides electroporation, which has been used with some success to load exosomes with RNA, it has been technically challenging to load the exosomes with different payloads.[165] Moreover, there are no satisfactory scalable procedures for exosome production and their separation from other, very similar, extracellular vesicles, which may lead to adverse effects. Finally, exosomes and other vesicles that bud directly from cell membranes, albeit under certain conditions and from specialized regions, do not recapitulate the chemical composition and protein repertoire of cytoplasmic membranes and may lead to unknown consequences. Accordingly, the clinical application of exosomes will require further basic research into their interactions *in vivo* and the development of currently unavailable technical skills.

8.4 Cellular Membrane-based Drug Delivery Vehicles

Many of the problems associated with using whole cells as drug delivery devices can be attributed to the fact that the cells are still able to differentiate and/or actually promote pathologies such as cancer and metastasis. Moreover, non-autologous cells pose many immunological problems, whereas the use of autologous cells is not always practical depending on the quantity desired, isolation procedure, viability and proliferation, and most importantly, patient disease stage. To circumvent these shortcomings, researchers have developed nano-sized cellular membrane vehicles

(CMVs), which are dissimilar to exosomes and other naturally-shed extracellular vesicles. They are physically reconstructed, nano-sized vesicles from the whole membranes of mammalian cells. They are devoid of their intracellular components, but still maintain the majority of their membrane proteins and lipids, which can be used for active or passive targeting, in addition to the EPR effect potentiated by their nano-size. Until recently, CMVs have only been used in cell membrane studies and cancer immunotherapy,[166,167] and not for physically encapsulating drugs or completely coating nanoparticles for targeted drug delivery. CMVs that have been used as DDVs include RBCs,[168–200] WBCs,[201] cancer cells[202] and MSC membranes.[166,167]

8.4.1 *RBC Cellular Membrane Vehicles*

There are two types of RBC-based CMVs. The first is engineered directly from erythrocyte ghosts and the drug is conjugated to, or encapsulated within, the membrane surface.[168–181] The second involves NP encapsulation of the drug prior to RBC-membrane coating.[182–199] Both devices have diameters between 100–200 nm, which allow for EPR-mediated passive targeting to tissues that have leaky vasculature.[188]

8.4.1.1 *Nanoerythrocytes/nanoerythrosomes*

Gaudreault *et al.* developed a method to prepare nano-sized vesicles from the extrusion of RBC ghosts, and patented these vesicles as nanoerythrosomes (nEryts).[168] RBC ghosts were first created by full hypotonic lysis, where all hemoglobin and other cellular constituents were eliminated.[169–171] The RBC ghosts were then extruded through polycarbonate membrane filters, causing them to break down into smaller, spherical vesicles with diameters of approximately 100–200 nm.[168–172] From one red blood cell, 4,000–6,000 nEryts can theoretically be formed, with a total surface area that is nearly 80-fold higher than their parent cells.[172] nEryts can also be engineered from ghost cells using sonication,[173–176] which converts the ghosts into smaller vesicles using a dismembrator, or electrical breakdown (EB), which makes use an electrical potential.[177] However, extrusion yields nEryts of more uniform size. Additionally, the heat generated through sonication and EB can cause membrane damage if not controlled. After sizing, drugs can be conjugated to the nEryt membrane using a chemical cross linker such as glutaraldehyde. To compare the sizing methods and their effect on drug conjugation, an anticancer drug methotrexate (MTX) was conjugated to the membrane after extrusion, sonication and EB (Table 8.4).

Table 8.4:   Comparison of sizing methods for nEryts and the effect of the sizing method on drug conjugation.

Method	Yield %	Size (nm)	Description	Drug Conc. (mg/ml)	Max. Conj. (ng/μg protein)
Extrusion	80.8	130±26	8 extrusions, 0.4 μm filter	1.25	175
Sonication	76.6	170±30	50 W, 3 min, 4°C	2.5	175
EB	76.42	190±25	2K V/cm, 200 μs, 37°C	2.0	160

MTX-nEryts showed higher cytotoxicity and antineoplastic activity than free drug *in vitro* using leukemia cells (L-1210). An effort was made to determine the mechanism of action, and it was established that MTX-nEryts neither diffused through nor entered the cell membrane by endocytosis. Instead, MTX-nEryts are rapidly absorbed onto the cell membrane and hydrolysis of the glutaraldehyde linking arms slowly releases free MTX, which can then penetrate into the cells and produce a higher concentration of the free drug at the cell vicinity over a longer period of time.[171] *In vivo* studies confirmed that nEryts prolonged the release of MTX over free drug, and preferentially localized within RES organs, but with no significant damage.[177]

Spectroscopic characterization of empty nEryts (without drug) was used to study the effect of polyethyleneglycol (PEG) bound to the erythrocyte membrane.[172] PEG is a nontoxic, synthetic polymer that is grafted onto the surface of liposomes and proteins to improve their stealth properties and reduce their aggregation.[203] PEG (2,000 and 5,000 g/mol) was covalently bound to nEryts via lysine residues on cell surface proteins. The attachment of PEG to the nEryt membrane reduced attractive forces and increased repulsive forces (such as steric and hydration) between the vesicles. Thus, protein aggregation was completely prevented up to temperatures of 50°C, whereas nEryt without PEG showed irreversible protein aggregation above 37°C.

Other therapeutic drugs that have been conjugated to nEryts include daunorubicin (DNR),[168–171] doxorubicin (DOX),[176,200] insulin,[173–175] pyrimethamine (PMA),[178,179] mitomycin-C, hydroxyurea and 6-mercaptopurine.[177] DNR, a common anticancer drug, was covalently attached to the nEryt membrane after extrusion.[168–171] *In vivo*, DNR-nEryts displayed a longer half-life, higher antineo-plastic activity, and a longer mean survival time in mice than free DNR.[168] However, the method used to purify the DNR-nEryts significantly affected their biodistribution.[169] Dialysis purification followed by IV administration led to their rapid removal from blood circulation and uptake mainly by the liver and spleen, whereas IP administration showed marked activity in the inguinal lymph nodes 2 h post-injection. When purified by centrifugation, they accumulated within the lungs, which was attributed to particle aggregation. Further investigation showed their average diameter was 10-fold greater than vesicles purified by dialysis.

Using probe sonication, Al-Achi *et al.* engineered nEryts to deliver DOX[176,200] and insulin.[173–175] DOX is well known for its anticancer potential; however, it is associated with serious cardiotoxicity that can lead to congestive heart failure. DOX-nEryts were compared with DOX-erythrocyte ghosts (DOX-EGs) and free DOX. *In vivo*, DOX-nEryts and DOX-EGs remained longer in the central compartment than free DOX. The concentration of DOX in the heart from DOX-nEryts was undetectable, whereas the heart had the highest concentration of all the organs for both DOX-EGs and free DOX. DOX-EGs displayed highest uptake by the spleen, which indicates that EGs are recognized by the body as damaged or dead erythrocytes, whereas nEryts are not. Within the liver, DOX-nEryts and DOX-EGs displayed higher uptake than free DOX, due to the fact that particles of smaller sizes are preferentially taken up by the RES, as are EGs.

Insulin-nEryts (I-nEryts) were created for intraduodenal[174] and oral[175] administration, because the usual subcutaneous route provides easy vascular access but slow and variable insulin absorption,[204] while the intravenous route gives rapid onset but short duration.[205] Oral delivery is complicated by digestive enzymes that can destroy peptides. Oral delivery with synthetic liposomes enhanced insulin absorption, but overall absorption remained low, the insulin was partially degraded, and no hypoglycemic effect was observed.[206] Unlike DOX-nEryts, I-nEryts were found to both encapsulate and bind to insulin,[173] and I-nEryts led to a significant decrease in glucose concentration following both intraduodenal and oral administration.[174,175]

The anti-malarial drug PMA was chemically conjugated to nEryts by Jain *et al.* after both extrusion[179] and sonication,[178] which led to a significant improvement in drug half-life.[179] Using dialysis membranes, nearly all of the free drug was released over 20 h, whereas only one quarter of drug was released from PMA-nEryts. *In vivo*, PMA-nEryts showed controlled drug release and lower accumulation in the kidneys, as compared to free PMA. However, PMA-nEryts showed higher accumulation in the liver, which may be the result of removal by Kupffer cells.

Aside from therapeutic drugs, nEryts have been used to encapsulate a near infrared (NIR) chromophore, indocyanine green (ICG),[180] to enhance the local optical absorption of a target, e.g. abnormal vasculature or a tumor mass. Currently, ICG is the only NIR dye that has been approved by the FDA for cardiocirculatory measurements, ophthalmological imaging and liver function tests.[207–209] ICG has also been used to map the sentinel lymph nodes in several cancers[210–213] and for other phototherapeutic applications.[214–218] ICG has a very short half-life (2–4 min); incorporating it within nano-sized vesicles increased its half-life to nearly 90 min.[219] Anvari *et al.* have successfully encapsulated ICG within nEryts, using the extrusion method, and found that the concentration that would typically result in fluorescence quenching in its free form could be encapsulated within nEryts without quenching.[180]

More recently, nEryts were used to encapsulate a small molecular weight drug, fasudil, and deliver it as an inhalation carrier.[181] Fasudil is an investigational drug used to treat pulmonary arterial hypertension (PAH),[220] and a rho-kinase (ROCK) inhibitor that works as a potent vasodilator. The drug was loaded prior to resealing of the ghost cells, and three sizing methods: bath sonication, probe sonication and extrusion, were used after drug loading.[181] Sonication gave rise to larger particles and reduced entrapment efficiency, whereas extrusion gave rise to smaller homogeneous particles and had a minimal effect on entrapment. F-nEryts showed the ability to withstand the force of nebulization and maintain a slow release profile. After 2 h, all of the free drug was released *in vitro*, whereas only a third was released from F-nEryts, and continuously over 48 h. nEryts were present in the cytoplasm of pulmonary arterial endothelial/smooth muscle cells (PA-E/SMCs), suggesting that they were taken up by vascular cells and can cross the air–blood barrier. nEryts produced a therapeutic effect in SMCs through ROCK inhibition and reduced mean PAH. Additionally, due to their small size (<250 nm), they can escape the respiratory clearance mechanism and extend drug residence time in the lungs. *In vivo* studies revealed a 6–8 fold increase in the half-life of F-nEryts over free drug administered by IV or IT, and no extensive lung damage was observed.

F-nEryts were also conjugated with a homing peptide, CARSKNKDC (CAR),[221] which is known to specifically accumulate in PAH lesions,[222] exhibit cell penetrating properties, and provide a bystander effect when administered with other vasodilators.[223] *In vitro* cellular uptake studies demonstrated that CAR-F-nEryts internalized more efficiently (~1.5 fold) and showed greater lung targeting (~2 fold) than F-nEryts.[221] The peptide reduces the frequency of administration and minimizes fluctuations in mean system arterial pressure that lead to serious side effects such as syncope and cardiovascular collapse. CAR-F-nEryts also demonstrated stronger and more extended vasodilatory duration and higher accumulation in PAH sites, which resulted in a significant reduction in mean pulmonary arterial pressure. Overall, these studies demonstrate that nEryts are capable of encapsulating and binding both small and large molecular weight drugs, maintaining controlled and sustained release profiles, and staying nontoxic and non-immunogenic (when autologous).

8.4.1.2 *RBC membrane coated nanoparticles*

RBC membrane coated nanoparticles (RBC NPs) have received a lot of attention in recent literature.[182–198] These vehicles, first reported by Zhang *et al.* in 2011, were created by extruding poly(lactic-co-glycolic acid) (PLGA) NPs with RBC ghosts.[187] RBC ghosts were bath sonicated, washed, extruded serially through 400 nm and 100 nm polycarbonate porous membranes, and fused with PLGA NPs through additional extrusion.[186] Experimentally, it was determined that 100 μL of mouse blood provided sufficient RBC membrane material to coat 1 mg of PLGA NPs, provide stability and prevent PLGA aggregation.[194] Mechanical force of the extrusion method allowed for sub-100-nm PLGA NPs to cross lipid bilayers and fuse with the purified RBC membrane. The bilayer structure was retained throughout the procedure, thus minimizing damage to membrane proteins. RBC NPs were sterically stabilized and did not aggregate due to hydrophilic surface glycans.[224]

The core-shell structure of RBC NPs, consisting of a 70 nm PLGA core and a 7–8 nm RBC membrane shell, remained intact after cellular internalization.[225] Membrane proteins were mostly retained throughout the procedure, and the membrane remained in the right-side-out orientation.[191] The negative surface charge of the PLGA NPs was determined to be the key element in providing proper core-shell formation, due to the negatively-charged extracellular membrane. Negatively-charged PLGA cores were fully covered following extrusion, whereas positively-charged PLGA cores clogged the extruder and showed dense aggregates. By conjugating a ssDNA probe to the RBC membrane prior to cell lysis, Cheng *et al.* quantified the cell membrane orientation of their RBC NPs, which were created by probe sonication of RBC membranes followed by overnight incubation with PLGA NPs.[197] RBC NPs exhibited a higher right-side-out membrane orientation (84%) than sonicated RBC membrane vesicles (RBC MVs, 70%). Thus, this confirms that the repulsive interaction between the membrane and the PLGA NPs provides proper coating. Both RBC MVs and NPs show higher permeability than RBCs and liposomes, due to the packing of lipids being less compact. Their permeability can be controlled by adjusting the lipid composition in the membrane formulation, thus improving control over the release kinetics of encapsulated drugs.[197] Gold nanoparticles (AuNPs) were also used successfully as the core material for these devices,[186,196] and demonstrated lower uptake by macrophages than naked AuNPs.[186] Additionally, AuNPs generally have a high affinity for thiolated compounds, and the RBC membrane effectively shielded the AuNPs from this interaction.

Researchers have also functionalized the RBC membrane for site-specific targeting.[184] Zhang *et al.* conjugated folate, a small molecule, and AS1411, a nucleolin-targeting aptamer, to the membrane of RBC NPs. Several types of cancers overexpress folate receptors, which has led researchers to create nanocarriers functionalized with folate.[226,227] When incubated with a KB cell line, a model cancer cell line that overexpresses the folate receptor, folate-RBC NPs showed an 8-fold increase in cellular uptake as compared to unmodified RBC NPs. Folate-RBC NPs incubated with A549 cells, a cancer cell line that does not overexpress folate receptors, showed no increased uptake. AS1411-RBC NPs demonstrated similar results, with a 2-fold increase in cellular uptake when compared to the unmodified RBC NPs.

RBC NPs were specifically studied for their ability to function as a biomimetic nanosponge (NSP) that absorbs pore forming toxins (PFTs),[190] one of the most common protein toxins found in nature.[228] PFTs lyse cellular membranes, damaging their permeability. Bacteria such as *Escherichia coli*, *Helicobacter pylori* and *Staphylococcus aureus* use PFTs as a major virulence mechanism. Studies have shown that inhibiting pore forming α-toxin can reduce the severity of *Staphylococcus aureus* infections.[229] After α-toxin exposure, RBCs incubated with the NSP remained intact, whereas RBCs incubated with PEGylated PLGA NPs, PEGylated liposomes, and empty RBC membranes exhibited more than 90% hemolysis.[190] Although both the NSP and RBC membranes retained nearly all of the toxins unlike other formulations that showed negligible retention, only the RBC membranes showed hemolytic behavior, which demonstrates the necessity of the PLGA core. *In vivo* studies in mice showed that an injection of free α-toxin caused necrosis of skin tissue, apoptosis, inflammatory infiltrate of neutrophils with dermal edema, and damage to underlying muscle tissue. However, no damage was observed when the NSP was injected with α-toxin. After a lethal α-toxin injection, the mice that did not receive any treatment exhibited 100% mortality within 6 h, the mice treated with NSP preinoculation exhibited only 11% mortality, and the mice treated with NSP postinoculation exhibited 56% mortality.

The NSP was used to create a NP-detained toxin for antitoxin vaccination.[195] Generally, antitoxin vaccines used to treat and prevent bacterial infections are based on inactivated bacterial toxins,[230] which is complicated by the need to maintain antigenic presentation of inactivated toxins in addition to a safe and non-toxic design. Thus, the RBC NSP was proposed to safely deliver non-disrupted PFTs for immune processing.[195] Staphylococcal α-haemolysin (Hla) was used as a model toxin and the device was referred to as nanotoxoid(Hla). Nanotoxoid(Hla) allowed for uptake of toxins by immune cells, and once Hla was engulfed into the digestive endolysosomal compartment, NP-facilitated cellular endocytosis prevented the toxin from perforating the cellular membrane, which allowed the non-disrupted toxin to be delivered for immune processing. Mice injected with nanotoxoid(Hla) exhibited an intact epithelial structure with no cellular apoptosis outside of hair follicles, whereas untreated Hla caused significant apoptosis and skin lesions, and results with heat-treated Hla were time dependent. Nanotoxoid(Hla) also demonstrated considerably higher Hla specific antibody titers as compared to heat-treated Hla. There was no mortality with mice that were vaccinated with nanotoxoid(Hla), whereas all unvaccinated mice died.

RBC NSPs were also used to manage type II immune hypersensitivity reactions, which are caused by pathological antibodies targeting self-antigens that are either naturally occurring or due to exposure to an exogenous substance present on the cellular exterior or ECM.[231] Many autoimmune diseases fall within this category, such as myasthenia gravis, pernicious anemia, Graves' disease, autoimmune hemolytic anemia (AIHA) and immune thrombocytopenia.[232] Administration of a new drug or certain infections can also lead to this reaction. Therapies are variable and contain a high risk for adverse side effects.[233] Using a model of antibody-induced anemia (AIA), which can be caused by AIHA or be induced by drugs, the ability for RBC NSPs to clear pathological antibodies was evaluated.[231] With AIA, auto-antibodies attack surface antigens present on RBCs. Standard therapy includes steroids, cytotoxic drugs, blood transfusions and possibly splenectomy,[234,235] which carries the risks of severe infection,[236] hemolytic transfusion reactions, formation of alloantibodies, and iron toxicity.[237–239] RBC NSPs were used to intercept the auto-antibodies that would otherwise attack healthy RBCs, and *in vivo* results demonstrated a reduced antibody-mediated immune response, whereas an equivalent dose of PEG NP with analogous physicochemical properties failed to moderate the anti-RBC antibodies. Wang *et al.* encapsulated vancomycin (Van), a model antibiotic, within supramolecular gelatin nanoparticles (SGNPs) in the core of RBC NSPs.[199] SGNPs disassemble in the presence of matrix metalloproteinases (MMPs)[240] secreted by a broad spectrum of bacteria.[241]

Thus, RBC membranes act as detoxifiers by absorbing the exotoxins produced by bacteria, while the SGNP core is degraded by gelatinase that is overexpressed in the infection microenvironment, and Van is released to kill the local pathogenic bacteria. *In vitro* studies showed that RBC SGNPs were barely internalized by macrophage cells, and displayed negligible toxicity towards human embryonic kidney and hepatocyte cell lines. Additionally, in the presence of gelatinase, RBC SGNPs provided a sustained release profile as compared to naked SGNPs. Yang *et al.* created Au nanocages (AuNCs) coated with RBC membranes for photothermal therapy (PTT) cancer treatment.[196] PTT involves NIR on target tissues that contain photothermal conversion agents (PTCA) such as AuNC, which strongly absorb light and convert it to heat. Since cancer cells possess reduced heat tolerance,[242] they can be selectively destroyed at temperatures above hyperthermia.[243] *In vivo*, RBC-AuNCs showed an almost 10-fold increased circulation half-life and a nearly 2-fold greater tumor uptake than PVP-AuNCs.[196] The average tumor volume of mice treated with RBC-AuNCs decreased at a faster rate and reached a smaller final value than those treated with PVP-AuNCs. Additionally, mice treated with RBC-AuNCs exhibited a consistent increase in average body weight and 100% survival, whereas those treated with PVP-AuNCs exhibited a decrease in body weight and 80% survival.

The main attributes of RBC NPs are that a wide variety of therapeutic agents can be encapsulated within the PLGA core. However, a potential obstacle is that the membranes maintain specific antigens that make these systems only viable when the donor membrane matches that of the host. Thus, blood will have to be drawn directly from the host, or donor-specific blood will have to be collected. Due to a universal lack of donor blood, this can compromise the proposed delivery system. Furthermore, mostly proof of concept (POC) studies have been conducted. Additional studies demonstrating therapeutic loading with a desired target need to be carried out.

8.4.2   *Cancer Cell Membrane Coated Nanoparticles*

Zhang *et al.* have also used cancer cell membranes to coat polymeric NPs (CCNPs).[202] CCNPs present membrane-bound tumor-associated antigens and when combined with immunological adjuvants, can be delivered to antigen presenting cells to promote an antitumor-specific immune response for vaccine applications.[244] Additionally, they possess the same cell adhesion molecules as their source cells, allowing them to exhibit source cell-specific targeting based on the natural homotypic binding mechanism frequently observed among tumor cells.[245,246] Using extrusion methods, spherical CCNPs were derived from B16-F10 mouse melanoma cells with diameters of about 110 nm.[202] *In vitro* cellular studies demonstrated that the cells remained intact upon endocytosis by bone marrow-derived mouse dendritic cells. CCNPs were incorporated with monophosphoryl lipid A (MPLA) to help deliver tumor-associated antigens and induce dendritic cell maturation. The cytokine interferon-gamma (IFN-γ), an indicator of cytotoxic T-lymphocyte stimulation, was quantified and showed MLPA-CCNPs could successfully elicit an antigen-specific response. CCNPs also exhibited increased cellular uptake when incubated with MDA-MB-435 cancer cells, as compared to both naked PLGA NP and RBC NPs.

8.4.3 *Leukocyte Membrane Coated Nanoparticles*

Tasciotti *et al.* used leukocytes to coat nanoporous silicon (NPS), and demonstrated that these leukolike vectors (LLVs) could successfully evade the immune system, cross biological barriers, and localize in target tissues.[201] Freshly harvested leukocytes were purified by ultracentrifugation using a discontinuous sucrose density gradient, and reconstituted as proteo-lipid patches. Electrostatic and hydrophobic interactions between the negatively-charged proteo-lipid patches and positively-charged NPS drove self-assembly, and the proteo-lipid patches fully covered the NPS surface, leaving critical transmembrane proteins in the same orientation as parent cells. *In vitro* studies were used to examine the ability of LLVs to inhibit particle internalization by murine J774 macrophages and human THP-1 phagocytic cells. When the donor membrane matched that of the host phagocytic cells, LLVs significantly decreased particle uptake; when the donor and host cells were mismatched, no significant decrease was observed. Previous studies have reported that early NPS internalization by HUVECs was characterized by the formation of filopodia "cages" that surrounded the particles and locked them against the cell surface, but when NPS was treated with LLVs, filopodia formation did not occur.[201] Additionally, LLV coating doubled the binding potential of NPS to HUVEC cells, and prevented lysosomal sequestration. LLVs were in direct contact with the cell cytoplasm, while naked NPS remained trapped inside the endolysosomal compartment and transferred to the perinuclear region. Under inflamed conditions, membrane coating increased particle transportation through an endothelium monolayer. DOX was used to evaluate the therapeutic potential of LLVs and it was determined that LLVs can successfully transport the DOX payload through the endothelium and release it into the lower chamber. LLVs have demonstrated the potential to be used as therapeutic delivery devices that can reduce opsonization, delay phagocytic uptake, bind to inflamed endothelium, facilitate transport across the endothelial layer while avoiding the lysosomal pathway, enhance particle circulation time, and improve tumoritropic accumulation.

8.4.4 *Nanoghosts*

Although all the previous CMVs have displayed tremendous potential in the field of drug delivery, there is a common and significant obstacle to their clinical success. Excluding CCNPs and CMVs conjugated with active targeting ligands, these cells rely primarily on the EPR effect to target desired organs, thus limiting the targeting potential of these vehicles. However, Machluf *et al.* developed two types of natural active targeting CMVs in 2011,[167] and patented these CMVs as nanoghosts

(NGs).[166,167] The first type of NG is derived from the membranes of cells expressing CCR5, contains a cytotoxic drug that can be administered along with soluble CD4, and was used to target, fuse and destroy HIV-infected cells.[167] The second type of NG is based on the cytoplasmic membrane of MSCs, encompasses cell surface molecules and preserves the active targeting mechanism of the cells from which they are made, and has demonstrated significant potential in cancer therapy.[166]

8.4.4.1 Nanoghosts-CCR5

Anti-retroviral therapies (ARTs) are a prominent treatment regime for HIV/AIDS therapy.[247–249] They slow disease progression by targeting the essential enzymes and proteins of the virus life cycle, but cannot cure it. HIV attaches to host cells by the binding of viral gp120 to host CD40, followed by cellular co-receptor CCR5 and/or CXCR4.[249] The host cell is anchored by gp41, which leads to viral-cell membrane fusion and insertion of the viral nucleocapsid into the host cell. This process makes it possible to develop a targeted delivery system against gp120 and gp41 using CD4, CCR5 and/or CXCR4 as homing/targeting ligands. Thus, cellular membranes expressing CCR5 were used to develop CCR5-NGs that also encapsulated a cytotoxic drug. When this formulation was administered with soluble CD4 it was shown to target, fuse and destroy HIV-infected cells that express the ENV gene and display bioactive gp12/gp41 on their surface.[167] This method eliminates the need for elaborate purification and reincorporation of CCR5 that is necessary for synthetic liposomes.[250]

Cytoplasmic membranes from Cf2th/CCR5 (C9) cells were selected for these CCR5-NGs, and non-human Cf2Th cells (canine origin) were selected as a source of recombinant human CCR5, so that any potential effect by the membrane components of gp120-expressing cells (most abundant cellular marker for HIV) on specific targeting would be eliminated. EDTA was encapsulated since it features high inhibitory effects over intracellular enzymes in HIV-susceptible cells, has low acute toxicity, has a low octanol-water partition coefficient, and is impermeable to the NG membrane. To encapsulate EDTA, ghost cells were resuspended in a sucrose/TM buffer solution, sonicated, and extruded.

Two ENV-expressing model cell lines that are based on BHK and Namalwa cells were used for *in vitro* targeting studies. Both cell lines were engineered to express gp120 and gp41 of the dual tropic HIV-1 strain p89.6, which mimics HIV-infected cells. Since BHK cells are of non-human origin, any effect of human surface proteins that could possibly influence targeting was excluded. Namalwa cells are derived from human B-lymphocytes, and so ENV expression may be more efficient than if non-human BHK cells were used. However, it was demonstrated that both cell lines express the bioactive complex of gp120 and gp41 and can be used as targets to mimic the membrane of HIV-infected cells. The binding of CCR5-NGs to both BHK/ENV and Namalwa/ENV cells was significant, but specific for Namalwa/ENV cells. CCR5-NGs derived from suspension-growing Jurkat cells demonstrated distribution *in vivo* over a larger area and duration than those derived from adherent Cf2Th cells, which confirmed that adherent cells are "stickier" and contribute to non-specific binding. PEGylation was shown to make CCR5-NGs less adhesive and less prone to opsonization, which is known to increase biodistribution area and duration.[251] The effectiveness of EDTA for inhibiting intracellular enzymes in HIV-infected cells was tested using ultrasound-induced membrane permeabilization.[167] It was demonstrated that EDTA exhibited cytotoxic effects on a variety of cells, but only with ultrasound-induced membrane permeabilization, and not when solubilized in cell culture media. When EDTA was encapsulated within CCR5-NGs, a nearly 60% cytotoxicity was seen. These results demonstrate that not only is the release of EDTA responsible for the cytotoxic effect, but also the disruption of the cellular membrane (post liposome fusion), which is observed in the literature.[252] These studies demonstrate the potential for personalized HIV/AIDS treatments, since CD4+/CXCR4+and/or CD4+/CCR5+ cells can be harvested from the patient during the asymptomatic phase of HIV and preserved for NG preparation during the symptomatic phase.[167] Additionally, these studies demonstrate the possibility of treating HIV by viral reservoir depravation without using any antiretrovirals. CCR5-NGs can specifically bind to and fuse with target cells, deliver their cargo, and destroy the target cells, all while sparing "healthy" control cells and minimizing adverse side effects.

8.4.4.2　*Nanoghosts-MSCs*

Another obstacle faced by all previously mentioned CMVs is that they must either be autologous or a specific donor match. As there is already limited donor blood available, and if the patient is gravely ill, then autologous cells may not be a viable option; research directed towards non-autologous cells for CMVs may provide an alternative solution. NGs based on the cytoplasmic membrane of MSCs were demonstrated to encompass cell surface molecules and preserve the active targeting mechanism of the cells from which they were made, and did not express specific antigens that require a donor match. Their unique targeting ability was demonstrated with many types of solid tumors (lung, breast, colon, ovarian, etc.), and may be attributed to the fact that tumors, much like "wounds that never heal", involve angiogenesis, inflammation and rearrangement of the molecular and cellular matrix surrounding the tumor, all of which require MSC support. Isolated membrane fractions of tumor cells appeared to contain more potent MSC attractants than cytoplasmic fractions of the same cells, even though the specific receptors-ligand pairs contributing to such physical interactions have yet to be fully identified.[253] These attributes permit MSCs to serve as an excellent foundation for the production of NGs, or as cancer-targeting drug delivery vesicles for cancer-related applications, including the early diagnosis of cancer when it would otherwise be undetectable. Additionally, NGs are easier and less costly to produce than other actively-targeted pharmaceutical preparations. Encompassing MSC surface proteins and utilizing their unique targeting capabilities, NGs may be loaded with various drugs, nucleic acids or contrast agents, and can be targeted against multiple cancers even in the early stages. The well-documented low immunogenic nature of the largely available MSCs assures that NGs can be readily prepared as a safe-to-use, off-the-shelf product. Furthermore, the mechanisms associated with MSC targeting, and the role of MSCs in cancer progression, can be determined using these NGs. Specific examples of MSC-based NGs follow.

As an initial POC study, NGs were prepared from human and rat MSCs (hMSCs and rMSCs), as well as human smooth muscle cells (hSMCs) as the non-mesenchymal control.[166] sTRAIL was selected as the encapsulated drug since its free form exhibits a short half-life and high hepatoxicity, which limit its clinical use. Additionally, when sTRAIL was administered in controlled release formulations, no effect was seen without the combination of additional drugs.[254] NGs were prepared by homogenization, sonication and extrusion methods mentioned in the previous section (Figure 8.1).[166] Extrusion was performed in a medium containing sTRAIL for drug encapsulation to take place. The final vesicles were obtained by ultracentrifugation and had an average diameter of approximately 180 nm, with negative zeta

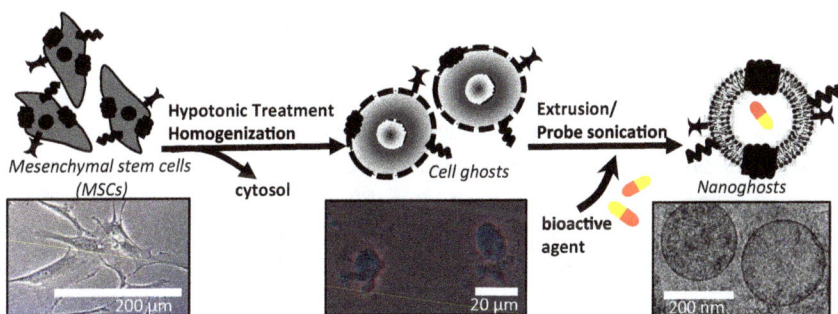

Figure 8.1:   Protocol: From mesenchymal stem cells to nanoghosts.

potential. The encapsulation efficiency of sTRAIL was 30%, and NGs exhibited a burst release profile, with approximately 70% of sTRAIL rapidly released. Flow cytometric analysis showed all hMSC surface markers were retained on the surface of NGs. *In vitro* studies were used to determine the selectivity of hMSC-NG binding to specific targets (PC3 and MCF7 cells) in comparison to nonspecific targets (baby hamster kidney cells, BHK cells and hSMCs). A time-dependent selectivity of hMSC-NGs was seen with PC3 and MCF7 tumor cells when compared with nonspecific target cells (baby hamster kidney cells, BHK cells and hSMCs), emphasizing that hMSC-NG targeting is tumor-specific and not species-specific, and one that is active and not passive.

In vivo mice studies showed significant accumulation of NGs within the tumor and liver after 24 h with both IV and IP administration, and only within the tumor after 1 week (IP administration). Other nanovesicle formulations have also demonstrated a more targeted biodistribution through IP.[255] No histological abnormalities or heptatotoxicity were detected within the liver.[166] A single administration of sTRAIL-hMSC-NGs via the IP route was sufficient in promoting tumor inhibition as compared to mice that were not treated, treated with empty NGs, or given free sTRAIL. Similar results were demonstrated with sTRAIL-rMSC-NGs with the exception of the last two time points (hMSC-NGs produced a larger inhibition effect), whereas sTRAIL-hSMC-NGs and synthetic liposomes containing sTRAIL did not inhibit tumor growth. The tumors were harvested and analyzed for apoptotic cells; sTRAIL-NGs showed a significantly higher percentage of apoptosis as compared to non-treated mice, mice treated with empty NGs, and mice given free sTRAIL. A higher reduction in tumor vascularization and proliferation was also seen with sTRAIL-NGs, as compared to the other formulations. PEGylation of hMSC-NGs was investigated for its ability to reduce opsonin-dependent uptake and increase circulation time. Although PEG-hMSC-NGs appeared to partially mask two of the four MSC surface markers tested, the remaining markers were still found in substantial quantities and in native confirmations, which was sufficient to support active targeting. Membrane asymmetry was also confirmed by observing that the expression of MSC markers on the surface of NGs were correctly oriented.

This initial POC study on the effectiveness of hMSC-NGs as a drug carrier clearly demonstrated that NGs act by both disrupting the target cell membranes, leading to some apoptosis, and interacting with the tumor microenvironment; the entire effect of free sTRAIL, on the other hand, is likely a result of apoptosis.[256] The anticancer effect displayed by empty hMSC-NGs can be attributed to interactions with different components of the tumor microenvironment such as angiogenic blood vessels and inflammatory cells, and the transition by MSCs into tumor-associated fibroblasts.[257,258] It was also proposed that NGs may compete with endogenous MSCs and interfere with the tumor support they provide. Active targeting was confirmed over EPR-based passive targeting because hSMC-NGs and hMSC-NGs both have similar size and physical properties but hSMC-NGs failed to target the tumor.[166] Additionally, synthetic liposomes that are widely investigated as drug carriers for various cancers[259] demonstrated no effect when administered under similar conditions, suggesting that passive targeting alone is insufficient. Following this POC study, additional hMSC-NGs were derived to further understand their physical characteristics, targeting potential, and specific drug delivery capability.

The shelf-life stability of hMSC-NGs was assessed in PBS and at different temperatures (4°C, −20°C and −196°C) for several weeks by observing their size (DLS), charge (Zeta potential) and morphology (cryo-TEM). NGs aggregated within several hours at room temperature, but remained stable at 4°C for over 10 days, making them suitable for short-term storage. As for long-term storage, NGs were found to be most stable in PBS at −20°C, where they maintained the same charge and size for more than two months.

The quantity of PEG necessary for the device and the kinetics of the PEGylation reaction were evaluated to determine the optimal PEGylation conditions that prevent particle aggregation and enhance targeting. Different quantities of PEG-5000 were reacted with fluorescently-labeled hMSC-NGs for various times, and the resulting fluorescent PEG-NGs were incubated *in vitro* with A549, PC3 and PC2 cancer cells to observe the effect of PEGylation on targeting. FACS analysis was used to observe the targeting results, and it was demonstrated that reacting 0.5 mg PEG for every 1 mg protein, as determined by the Bradford Assay, for 1 h was ideal. Below 0.5 mg PEG, or if the reaction time was less than 1 h, targeting was decreased as a result of particle aggregation. Above 0.5 mg PEG, or for reactions longer than 1 h, targeting was decreased as a result of the MSC-surface markers being masked, or over-PEGylation. A quantifiable reaction yield still needs to be determined, and results need to be verified *in vivo*.

Studying the interactions of hMSC-NGs with cancer cells revealed that NGs participate in cell surface interactions, which leads to internalization by cancer cells. Cellular interactions with hMSC-NGs include fusion and energy-dependent endocytosis. Endocytosis studies using inhibitors revealed that all three major pathways (lipid raft/caveolae, clathrin, and macropinocytosis or phagocytosis) are involved in the cellular uptake of hMSC-NGs. *In vitro* studies showed that integrins, specifically integrin $\beta 1$ and intra-cellular adhesion molecule (ICAM), have a role in cancer cell-hMSC-NG interactions. hMSC-NG interactions with cancer cells and cells of the tumor niche were also affected by stimulation of hMSCs prior to

NG production. Cancer cells and macrophages demonstrated increased uptake of hMSC-NGs when hMSCs were stimulated by cancer-derived conditioned growth media, while endothelial cells showed the highest uptake of NGs produced from pro-inflammatory cytokine (TNF-α, IL-1β)-stimulated hMSCs. These results suggest that the targeting abilities of hMSC-NGs can be programmed by tuning the cultivation conditions of hMSCs.

The ability of hMSC-NGs to target glial cells within the brain was investigated for the possible treatment of Alzheimer's and Parkinson's diseases, brain tumors, and multiple sclerosis, since these diseases remain severely undertreated due to the high selectivity of the blood brain barrier (BBB).[2] The BBB, which is formed by impermeable tight junctions (TJs) between brain capillary endothelial cells (BCEC), serves as an active and selective barrier regulating the homeostasis of the brain and protecting it from toxic substances. To evaluate the targeting ability of NGs towards glial cells, fluorescently-labeled NGs were incubated for up to one hour with rat glial cells that were later analyzed by flow cytometry for NG binding. NG binding to microglia cells was much faster than to astrocytes, albeit the final proportion of NG-bound astrocytes was higher. In addition, biodistribution studies with NGs were performed in mice. Fluorescently-labeled NGs were IV injected into nude mice, and 24 h later, the mice were sacrificed and their brains were harvested and sectioned to the left and right hemispheres and the cerebellum. All of the brain sections, which contained the same two cell populations, gave similar results: NG binding to one of the populations was much higher than to the other population. hMSC-NGs were also studied for their ability to selectively target metastatic non-small cell lung carcinoma (NSCLC, Figure 8.2), the most common lung malignancy that also holds the highest mortality rate of all cancers.[260,261] NGs were encapsulated with Cisplatin, a platinum-based anticancer drug that is widely used in the clinic to treat NSCLC, but causes severe side effects in its free form. *In vitro*, Cisplatin-NGs demonstrated an improved therapeutic index toward the NSCLC cell line (A549) as compared to the free drug. An *in vivo* metastatic NSCLC model in athymic nude mice was created by IV injection of A549 cells and by monitoring the progress of metastases using MRI imaging. Future studies will be carried out to evaluate the efficacy of cisplatin-NGs using this model.

The delivery of DNA by nEryts for cancer therapy was evaluated. To improve DNA loading into NGs, plasmid DNA encoding for GFP (pGFP) was complexed with linear polyethylenimine (PEI, 25 kDa) at different nitrogen-to-phosphate (N:P) ratios ranging from 5:1 to 20:1. NGs were passively loaded with PEI:DNA complexes during sonication of the ghosts, and their loading efficiencies were determined using qPCR. An N:P ratio of 15:1 yielded the highest loading efficiency of more

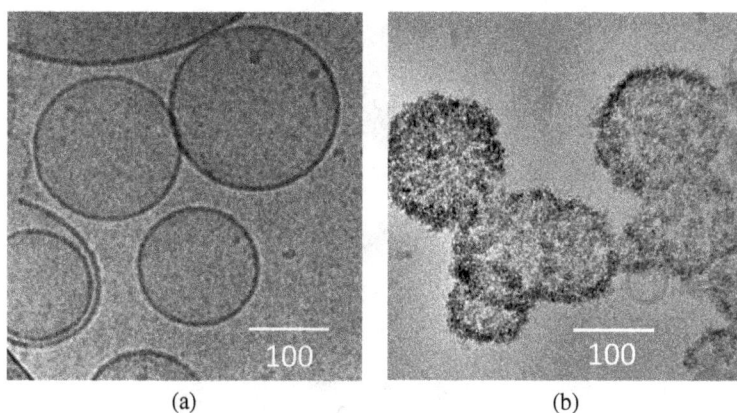

Figure 8.2:   Cryo–TEM imaging of (a) empty NGs and (b) NGs loaded with Cisplatin.

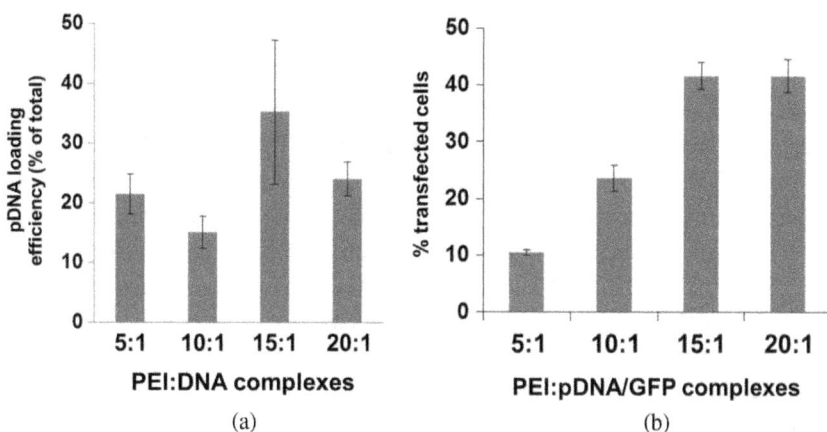

Figure 8.3:   (a) pDNA loading efficiency and (b) transfection efficiency.

than 30% [Figure 8.3(a)]. When the amount of DNA was held constant, NGs loaded with different PEI:pGFP ratios were used to transfect PC3 cells that were analyzed by FACS for GFP expression 24 h post transfection. An N:P ratio of over 15:1 gave the highest transfection of almost 40% of the cells [Figure 8.3(b)].

hMSC-NGs have also been exploited for their ability to home to sites of inflammation and specifically target anti-inflammatory drugs to inflamed tissue. To investigate the interaction of NGs with macrophages, fluorescently-labeled NGs were incubated with THP-1 cells for 3 h. At the end of the incubation, NGs were clearly seen inside the cells (Figure 8.4). The entrapment of anti-inflammatory

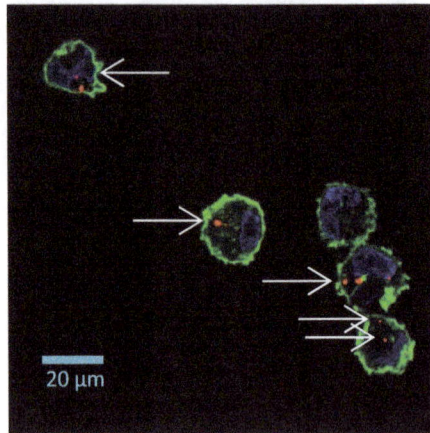

Figure 8.4: THP-1 cells and NGs; DID (red, NGs), Hoechst (blue, nucleus), phalloidin (green, actin filaments).

drugs such as corticosteroids prevents off-target toxicity, which will decrease side effects and increase therapeutic efficiency. Empty-hMSC-NGs and NGs loaded with prednisolone-sodium phosphate (PLP) were injected using IV or IP administration into a rat model of ischemia-reperfusion, and were detected 24 h post injection in the inflamed myocardium (communicated data). *In vitro* studies were also performed to evaluate the interaction of NGs with immune cells. Empty-NGs demonstrated the ability to interact with macrophages, particularly after stimulation. For example, when incubated with LPS-stimulated macrophages, empty-NGs were shown to affect the mRNA levels of pro-inflammatory cytokines. PLP-NGs showed a similar effect on nitric oxide levels produced by LPS-stimulated macrophages to that of the free drug.

References

1. Latorre-Romero, C. *et al.* Using living cells to transport therapeutic genes for cancer treatment. *Clinical & Translational Oncology* 13, 10–17 (2011).
2. Altaner, C. Prodrug cancer gene therapy. *Cancer Letters* 270, 191–201 (2008).
3. Willmon, C. *et al.* Cell carriers for oncolytic viruses: Fed Ex for cancer therapy. *Molecular Therapy* 17, 1667–1676 (2009).
4. Balyasnikova, I.V., Ferguson, S.D., Sengupta, S., Han, Y., and Lesniak, M.S. Mesenchymal stem cells modified with a single-chain antibody against EGFRvIII successfully inhibit the growth of human xenograft malignant glioma. *Plos One* 5 (2010).

5. Lee, H.K. *et al.* Mesenchymal stem cells deliver synthetic microRNA mimics to glioma cells and glioma stem cells and inhibit their cell migration and self-renewal. *Oncotarget* 4, 346–361 (2013).

6. Sun, X.L. *et al.* Molecular targeting of malignant glioma cells with an EphA2-specific immunotoxin delivered by human bone marrow-derived mesenchymal stem cells. *Cancer Letters* 312, 168–177 (2011).

7. Auffinger, B. *et al.* Drug-loaded nanoparticle systems and adult stem cells: a potential marriage for the treatment of malignant glioma? *Oncotarget* 4, 378–396 (2013).

8. Porada, C.D., and Almeida-Porada, G. Mesenchymal stem cells as therapeutics and vehicles for gene and drug delivery. *Advanced Drug Delivery Reviews* 62, 1156–1166 (2010).

9. Pontikoglou, C., Deschaseaux, F., Sensebe, L., and Papadaki, H.A. Bone marrow mesenchymal stem cells: biological properties and their role in hematopoiesis and hematopoietic stem cell transplantation. *Stem Cell Reviews and Reports* 7, 569–589 (2011).

10. Wang, S., Qu, X., and Zhao, R.C. Clinical applications of mesenchymal stem cells. *Journal of Hematology & Oncology* 5, 19 (2012).

11. Kim, N., and Cho, S.G. Clinical applications of mesenchymal stem cells. *The Korean Journal of Internal Medicine* 28, 387–402 (2013).

12. Zhang, T. *et al.* Bone marrow-derived mesenchymal stem cells promote growth and angiogenesis of breast and prostate tumors. *Stem Cell Research & Therapy* 4, 70 (2013).

13. Hass, R., and Otte, A. Mesenchymal stem cells as all-round supporters in a normal and neoplastic microenvironment. *Cell Communication and Signaling: CCS* 10, 26 (2012).

14. Caplan, A.I. Adult mesenchymal stem cells for tissue engineering versus regenerative medicine. *Journal of Cellular Physiology* 213, 341–347 (2007).

15. Griffin, M.D. *et al.* Anti-donor immune responses elicited by allogeneic mesenchymal stem cells: what have we learned so far? *Immunology and Cell Biology* 91, 40–51 (2013).

16. Ankrum, J., and Karp, J.M. Mesenchymal stem cell therapy: two steps forward, one step back. *Trends in Molecular Medicine* 16, 203–209 (2010).

17. Beggs, K.J. *et al.* Immunologic consequences of multiple, high-dose administration of allogeneic mesenchymal stem cells to baboons. *Cell Transplantation* 15, 711–721 (2006).

18. Roorda, B.D., Elst, A., Boer, T.G., Kamps, W.A., and de Bont, E.S. Mesenchymal stem cells contribute to tumor cell proliferation by direct cell-cell contact interactions. *Cancer Investigation* 28, 526–534 (2010).

19. Khakoo, A.Y. *et al.* Human mesenchymal stem cells exert potent antitumorigenic effects in a model of Kaposi's sarcoma. *The Journal of Experimental Medicine* 203, 1235–1247 (2006).

20. Roorda, B.D., ter Elst, A., Kamps, W.A., and de Bont, E.S. Bone marrow-derived cells and tumor growth: contribution of bone marrow-derived cells to tumor microenvironments with special focus on mesenchymal stem cells. *Critical Reviews in Oncology/Hematology* 69, 187–198 (2009).

21. Bergfeld, S.A., and DeClerck, Y.A. Bone marrow-derived mesenchymal stem cells and the tumor microenvironment. *Cancer Metastasis Reviews* 29, 249–261 (2010).

22. Martin, F.T. *et al.* Potential role of mesenchymal stem cells (MSCs) in the breast tumour microenvironment: stimulation of epithelial to mesenchymal transition (EMT). *Breast Cancer Research and Treatment* 124, 317–326 (2010).

23. Kauser, K., and Zeiher, A.-M. Bone marrow-derived progenitors, Vol. 180. (Springer, 2007).
24. Traggiai, E. *et al.* Bone marrow-derived mesenchymal stem cells induce both polyclonal expansion and differentiation of B cells isolated from healthy donors and systemic lupus erythematosus patients. *Stem Cells* 26, 562–569 (2008).
25. Ren, G. *et al.* Mesenchymal stem cell-mediated immunosuppression occurs via concerted action of chemokines and nitric oxide. *Cell Stem Cell* 2, 141–150 (2008).
26. Binello, E., and Germano, I.M. Stem cells as therapeutic vehicles for the treatment of high-grade gliomas. *Neuro-Oncology* 14, 256–265 (2012).
27. Albarenque, S.M., Zwacka, R.M., and Mohr, A. Both human and mouse mesenchymal stem cells promote breast cancer metastasis. *Stem Cell Research* 7, 163–171 (2011).
28. Akimoto, K. *et al.* Umbilical cord blood-derived mesenchymal stem cells inhibit, but adipose tissue-derived mesenchymal stem cells promote, glioblastoma multiforme proliferation. *Stem Cells and Development* 22, 1370–1386 (2013).
29. Zhu, W. *et al.* Mesenchymal stem cells derived from bone marrow favor tumor cell growth in vivo. *Experimental and Molecular Pathology* 80, 267–274 (2006).
30. Yu, J.M., Jun, E.S., Bae, Y.C., and Jung, J.S. Mesenchymal stem cells derived from human adipose tissues favor tumor cell growth in vivo. *Stem Cells and Development* 17, 463–473 (2008).
31. Pasquet, M. *et al.* Hospicells (ascites-derived stromal cells) promote tumorigenicity and angiogenesis. *International Journal of Cancer* 126, 2090–2101 (2010).
32. Roodhart, J.M.L. *et al.* Mesenchymal stem cells induce resistance to chemotherapy through the release of platinum-induced fatty acids. *Cancer Cell* 20, 370–383 (2011).
33. Teng, I.W. *et al.* Targeted methylation of two tumor suppressor genes is sufficient to transform mesenchymal stem cells into cancer stem/initiating cells. *Cancer Research* 71, 4653–4663 (2011).
34. Gao, Z.B., Zhang, L.N., Hu, J., and Sun, Y.J. Mesenchymal stem cells: a potential targeted-delivery vehicle for anti-cancer drug, loaded nanoparticles. *Nanomedicine-Nanotechnology Biology and Medicine* 9, 174–184 (2013).
35. Bao, Q. *et al.* Mesenchymal stem cell-based tumor-targeted gene therapy in gastrointestinal cancer. *Stem Cells and Development* 21, 2355–2363 (2012).
36. Kolluri, K.K., Laurent, G.J., and Janes, S.M. Mesenchymal stem cells as vectors for lung cancer therapy. *Respiration* 85, 443–451 (2013).
37. Gjorgieva, D., Zaidman, N., and Bosnakovski, D. Mesenchymal stem cells for anti-cancer drug delivery. *Recent Patents on Anti-Cancer Drug Discovery* 8, 310–318 (2013).
38. Li, L. *et al.* Human mesenchymal stem cells with adenovirus-mediated TRAIL gene transduction have antitumor effects on esophageal cancer cell line Eca-109. *Acta Biochimica Et Biophysica Sinica* 46, 471–476 (2014).
39. Lee, H.J. *et al.* A Therapeutic strategy for metastatic malignant fibrous histiocytoma through mesenchymal stromal cell-mediated TRAIL production. *Annals of Surgery* 257, 952–960 (2013).
40. Zhang, B. *et al.* The inhibitory effect of MSCs expressing TRAIL as a cellular delivery vehicle in combination with cisplatin on hepatocellular carcinoma. *Cancer Biology & Therapy* 13, 1175–1184 (2012).

41. Nesterenko, I., Wanningen, S., Bagci-Onder, T., Anderegg, M., and Shah, K. Evaluating the effect of therapeutic stem cells on TRAIL resistant and sensitive medulloblastomas. *Plos One* 7 (2012).

42. Karshieva, S.S., Krasikova, L.S., and Belyavskii, A.V. Mesenchymal stem cells as tool for antitumor therapy. *Molecular Biology* 47, 45–54 (2013).

43. Dembinski, J.L. *et al.* Tumor stroma engraftment of gene-modified mesenchymal stem cells as anti-tumor therapy against ovarian cancer. *Cytotherapy* 15, 20–32 (2013).

44. Drela, K., Siedlecka, P., Sarnowska, A., and Domanska-Janik, K. Human mesenchymal stem cells in the treatment of neurological diseases. *Acta Neurobiologiae Experimentalis* 73, 38–56 (2013).

45. Yang, X., Du, J., Xu, X., Xu, C., and Song, W. IFN-gamma-secreting-mesenchymal stem cells exert an antitumor effect *in vivo* via the TRAIL pathway. *Journal of Immunology Research*, 9 (2014).

46. Martini, M. *et al.* IFN-gamma-mediated upmodulation of MHC class I expression activates tumor-specific immune response in a mouse model of prostate cancer. *Vaccine* 28, 3548–3557 (2010).

47. Hodgkinson, C.P., Gomez, J.A., Mirotsou, M., and Dzau, V.J. Genetic engineering of mesenchymal stem cells and its application in human disease therapy. *Human Gene Therapy* 21, 1513–1526 (2010).

48. Payne, N.L. *et al.* Early intervention with gene-modified mesenchymal stem cells overexpressing interleukin-4 enhances anti-inflammatory responses and functional recovery in experimental autoimmune demyelination. *Cell Adhesion & Migration* 6, 179–189 (2012).

49. Payne, N.L. *et al.* Human adipose-derived mesenchymal stem cells engineered to secrete IL-10 inhibit APC function and limit CNS autoimmunity. *Brain Behavior and Immunity* 30, 103–114 (2013).

50. Young, J.S., Kim, J.W., Ahmed, A.U., and Lesniak, M.S. Therapeutic cell carriers: a potential road to cure glioma. *Expert Review of Neurotherapeutics* 14, 651–660 (2014).

51. Chen, X. *et al.* A tumor-selective biotherapy with prolonged impact on established metastases based on cytokine gene-engineered MSCs. *Molecular Therapy* 16, 749–756 (2008).

52. Zhang, Y.X. *et al.* Gene therapy of ovarian cancer using IL-21-secreting human umbilical cord mesenchymal stem cells in nude mice. *Journal of Ovarian Research* 7, 10 (2014).

53. Becerra, S.P., and Notario, V. The effects of PEDF on cancer biology: mechanisms of action and therapeutic potential. *Nature Reviews Cancer* 13, 258–271 (2013).

54. Li, Q. *et al.* Mesenchymal stem cells from human fat engineered to secrete BMP4 are nononcogenic, suppress brain cancer, and prolong survival. *Clinical Cancer Research* 20, 2375–2387 (2014).

55. Zheng, L. *et al.* Antitumor activities of human placenta-derived mesenchymal stem cells expressing endostatin on ovarian cancer. *Plos One* 7, 10 (2012).

56. Sun, Y.P. *et al.* Effect of NK4 Transduction in bone marrow-derived mesenchymal stem cells on biological characteristics of pancreatic cancer cells. *International Journal of Molecular Sciences* 15, 3729–3745 (2014).

57. Wang, N. *et al.* Autologous bone marrow stromal cells genetically engineered to secrete an IGF-I receptor decoy prevent the growth of liver metastases. *Molecular Therapy* 17, 1241–1249 (2009).

58. Eliopoulos, N. *et al.* Erythropoietin gene-enhanced marrow mesenchymal stromal cells decrease cisplatin-induced kidney injury and improve survival of allogeneic mice. *Molecular Therapy* 19, 2072–2083 (2011).

59. Mok, P.L., Cheong, S.K., Leong, C.F., Chua, K.H., and Ainoon, O. Human mesenchymal stromal cells could deliver erythropoietin and migrate to the basal layer of hair shaft when subcutaneously implanted in a murine model. *Tissue & Cell* 44, 249–256 (2012).

60. Shahrokhi, S., Daneshmandi, S., and Menaa, F. Tumor necrosis factor-alpha/cd40 ligand-engineered mesenchymal stem cells greatly enhanced the antitumor immune response and lifespan in mice. *Human Gene Therapy* 25, 240–253 (2014).

61. Cawthorn, W.P., Scheller, E.L., and MacDougald, O.A. Adipose tissue stem cells: the great WAT hope. *Trends in Endocrinology and Metabolism* 23, 270–277 (2012).

62. Kikuchi, H. *et al.* Therapeutic potential of transgenic mesenchymal stem cells engineered to mediate anti-high mobility group box 1 activity: targeting of colon cancer. *Journal of Surgical Research* 190, 134–143 (2014).

63. Amara, I., Touati, W., Beaune, P., and de Waziers, I. Mesenchymal stem cells as cellular vehicles for prodrug gene therapy against tumors. *Biochimie* 105C, 4–11 (2014).

64. Zhu, X.H. *et al.* Gene therapy of gastric cancer using LIGHT-secreting human umbilical cord blood-derived mesenchymal stem cells. *Gastric Cancer* 16, 155–166 (2013).

65. Rahmati, S., Alijani, N., and Kadivar, M. *In vitro* generation of glucose-responsive insulin producing cells using lentiviral based pdx-1 gene transduction of mouse (C57BL/6) mesenchymal stem cells. *Biochemical and Biophysical Research Communications* 437, 413–419 (2013).

66. Zhong, Y.S. *et al.* Generation of a human bone marrow-derived mesenchymal stem cell line expressing and secreting high levels of bioactive alpha-melanocyte-stimulating hormone. *Journal of Biochemistry* 153, 371–379 (2013).

67. Hu, J.W. *et al.* Infusion of Trx-1-Overexpressing hucMSC prolongs the survival of acutely irradiated NOD/SCID mice by decreasing excessive inflammatory injury. *Plos One* 8, 12 (2013).

68. Knoop, K. *et al.* Stromal targeting of sodium iodide symporter using mesenchymal stem cells allows enhanced imaging and therapy of hepatocellular carcinoma. *Human Gene Therapy* 24, 306–316 (2013).

69. Bak, X.Y. *et al.* Human embryonic stem cell-derived mesenchymal stem cells as cellular delivery vehicles for prodrug gene therapy of glioblastoma. *Human Gene Therapy* 22, 1365–1377 (2011).

70. Miletic, H. *et al.* Bystander killing of malignant glioma by bone marrow-derived tumor-infiltrating progenitor cells expressing a suicide gene. *Molecular Therapy* 15, 1373–1381 (2007).

71. Leng, L. *et al.* Molecular imaging for assessment of mesenchymal stem cells mediated breast cancer therapy. *Biomaterials* 35, 5162–5170 (2014).

72. Kang, N.H. *et al.* Human amniotic fluid-derived stem cells expressing cytosine deaminase and thymidine kinase inhibits the growth of breast cancer cells in cellular and xenograft mouse models. *Cancer Gene Therapy* 19, 412–419 (2012).

73. Jiang, J.Y. *et al.* A preliminary study on the construction of double suicide gene delivery vectors by mesenchymal stem cells and the *in vitro* inhibitory effects on SKOV3 cells. *Oncology Reports* 31, 781–787 (2014).

74. Xia, X. *et al.* Mesenchymal stem cells as carriers and amplifiers in CRAd delivery to tumors. *Molecular Cancer* 10, 12 (2011).

75. Brennen, W.N., Denmeade, S.R., and Isaacs, J.T. Mesenchymal stem cells as a vector for the inflammatory prostate microenvironment. *Endocrine-Related Cancer* 20, R269-R290 (2013).

76. Hsiao, W.C., Sung, S.Y., Liao, C.H., Wu, H.C., and Hsieh, C.L. Vitamin D-3-inducible mesenchymal stem cell-based delivery of conditionally replicating adenoviruses effectively targets renal cell carcinoma and inhibits tumor growth. *Molecular Pharmaceutics* 9, 1396–1408 (2012).

77. Bolontrade, M.F. *et al.* A specific subpopulation of mesenchymal stromal cell carriers overrides melanoma resistance to an oncolytic adenovirus. *Stem Cells and Development* 21, 2689–2702 (2012).

78. Stuckey, D.W., and Shah, K. Stem cell-based therapies for cancer treatment: separating hope from hype. *Nature Reviews Cancer* 14, 683–691 (2014).

79. Huang, X.L. *et al.* Mesenchymal stem cell-based cell engineering with multifunctional mesoporous silica nanoparticles for tumor delivery. *Biomaterials* 34, 1772–1780 (2013).

80. Cao, B., Yang, M., Zhu, Y., Qu, X., and Mao, C. Stem cells loaded with nanoparticles as a drug carrier for in vivo breast cancer therapy. *Advanced Materials* 26, 4627–4631 (2014).

81. Sisto, F. *et al.* Human mesenchymal stromal cells can uptake and release ciprofloxacin, acquiring in vitro anti-bacterial activity. *Cytotherapy* 16, 181–190 (2014).

82. Hu, C.C. *et al.* Human mesenchymal stem cells-like cells as cellular vehicles for delivery of immunotoxin *in vitro*. *Biotechnology Letters* 31, 181–189 (2009).

83. Mariotti, V., Greco, S., Mohan, R., Nahas, G., and Rameshwar, P. Stem cell in alternative treatments for brain tumors: potential for gene delivery. *Molecular and Cellular Therapies* 2, 1–10 (2014).

84. Muller, F.J., Snyder, E.Y., and Loring, J.F. Gene therapy: can neural stem cells deliver? *Nature Reviews Neuroscience* 7, 75–84 (2006).

85. Doetsch, F., Caillé, I., Lim, D.A., García-Verdugo, J.M., and Alvarez-Buylla, A. Subventricular zone astrocytes are neural stem cells in the adult mammalian brain. *Cell* 97, 703–716 (1999).

86. Aboody, K.S. *et al.* Neural stem cells display extensive tropism for pathology in adult brain: Evidence from intracranial gliomas. *Proceedings of the National Academy of Sciences of the United States of America* 97, 12846–12851 (2000).

87. Joo, K.M. *et al.* Human neural stem cells can target and deliver therapeutic genes to breast cancer brain metastases. *Molecular Therapy* 17, 570–575 (2009).

88. Aboody, K.S. *et al.* Development of a tumor-selective approach to treat metastatic cancer. *Plos One* 1 (2006).

89. Kim, S.K. *et al.* Human neural stem cells target experimental intracranial medulloblastoma and deliver a therapeutic gene leading totumor regression. *Clinical Cancer Research* 12, 5550–5556 (2006).

90. Sims, T.L. *et al.* Neural progenitor cell-mediated delivery of osteoprotegerin limits disease progression in a preclinical model of neuroblastoma bone metastasis. *Journal of Pediatric Surgery* 44, 204–211 (2009).

91. Ahmed, A.U. *et al*. A comparative study of neural and mesenchymal stem cell-based carriers for oncolytic adenovirus in a model of malignant glioma. *Molecular Pharmaceutics* 8, 1559–1572 (2011).

92. Altanerova, V. *et al*. Human adipose tissue-derived mesenchymal stem cells expressing yeast cytosinedeaminase: Uracil phosphoribosyltransferase inhibit intracerebral rat glioblastoma. *International Journal of Cancer* 130, 2455–2463 (2012).

93. Yang, J. *et al*. Tumor tropism of intravenously injected human-induced pluripotent stem cell-derived neural stem cells and their gene therapy application in a metastatic breast cancer model. *Stem Cells* 30, 1021–1029 (2012).

94. Sonabend, A.M. *et al*. Mesenchymal stem cells effectively deliver an oncolytic adenovirus to intracranial glioma. *Stem Cells* 26, 831–841 (2008).

95. Ryu, C.H. *et al*. Gene therapy of intracranial glioma using interleukin 12-secreting human umbilical cord blood-derived mesenchymal stem cells. *Human Gene Therapy* 22, 733–743 (2011).

96. Krystkowiak, P. *et al*. Alloimmunisation to donor antigens and immune rejection following foetal neural grafts to the brain in patients with Huntington's Disease. *Plos One* 2 (2007).

97. Ernst, N. *et al*. An improved, standardised protocol for the isolation, enrichment and targeted neural differentiation of Nestin plus progenitors from adult human dermis. *Experimental Dermatology* 19, 549–555 (2010).

98. Mitrecic, D., Gajovic, S., and Pochet, R. Toward the treatments with neural stem cells: experiences from amyotrophic lateral sclerosis. *Anatomical Record-Advances in Integrative Anatomy and Evolutionary Biology* 292, 1962–1967 (2009).

99. Casalbore, P. *et al*. Tumorigenic potential of olfactory bulb-derived human adult neural stem cells associates with activation of TERT and NOTCH1. *Plos One* 4 (2009).

100. Zhao, Y. *et al*. Targeted suicide gene therapy for glioma using human embryonic stem cell-derived neural stem cells genetically modified by baculoviral vectors. *Gene Therapy* 19, 189–200 (2012).

101. Lee, E.X.W. *et al*. Glioma gene therapy using induced pluripotent stem cell derived neural stem cells. *Molecular Pharmaceutics* 8, 1515–1524 (2011).

102. Parihar, A., Eubank, T.D., and Doseff, A.I. Monocytes and macrophages regulate immunity through dynamic networks of survival and cell death. *Journal of Innate Immunity* 2, 204–215 (2010).

103. Doshi, N. *et al*. Cell-based drug delivery devices using phagocytosis-resistant backpacks. *Advanced Materials* 23, H105–H109 (2011).

104. Davies, J., and Gordon, S. Isolation and culture of human macrophages, in *Basic Cell Culture Protocols*, Vol. 290, eds. C. Helgason and C. Miller, 105–116 (Humana Press, 2005).

105. Choi, M.R. *et al*. A cellular Trojan horse for delivery of therapeutic nanoparticles into tumors. *Nano Letters* 7, 3759–3765 (2007).

106. Koppensteiner, H., Brack-Werner, R., and Schindler, M. Macrophages and their relevance in Human Immunodeficiency Virus Type I infection. *Retrovirology* 9 (2012).

107. Feng, S., Cui, S.S., Jin, J., and Gu, Y.Q. Macrophage as cellular vehicles for delivery of nanoparticles. *Journal of Innovative Optical Health Sciences* 7, 7 (2014).

108. Dou, H. *et al*. Development of a macrophage-based nanoparticle platform for antiretroviral drug delivery. *Blood* 108, 2827–2835 (2006).

109. Dou, H.Y. *et al.* Macrophage delivery of nanoformulated antiretroviral drug to the brain in a murine model of NeuroAIDS. *Journal of Immunology* 183, 661–669 (2009).

110. Kelly, P.M.A., Davison, R.S., Bliss, E., and McGee, J.O. Macrophages in human-breast disease — a quantitative immunohistochemical study. *British Journal of Cancer* 57, 174–177 (1988).

111. Tao, Y.H., Ning, M.M., and Dou, H.Y. A novel therapeutic system for malignant glioma: nanoformulation, pharmacokinetic, and anticancer properties of cell-nano-drug delivery. *Nanomedicine-Nanotechnology Biology and Medicine* 9, 222–232 (2013).

112. Hwang, S.-A., and Actor, J.K. in eLS (John Wiley & Sons, Ltd, 2001).

113. Spits, H. *et al.* Innate lymphoid cells — a proposal for uniform nomenclature. *Nature Reviews Immunology* 13, 145–149 (2013).

114. Janeway, C.A., and Bottomly, K. Signals and signs for lymphocyte-responses. *Cell* 76, 275–285 (1994).

115. June, C.H., Blazar, B.R., and Riley, J.L. Engineering lymphocyte subsets: tools, trials and tribulations. *Nature Reviews Immunology* 9, 704–716 (2009).

116. Dudley, M.E., and Rosenberg, S.A. Adoptive cell transfer therapy. *Seminars in Oncology* 34, 524–531 (2007).

117. Kottke, T. *et al.* The perforin-dependent immunological synapse allows T cell activation-dependent tumor targeting by MLV vector particles. *Gene Therapy* 13, 1166–1177 (2006).

118. Lamers, C.H.J. *et al.* Treatment of metastatic renal cell carcinoma with CAIX CAR-engineered T cells: clinical evaluation and management of on-target toxicity. *Molecular Therapy* 21, 904–912 (2013).

119. Lanitis, E. *et al.* Redirected antitumor activity of primary human lymphocytes trans-duced with a fully human anti-mesothelin chimeric receptor. *Molecular Therapy* 20, 633–643 (2012).

120. Lieberman, J. The ABCs of granule-mediated cytotoxicity: new weapons in the arsenal. *Nature Reviews Immunology* 3, 361–370 (2003).

121. Bots, M. *et al.* SPI-CI and SPI-6 cooperate in the protection from effector cell-mediated cytotoxicity. *Blood* 105, 1153–1161 (2005).

122. Chester, J. *et al.* Tumor antigen-specific induction of transcriptionally targeted retrovi-ral vectors from chimeric immune receptor-modified T cells. *Nature Biotechnology* 20, 256–263 (2002).

123. Thorne, S.H., and Contag, C.H. Combining immune cell and viral therapy for the treatment of cancer. *Cellular and Molecular Life Sciences* 64, 1449–1451 (2007).

124. Sumpio, B.E., Riley, J.T., and Dardik, A. Cells in focus: endothelial cell. *International Journal of Biochemistry & Cell Biology* 34, 1508–1512 (2002).

125. Alberts, B., Johnson, A., and Lewis, J. in Molecular Biology of the Cell, Edn. 4th (Garland Science, New York; 2002).

126. Dudek, A.Z. Endothelial lineage cell as a vehicle for systemic delivery of cancer gene therapy. *Translational Research* 156, 136–146 (2010).

127. Lucas, T. *et al.* Adenoviral-mediated endothelial precursor cell delivery of soluble cd115 suppresses human prostate cancer xenograft growth in mice. *Stem Cells* 27, 2342–2352 (2009).

128. Muta, M. *et al.* Impact of vasculogenesis on solid tumor growth in a rat model. *Oncology Reports* 10, 1213–1218 (2003).

129. Rancourt, C. *et al.* Endothelial cell vehicles for delivery of cytotoxic genes as a gene therapy approach for carcinoma of the ovary. *Clinical Cancer Research* 4, 265–270 (1998).

130. Wei, J., Jarmy, G., Genuneit, J., Debatin, K.M., and Beltinger, C. Human blood late outgrowth endothelial cells for gene therapy of cancer: determinants of efficacy. *Gene Therapy* 14, 344–356 (2007).

131. Hamanishi, J. *et al.* Activated local immunity by CC chemokine ligand 19-transduced embryonic endothelial progenitor cells suppresses metastasis of murine ovarian cancer. *Stem Cells* 28, 164–173 (2010).

132. Wei, J.W. *et al.* Embryonic endothelial progenitor cells armed with a suicide gene target hypoxic lung metastases after intravenous delivery. *Cancer Cell* 5, 477–488 (2004).

133. Tura, O. *et al.* Late outgrowth endothelial cells resemble mature endothelial cells and are not derived from bone marrow. *Stem Cells* 31, 338–348 (2013).

134. Hamada, K. *et al.* Carrier cell-mediated delivery of a replication-competent adenovirus for cancer gene therapy. *Molecular Therapy* 15, 1121–1128 (2007).

135. Iguchi, K. *et al.* Efficient antitumor effects of carrier cells loaded with a fiber-substituted conditionally replicating adenovirus on CAR-negative tumor cells. *Cancer Gene Therapy* 19, 118–125 (2012).

136. Zhang, T. *et al.* Gene therapy for oral squamous cell carcinoma with IAI.3B promoter-driven oncolytic adenovirus-infected carrier cells. *Oncology Reports* 25, 795–802 (2011).

137. Power, A.T. *et al.* Carrier cell-based delivery of an oncolytic virus circumvents antiviral immunity. *Molecular Therapy* 15, 123–130 (2007).

138. Harrington, K. *et al.* Cells as vehicles for cancer gene therapy: The missing link between targeted vectors and systemic delivery? *Human Gene Therapy* 13, 1263–1280 (2002).

139. Gothoskar, A.V. Resealed erythrocytes: a review. *Pharmaceutical Technology*, 140–158 (2004).

140. Hamidi, M., and Tajerzadeh, H. Carrier erythrocytes: an overview. *Drug Delivery* 10, 9–20 (2003).

141. Patel, P.D., Dand, N., Hirlekar, R.S., and Kadam, V.J. Drug loaded erythrocytes: as novel drug delivery system. *Current Pharmaceutical Design* 14, 63–70 (2008).

142. Hamidi, M., Zarrin, A., Foroozesh, M., and Mohammadi-Samani, S. Applications of carrier erythrocytes in delivery of biopharmaceuticals. *Journal of Controlled Release* 118, 145–160 (2007).

143. Banker, G.S., and Rhodes, C.T. Modern Pharmaceutics, 4th Edition. (New York; 2002).

144. Ihler, G.M., Glew, R.H., and Schnure, F.W. Enzyme loading of erythrocytes. *Proceedings of the National Academy of Sciences of the United States of America* 70, 2663–2666 (1973).

145. Zimmerman, U. Jahresbericht der kernforschungsanlage Jülich GmbH. *Nuclear Research Center Julich*, 55–58 (1973).

146. Rossi, L. *et al.* Erythrocyte-based drug delivery. *Expert Opinion on Drug Delivery* 2, 311–322 (2005).

147. Anselmo, A.C. *et al.* Delivering nanoparticles to lungs while avoiding liver and spleen through adsorption on red blood cells. *Acs Nano* 7, 11129–11137 (2013).

148. Fraternale, A. *et al.* Macrophage protection by addition of glutathione (GSH)-loaded erythrocytes to AZT and DDI in a murine AIDS model. *Antiviral Research* 56, 263–272 (2002).

149. Zimmerman, U. in Targeted drugs. (ed. E.P. Goldberg) 153–200 (John Wiley & Sons, New York, New York; 1983).

150. Price, R.J., Skyba, D.M., Kaul, S., and Skalak, T.C. Delivery of colloidal, particles and red blood cells to tissue through microvessel ruptures created by targeted microbubble destruction with ultrasound. *Circulation* 98, 1264–1267 (1998).
151. Mukthavaram, R., Shi, G.X., Kesari, S., and Simberg, D. Targeting and depletion of circulating leukocytes and cancer cells by lipophilic antibody-modified erythrocytes. *Journal of Controlled Release* 183, 146–153 (2014).
152. Pragya, V.R. Resealed erythrocytes: a promising drug carrier. *International Journal of Pharmacy and Pharmaceutical Sciences* 4, 75–82 (2012).
153. Yuan, S.H., Ge, W.H., Huo, J., and Wang, X.H. Slow release properties and liver-targeting characteristics of methotrexate erythrocyte carriers. *Fundamental & Clinical Pharmacology* 23, 189–196 (2009).
154. Ravindra, K. *et al.* Encapsulated erythrocytes for novel drug delivery system. *Journal of Drug Delivery & Therapeutics* 2, 61–67 (2012).
155. Hamidi, M., Rafiei, P., Azadi, A., and Mohammadi-Samani, S. Encapsulation of valproate-loaded hydrogel nanoparticles in intact human erythrocytes: a novel nano-cell composite for drug delivery. *Journal of Pharmaceutical Sciences* 100, 1702–1711 (2011).
156. Chambers, E., and Mitragotri, S. Prolonged circulation of large polymeric nanoparticles by non-covalent adsorption on erythrocytes. *Journal of Controlled Release* 100, 111–119 (2004).
157. Chambers, E., and Mitragotri, S. Long circulating nanoparticles via adhesion on red blood cells: Mechanism and extended circulation. *Experimental Biology and Medicine* 232, 958–966 (2007).
158. Marczak, A., and Bukowska, B. ROS production and their influence on the cellular antioxidative system in human erythrocytes incubated with daunorubicin and glutaraldehyde. *Environmental Toxicology and Pharmacology* 36, 171–181 (2013).
159. Thery, C., Ostrowski, M., and Segura, E. Membrane vesicles as conveyors of immune responses. *Nature Reviews Immunology* 9, 581–593 (2009).
160. O'Loughlin, A.J., Woffindale, C.A., and Wood, M.J.A. Exosomes and the emerging field of exosome-based gene therapy. *Current Gene Therapy* 12, 262–274 (2012).
161. Jang, S.C. *et al.* Bioinspired exosome-mimetic nanovesicles for targeted delivery of chemotherapeutics to malignant tumors. *ACS Nano* 7, 7698–7710 (2013).
162. Jeong, D. *et al.* Nanovesicles engineered from ES cells for enhanced cell proliferation. *Biomaterials* 35, 9302–9310 (2014).
163. Jo, W. *et al.* Microfluidic fabrication of cell-derived nanovesicles as endogenous RNA carriers. *Lab on a Chip* 14, 1261–1269 (2014).
164. Jo, W. *et al.* Large-scale generation of cell-derived nanovesicles. *Nanoscale* 6, 12056–12064 (2014).
165. Hood, J.L., Scott, M.J., and Wickline, S.A. Maximizing exosome colloidal stability following electroporation. *Analytical Biochemistry* 448, 41–49 (2014).
166. Furman, N.E.T. *et al.* Reconstructed stem cell nanoghosts: a natural tumor targeting platform. *Nano Letters* 13, 3248–3255 (2013).
167. Bronshtein, T., Toledano, N., Danino, D., Pollack, S., and Machluf, M. Cell derived liposomes expressing CCR5 as a new targeted drug-delivery system for HIV infected cells. *Journal of Controlled Release* 151, 139–148 (2011).

168. Lejeune, A. *et al.* Nanoerythrosome, a new derivative of erythrocyte ghost — preparation and antineoplastic potential as drug carrier for daunorubicin. *Anticancer Research* 14, 915–919 (1994).
169. Desilets, J., Lejeune, A., Mercer, J., and Gicquaud, C. Nanoerythrosomes, a new derivative of erythrocyte ghost: IV. Fate of reinjected nanoerythrosomes. *Anticancer Research* 21, 1741–1747 (2001).
170. Lejeune, A., Poyet, P., Gaudreault, R.C., and Gicquaud, C. Nanoerythrosomes, a new derivative of erythrocyte ghost: III. Is phagocytosis involved in the mechanism of action? *Anticancer Research* 17, 3599–3603 (1997).
171. Moorjani, M. *et al.* Nanoerythrosomes, a new derivative of erythrocyte ghost: II. Identification of the mechanism of action. *Anticancer Research* 16, 2831–2836 (1996).
172. Pouliot, R. *et al.* Spectroscopic characterization of nanoerythrosomes in the absence and presence of conjugated polyethyleneglycols: an FTIR and P-31-NMR study. *Biochimica Et Biophysica Acta-Biomembranes* 1564, 317–324 (2002).
173. Alachi, A., and Greenwood, R. Human insulin binding to erythrocyte-membrane. *Drug Development and Industrial Pharmacy* 19, 673–684 (1993).
174. Alachi, A., and Greenwood, R. Intraduodenal administration of biocarrier-insulin systems. *Drug Development and Industrial Pharmacy* 19, 1303–1315 (1993).
175. Al-Achi, A., and Greenwood, R. Erythrocytes as oral delivery systems for human insulin. *Drug Development and Industrial Pharmacy* 24, 67–72 (1998).
176. Alachi, A., and Boroujerdi, M. Pharmacokinetics and tissue uptake of doxorubicin associated with erythrocyte-membrane — erythrocyte-ghosts vs erythrocyte-vesicles. *Drug Development and Industrial Pharmacy* 16, 2199–2219 (1990).
177. Jain, S., and Jain, N.K. (Plenum Publishers, New York, New York; 2003).
178. Agnihotri, J., and Jain, N.K. Biodegradable long circulating cellular carrier for antimalarial drug pyrimethamine. *Artificial Cells Nanomedicine and Biotechnology* 41, 309–314 (2013).
179. Agnihotri, J., Gajbhiye, V., and Jain, N. Engineered cellular carrier nanoerythrosomes as potential targeting vectors for anti-malarial drug. *Asian Journal of Pharmaceutics* 4, 116–120 (2010).
180. Bahmani, B., Bacon, D., and Anvari, B. Erythrocyte-derived photo-theranostic agents: hybrid nano-vesicles containing indocyanine green for near infrared imaging and therapeutic applications. *Scientific Reports* 3, 7 (2013).
181. Gupta, N., Patel, B., and Ahsan, F. Nano-engineered erythrocyte ghosts as inhalational carriers for delivery of fasudil: preparation and characterization. *Pharmaceutical Research* 31, 1553–1565 (2014).
182. Aryal, S. *et al.* Erythrocyte membrane-cloaked polymeric nanoparticles for controlled drug loading and release. *Nanomedicine* 8, 1271–1280 (2013).
183. Fang, R.H., Hu, C.M.J., and Zhang, L.F. Nanoparticles disguised as red blood cells to evade the immune system. *Expert Opinion on Biological Therapy* 12, 385–389 (2012).
184. Fang, R.N.H. *et al.* Lipid-insertion enables targeting functionalization of erythrocyte membrane-cloaked nanoparticles. *Nanoscale* 5, 8884–8888 (2013).
185. Gao, W.W., Hu, C.M.J., Fang, R.H., and Zhang, L.F. Liposome-like nanostructures for drug delivery. *Journal of Materials Chemistry B* 1, 6569–6585 (2013).
186. Gao, W.W. *et al.* Surface functionalization of gold nanoparticles with red blood cell membranes. *Advanced Materials* 25, 3549–3553 (2013).

187. Hu, C.M.J. *et al.* Erythrocyte membrane-camouflaged polymeric nanoparticles as a biomimetic delivery platform. *Proceedings of the National Academy of Sciences of the United States of America* 108, 10980–10985 (2011).
188. Hu, C.M.J., Fang, R.H., and Zhang, L.F. Erythrocyte-inspired delivery systems. *Advanced Healthcare Materials* 1, 537–547 (2012).
189. Hu, C.M.J., Fang, R., and Zhang, L.F. Red blood cell membrane coated polymeric nanoparticles p p. *Abstracts of Papers of the American Chemical Society* 243, 1 (2012).
190. Hu, C.M.J., Fang, R.H., Copp, J., Luk, B.T., and Zhang, L.F. A biomimetic nanosponge that absorbs pore-forming toxins. *Nature Nanotechnology* 8, 336–340 (2013).
191. Hu, C.M.J. *et al.* 'Marker-of-self' functionalization of nanoscale particles through a top-down cellular membrane coating approach. *Nanoscale* 5, 2664–2668 (2013).
192. Hu, C.M., Fang, R., Luk, B., and Zhang, L.F. Synthetic colloidal particles cloaked by natural RBC membranes as biomimetic delivery vehicles. *Abstracts of Papers of the American Chemical Society* 245, 1 (2013).
193. Hu, C.M.J., Fang, R.H., Luk, B.T., and Zhang, L.F. Polymeric nanotherapeutics: clinical development and advances in stealth functionalization strategies. *Nanoscale* 6, 65–75 (2014).
194. Luk, B.T. *et al.* Interfacial interactions between natural RBC membranes and synthetic polymeric nanoparticles. *Nanoscale* 6, 2730–2737 (2014).
195. Hu, C.M.J., Fang, R.H., Luk, B.T., and Zhang, L.F. Nanoparticle-detained toxins for safe and effective vaccination. *Nature Nanotechnology* 8, 933–938 (2013).
196. Piao, J.-G. *et al.* Erythrocyte membrane is an alternative coating to polyethylene glycol for prolonging the circulation lifetime of gold nanocages for photothermal therapy. *Acs Nano* 8, 10414–10425 (2014).
197. Fan, Z., Zhou, H., Li, P.Y., Speer, J.E., and Cheng, H. Structural elucidation of cell membrane-derived nanoparticles using molecular probes. *Journal of Materials Chemistry B* 2, 8231–8238 (2014).
198. Hu, C.M.J., and Zhang, L.F. Nanotoxoid vaccines. *Nano Today* 9, 401–404 (2014).
199. Li, L.L. *et al.* Core-shell supramolecular gelatin nanoparticles for adaptive and "on-demand" antibiotic delivery. *Acs Nano* 8, 4975–4983 (2014).
200. Alachi, A., and Boroujerdi, M. Adsorption-isotherm for doxorubicin on erythrocyte-membrane. *Drug Development and Industrial Pharmacy* 16, 1325–1338 (1990).
201. Parodi, A. *et al.* Synthetic nanoparticles functionalized with biomimetic leukocyte membranes possess cell-like functions. *Nature Nanotechnology* 8, 61–68 (2013).
202. Fang, R.H. *et al.* Cancer cell membrane-coated nanoparticles for anticancer vaccination and drug delivery. *Nano Letters* 14, 2181–2188 (2014).
203. Pasut, G., and Veronese, F.M. State of the art in PEGylation: the great versatility achieved after forty years of research. *Journal of Controlled Release* 161, 461–472 (2012).
204. Bottermann, P., Gyaram, H., Wahl, K., Ermler, R., and Lebender, A. Pharmacokinetics of biosynthetic human insulin and characteristics of its effect. *Diabetes Care* 4, 168–169 (1981).
205. Salzman, R. *et al.* Intranasal aerosolized insulin — mixed-meal studies and long-term use in type-I diabetes. *New England Journal of Medicine* 312, 1078–1084 (1985).
206. Schilling, R.J., and Mitra, A.K. Intestinal mucosal transport of insulin. *International Journal of Pharmaceutics* 62, 53–64 (1990).

207. Yannuzzi, L.A. Indocyanine green angiography: a perspective on use in the clinical setting. *American Journal of Ophthalmology* 151, 745–751 (2011).
208. Caesar, J., Shaldon, S., Chiandussi, L., Guevara, L., and Sherlock, S. The use of indocyanine green in the measurement of hepatic blood flow and as a test of hepatic function. *Clin Sci* 21, 43–57 (1961).
209. Fox, I.J., and Wood, E.H. Applications of dilution curves recorded from the right side of the heart or venous circulation with the aid of a new indicator dye. *Proc Staff Meet Mayo Clin* 32, 541–550 (1957).
210. van der Vorst, J.R. *et al.* Dose optimization for near-infrared fluorescence sentinel lymph node mapping in patients with melanoma. *British Journal of Dermatology* 168, 93–98 (2013).
211. Jeschke, S. *et al.* Visualisation of the lymph node pathway in real time by laparoscopic radioisotope- and fluorescence-guided sentinel lymph node dissection in prostate cancer staging. *Urology* 80, 1080–1086 (2012).
212. Sevick-Muraca, E.M. *et al.* Imaging of lymph flow in breast cancer patients after microdose administration of a near-infrared flurophore: Feasibility study. *Radiology* 246, 734–741 (2008).
213. Crane, L.M.A. *et al.* Intraoperative near-infrared fluorescence imaging for sentinel lymph node detection in vulvar cancer: First clinical results. *Gynecologic Oncology* 120, 291–295 (2011).
214. Smretschnig, E. *et al.* Half-fluence photodynamic therapy in chronic central serous chorioretinopathy. *Retina-the Journal of Retinal and Vitreous Diseases* 33, 316–323 (2013).
215. Yannuzzi, L.A. *et al.* Indocyanine green angiography-guided photodynamic therapy for treatment of chronic central serous chorioretinopathy — A pilot study. *Retina-the Journal of Retinal and Vitreous Diseases* 23, 288–298 (2003).
216. Liggett, P.E., Lavaque, A.J., Chaudhry, N.A., Jablon, E.P., and Quiroz-Mercado, H. Preliminary results of combined simultaneous transpupillary thermotherapy and ICG-based photodynamic therapy for choroidal melanoma. *Ophthalmic Surgery Lasers & Imaging* 36, 463–470 (2005).
217. Costa, R.A. *et al.* Indocyanine green-mediated photothrombosis as a new technique of treatment for persistent central serous chorioretinopathy. *Current Eye Research* 25, 287–297 (2002).
218. Klein, A. *et al.* Indocyanine green-augmented diode laser treatment of port-wine stains: clinical and histological evidence for a new treatment option from a randomized controlled trial. *British Journal of Dermatology* 167, 333–342 (2012).
219. Yaseen, M.A., Yu, J., Wong, M.S., and Anvari, B. *In-vivo* fluorescence imaging of mammalian organs using charge-assembled mesocapsule constructs containing indocyanine green. *Optics Express* 16, 20577–20587 (2008).
220. Schwenke, D.O. *et al.* Role of Rho-kinase signaling and endothelial dysfunction in modulating blood flow distribution in pulmonary hypertension. *Journal of Applied Physiology* 110, 901–908 (2011).
221. Gupta, N., Patel, B., Nahar, K., and Ahsan, F. Cell permeable peptide conjugated nanoerythrosomes of fasudil prolong pulmonary arterial vasodilation in PAH rats. *Eur J Pharm Biopharm* 88, 1046–1055 (2014).

222. Urakami, T. *et al.* Peptide-directed highly selective targeting of pulmonary arterial hypertension. *American Journal of Pathology* 178, 2489–2495 (2011).

223. Toba, M. *et al.* A novel vascular homing peptide strategy to selectively enhance pulmonary drug efficacy in pulmonary arterial hypertension. *American Journal of Pathology* 184, 369–375 (2014).

224. Collins, B.E., and Paulson, J.C. Cell surface biology mediated by low affinity multivalent protein-glycan interactions. *Current Opinion in Chemical Biology* 8, 617–625 (2004).

225. Hochmuth, R.M., Evans, E.A., Wiles, H.C., and McCown, J.T. Mechanical measurement of red-cell membrane thickness. *Science* 220, 101–102 (1983).

226. Pan, X., and Lee, R.J. Tumour-selective drug delivery via folate receptor-targeted liposomes. *Expert Opinion on Drug Delivery* 1, 7–17 (2004).

227. Byrne, J.D., Betancourt, T., and Brannon-Peppas, L. Active targeting schemes for nanoparticle systems in cancer therapeutics. *Advanced Drug Delivery Reviews* 60, 1615–1626 (2008).

228. Rosado, C.J. *et al.* The MACPF/CDC family of pore-forming toxins. *Cellular Microbiology* 10, 1765–1774 (2008).

229. Shoham, M. Antivirulence agents against MRSA. *Future Medicinal Chemistry* 3, 775–777 (2011).

230. Kitchin, N.R.E. Review of diphtheria, tetanus and pertussis vaccines in clinical development. *Expert Review of Vaccines* 10, 605–615 (2011).

231. Copp, J.A. *et al.* Clearance of pathological antibodies using biomimetic nanoparticles. *Proceedings of the National Academy of Sciences of the United States of America* 111, 13481–13486 (2014).

232. Jacobson, D.L., Gange, S.J., Rose, N.R., and Graham, N.M.H. Epidemiology and estimated population burden of selected autoimmune diseases in the United States. *Clinical Immunology and Immunopathology* 84, 223–243 (1997).

233. Tabas, I., and Glass, C.K. Anti-inflammatory therapy in chronic disease: challenges and opportunities. *Science* 339, 166–172 (2013).

234. Lechner, K., and Jaeger, U. How I treat autoimmune hemolytic anemias in adults. *Blood* 116, 1831–1838 (2010).

235. Petz, L.D. A physician's guide to transfusion in autoimmune haemolytic anaemia. *British Journal of Haematology* 124, 712–716 (2004).

236. Kyaw, M.H. *et al.* Evaluation of severe infection and survival after splenectomy. *American Journal of Medicine* 119 (2006).

237. Ahrens, N., Pruss, A., Kahne, A., Kiesewetter, H., and Salama, A. Coexistence of autoantibodies and alloantibodies to red blood cells due to blood transfusion. *Transfusion* 47, 813–816 (2007).

238. Salama, A., Berghofer, H., and Muellereckhardt, C. Red-blood-cell transfusion in warm-type autoimmune hemolytic-anemia. *Lancet* 340, 1515–1517 (1992).

239. Shander, A., Cappellini, M.D., and Goodnough, L.T. Iron overload and toxicity: the hidden risk of multiple blood transfusions. *Vox Sanguinis* 97, 185–197 (2009).

240. Xu, J.-H. *et al.* Supramolecular gelatin nanoparticles as matrix metalloproteinase responsive cancer cell imaging probes. *Chemical Communications* 49, 4462–4464 (2013).

241. Forsyth, P.A. *et al.* Gelatinase-A (MMP-2), gelatinase-B (MMP-9) and membrane type matrix metalloproteinase-1 (MT1-MMP) are involved in different aspects of the pathophysiology of malignant gliomas. *British Journal of Cancer* 79, 1828–1835 (1999).

242. Svaasand, L.O., Gomer, C.J., and Morinelli, E. On the physical rationale of laser induced hyperthermia. *Lasers in Medical Science* 5, 121–128 (1990).

243. Landry, J., Samson, S., and Chretien, P. Hyperthermia-induced cell-death, thermo-tolerance, and heat-shock proteins in normal, respiration-deficient, and glycolysis-deficient Chinese-hamster cells. *Cancer Research* 46, 324–327 (1986).

244. Schwartzentruber, D.J. *et al.* gp100 peptide vaccine and interleukin-2 in patients with advanced melanoma. *New England Journal of Medicine* 364, 2119–2127 (2011).

245. Glinsky, V.V. *et al.* Intravascular metastatic cancer cell homotypic aggregation at the sites of primary attachment to the endothelium. *Cancer Research* 63, 3805–3811 (2003).

246. Khaldoyanidi, S.K. *et al.* MDA-MB-435 human breast carcinoma cell homo- and het-erotypic adhesion under flow conditions is mediated in part by Thomsen-Friedenreich antigen-galectin-3 interactions. *Journal of Biological Chemistry* 278, 4127–4134 (2003).

247. Doranz, B.J. *et al.* A dual-tropic primary HIV-1 isolate that uses fusin and the beta-chemokine receptors CKR-5, CKR-3, and CKR-2b as fusion cofactors. *Cell* 85, 1149–1158 (1996).

248. Rambaut, A., Posada, D., Crandall, K.A., and Holmes, E.C. The causes and conse-quences of HIV evolution. *Nature Reviews Genetics* 5, 52–61 (2004).

249. Sierra, S., Kupfer, B., and Kaiser, R. Basics of the virology of HIV-1 and its replication. *Journal of Clinical Virology* 34, 233–244 (2005).

250. Mirzabekov, T., Kontos, H., Farzan, M., Marasco, W., and Sodroski, J. Paramagnetic proteoliposomes containing a pure, native, and oriented seven-transmembrane segment protein, CCR5. *Nature Biotechnology* 18, 649–654 (2000).

251. Lentz, B.R. PEG as a tool to gain insight into membrane fusion. *European Biophysics Journal with Biophysics Letters* 36, 315–326 (2007).

252. Patil, S.D., Rhodes, D.G., and Burgess, D.J. Anionic liposomal delivery system for DNA transfection. *The AAPS journal* 6, e29-e29 (2004).

253. Hall, B., Andreeff, M., and Marini, F. The participation of mesenchymal stem cells in tumor stroma formation and their application as targeted-gene delivery vehicles. *Handb Exp Pharmacol*, 263–283 (2007).

254. Kim, T.H. *et al.* PEGylated TNF-related apoptosis-inducing ligand (TRAIL)-loaded sustained release PLGA microspheres for enhanced stability and antitumor activity. *Journal of Controlled Release* 150, 63–69 (2011).

255. Dadashzadeh, S., Mirahmadi, N., Babaei, M.H., and Vali, A.M. Peritoneal retention of liposomes: effects of lipid composition, PEG coating and liposome charge. *Journal of Controlled Release* 148, 177–186 (2010).

256. Shah, K., Tung, C.H., Breakefield, X.O., and Weissleder, R. *In vivo* imaging of S-TRAIL-mediated tumor regression and apoptosis. *Molecular Therapy* 11, 926–931 (2005).

257. Nel, A., Xia, T., Madler, L., and Li, N. Toxic potential of materials at the nanolevel. *Science* 311, 622–627 (2006).

258. Yang, F. *et al.* Liposome based delivery systems in pancreatic cancer treatment: From bench to bedside. *Cancer Treatment Reviews* 37, 633–642 (2011).

259. Guo, L. *et al.* TRAIL and doxorubicin combination enhances anti-glioblastoma effect based on passive tumor targeting of liposomes. *Journal of Controlled Release* 154, 93–102 (2011).

260. Onn, A. *et al.* Development of an orthotopic model to study the biology and therapy of primary human lung cancer in nude mice. *Clinical Cancer Research* 9, 5532–5539 (2003).
261. Eberhardt, W., Pottgen, C., and Stuschke, M. Chemoradiation paradigm for the treatment of lung cancer. *Nature Clinical Practice Oncology* 3, 188–199 (2006).

Nanoparticles for Improved Topical Drug Delivery for Skin Diseases

9

Razina Z. Seeni*, *Sangeetha Krishnamurthy** and *Juliana M. Chan*[†]

*School of Chemical and Biomedical Engineering,
Nanyang Technological University, Singapore 637457
[†] *School of Chemical and Biomedical Engineering and Lee Kong Chian*
School of Medicine, Nanyang Technological University, Singapore 637457

Abstract

Due to the ease of application and lower risk of systemic side effects, topical drug delivery is sometimes preferred to oral and parenteral delivery for the treatment of skin disease. However, the stratum corneum (SC) layer of the skin acts as a physical barrier to drug uptake. In addition, local adverse effects that result from the use of drugs such as topical corticosteroids may limit the long-term viability of topical treatments. In recent years, nanoparticle-based formulations have been gaining interest among researchers for beneficial properties such as site-specific targeting, controlled release and increased skin permeability. In this chapter, we discuss the use of nanoparticle-based drug delivery for the treatment of four common skin conditions: acne vulgaris, atopic dermatitis, psoriasis and fungal infections. We also discuss some of the challenges that need to be overcome for the successful application of nanotechnology in the clinic.

9.1 Introduction

In 1959, Nobel Laureate Richard Feynman predicted that we would one day be able to "swallow the doctor" to heal ailments from deep within our bodies.[1] In recent decades, the emergence of nanotechnology in medicine has brought with it many promising applications in disease diagnostics and therapeutics. Nanotechnology encompasses the use of nanoparticles with at least one dimension in the size range of 100 to 1,000 nanometers.[2] Today, a number of commercially-available products such as food, drugs and cosmetics already contain nanoparticles as additives, carriers and active agents.

Nanomedicine, a hybrid term coined from the use of nanotechnology in medicine, is a field that has gained significant interest among researchers in the translational research field. Due to a high surface-to-volume ratio derived from their small sizes, nanoparticles exhibit unique physicochemical properties that overcome the limitations of conventional therapies. Favorable characteristics of nanoparticle-based drug delivery systems include higher drug encapsulation, targeted action to specific sites, sustained drug release, and improved skin penetration for topical treatments. By reducing the overall drug dose and side effects, these properties may also help to increase overall patient compliance, making nanoparticles a very attractive drug delivery system for a wide number of clinical applications.[3]

In particular, dermatology is one such clinical field that may benefit richly from the unique properties of nanoparticle drug carriers. In the treatment of skin diseases, topical therapy offers several key advantages over conventional intravenous and oral delivery, such as accessibility, minimization of pain and avoidance of first metabolism reactions of the body.[4] A significant challenge to topical delivery, however, is the stratum corneum (SC) layer of the skin. While the SC provides a beneficial protective barrier against microbial and physical influences, the diffusion of drug molecules through the lipid bilayers of the SC remains a challenge despite the large surface area of the skin. In addition, treatment of inflammatory skin diseases requires active targeting of specific cell types which are present in the deeper skin layers.

Fortunately, nanomedicine researchers have discovered that nanoparticle properties such as size and lipophilicity can significantly influence the degree of skin penetration.[5] For these reasons, nanoparticles that are capable of solubilizing poorly water-soluble drugs and improving bioavailability are favored in topical skin treatments. In this chapter, we provide a detailed overview of the research taking place in nanomedicine to improve the topical uptake of drugs. Specifically, we highlight pre-clinical and clinical studies that have been carried out for the treatment of four skin conditions: acne vulgaris, atopic dermatitis, psoriasis and fungal infections.

9.2 The Anatomy of the Skin

While researchers have recognized the potential of using intact skin as the port of drug delivery to the human body, the skin poses a formidable barrier to the ingress of materials and only small quantities of a drug are able to penetrate over a period of time. In general, the desired outcome of topical delivery is to achieve penetration through the skin and retention of drugs in the skin layers without entering systemic circulation,[6] although in some cases systemic delivery may also be desired. Thus, in designing a suitable drug delivery system for topical application, one must first understand the anatomy of skin and its implications on the choice of drug and method of delivery.

As the largest organ in the human body, making up approximately $2\,m^2$ of surface area and 10% of body mass, the skin plays a vital role in protecting the human body against pathogens by creating a physical barrier from the external environment. Skin is made up of three basic layers: the epidermis, the dermis and subcutaneous tissue. The outermost stratified epidermis layer is mainly composed of keratinocytes and consists of five layers: SC, stratum lucidum, stratum granulosum, stratum spinosum and stratum basale. In addition, several appendages are present within these layers: hair follicles, sweat glands and nails. Other key cells present within the epidermis include melanocytes, Merkel cells and Langerhans cells, which play vital roles in the pigmentation, sensory and immune functions of the skin, respectively.[7]

The underlying dermis layer is made up of connective tissue and serves as a supporting matrix. It contains polysaccharides and proteins such as glycosamino-glycan and proteoglycan macromolecules that help to maintain skin hydration.[8] Aside from its remarkable capacity to retain water, the dermis also contains protein fibers such as collagen and elastin, which give skin its tensile strength, elasticity and pliability. Rich in blood capillaries, the dermis layer is highly vascularized with two intercommunicating plexuses: the superficial plexus and lower plexus. The lower plexus, which is located at the dermal-subcutaneous interface, is supplied by large blood vessels.[9] Beneath the dermis, the subcutaneous fat layer provides support for the epidermis and dermis and acts as a storage for fats and lipids.

9.2.1 *The Stratum Corneum and its Barrier Function*

The SC (Latin for "horny layer") represents the main resistance to skin permeation by preventing the passage of molecules larger than 500 kDa through the skin.[6] The epidermal barrier regulates transepidermal water loss and provides a selective permeability to exogenous and endogenous substances. Usually 15–20 μm thick, the SC has a peculiar structure commonly known as "bricks and mortar". It consists of corneocytes (bricks), which are surrounded by an extracellular milieu of lipids (mortar) such as ceramides, free fatty acids and cholesterol. The division of keratinocytes in the basal layer of the epidermis forms the spinous layer — the cells further differentiate into corneocytes and progress outwards forming the granular layer and the SC. In normal healthy skin, the progression of cells from the basal layer to the SC takes approximately 30 days. In diseased skin conditions such as psoriasis, however, the process is significantly accelerated, leading to a thicker SC layer.[7]

The multiple lamellar bilayers of corneocytes and lipids in the SC prevent excessive water loss and regulate water flux to maintain homeostasis. In addition, the SC allows only the passage of lipid-soluble molecules that are of low molecular weight.[10] In skin diseases such as atopic dermatitis (AD) and psoriasis, impaired epithelial architecture and altered physicochemical composition of the SC makes the barrier less efficient. In AD, it has been reported that a decrease in lipid ceramides and a higher proportion of packing of the lipids disrupts the formation of an effective skin barrier. In psoriatic plaques, it has been observed that altered ceramide composition and increased keratinocyte proliferation leads to the formation of an irregular horny layer.[5] Therefore, in some cases, the altered barrier nature of diseased skin may help in the delivery of nanoparticle drugs.

9.2.2 *Skin Penetration Pathways*

In addition to the cellular architecture of the skin, there is also a need to understand the transport mechanisms of nanoparticles into the skin. The absorption of active ingredients through the skin is primarily achieved by passive diffusion via three pathways: intercellular, transcellular and follicular routes. Physicochemical factors such as molecular weight, solubility and lipophilicity of the penetrating molecule influences the diffusion process and hence the route of penetration.[6]

The intercellular route is defined by the tortuous diffusion of molecules through interlamellar regions in the SC. Although the average thickness of the SC is about 20 μm, the diffusion length through this route is approximately 400 μm, suggesting the difficulty of permeation through intact SC.[11] Very often, it has been observed that lipophilic substances follow this particular route.

Another transport mechanism across the SC is the transcellular route, which allows molecules to pass directly through the intracellular matrix composed of keratinocytes with the shortest distance of travel. However, penetrating molecules experience significant resistance as they encounter the lipophilic membrane of each cell present within the SC.[12]

Regarded as the least important pathway in drug penetration — at least until recently — is the follicular route. Appendages such as hair follicles represent only 0.1–1% of the total skin surface area, but allow direct access to the epidermis and act as a depot for drug accumulation.[13] Interestingly, it has been observed that nanoparticles penetrate the hair follicle canal preferentially over the SC, and accumulate in the follicular infundibulum for ten times as long.[14] Studies have also shown that most nanoparticles do not cross the SC and that the transfollicular route may be the dominant pathway of nanoparticle entry into the skin.[4]

9.3 Topical Formulations for Skin Diseases

Researchers have tested various methods to increase the permeation of drug molecules through the skin. These physical and chemical penetration enhancers include electroporation, laser ablation and chemicals, all which may lead to side effects and reduced therapeutic practicality.[15] Current topical formulations such as creams and ointments often produce a highly concentrated layer of active ingredient on the skin, but due to poor skin penetration the amount of drug reaching the target site is often low and variable.

The two key characteristics of modern pharmacology is to achieve temporal and spatial delivery of the active ingredient(s).[12] Recent developments suggest that nanoparticle delivery systems may be useful in improving local and site-directed drug delivery to diseased skin cells. The small size of the nanocarriers allows for interaction at the sub-atomic level with the skin tissue, thus improving drug penetration and facilitating controlled release of the active ingredient(s). Besides ensuring direct contact with the SC and its associated appendages, nanoparticle encapsulation of the drug also shields the drug against chemical instability and degradation.[6]

9.3.1 Types of Nanoparticle Drug Carriers

To improve the efficacy and safety of topical treatment, the choice of drug carrier vehicle remains a crucial one. Hence, researchers have focused on different formulation strategies to increase drug absorption and reduce undesirable side effects, all without reducing drug efficacy. Figure 9.1 shows the different types of novel delivery vehicles that researchers have designed for drug delivery applications and which can be classified broadly as lipid-based and polymer-based carriers, each with their own advantages and limitations.

Liposomes are the most extensively studied lipid-based topical drug delivery systems. They are spherical bilayer vesicles made up of natural or synthetic phospholipids that can be used to encapsulate both hydrophilic and lipophilic drugs.[6,12]

Novel colloidal carrier

Lipid based carriers

Polymer based carriers

Vesicular | Particulate | Emulsion | Self assembled | Particulate | Capsular

Ethosome | SLN | Microemulsion/Nanoemulsion | Dendrimer | Nanosphere | Nanocapsule

Liposome | NLC | Multiple emulsion | Micelle | Microsphere | Microcapsule

Niosome | Liposphere | | Hydrogel

Transferosome

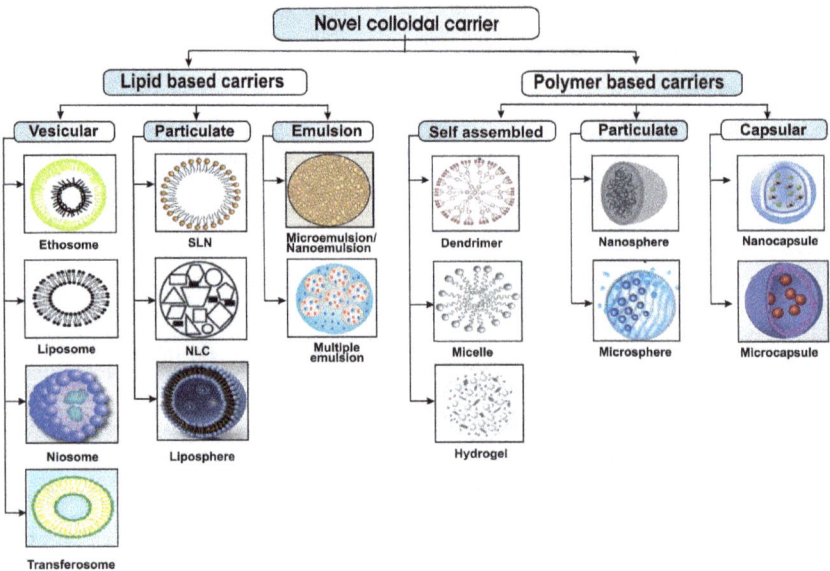

Figure 9.1: Various novel drug carriers.[16]

Liposomes have been shown to achieve significant penetration through the SC and enhance drug deposition in the skin. They have also been shown to be biodegradable and non-toxic, although there have been concerns over drug leakage from vesicles and their overall stability.[17]

Alternatives to liposomes include particulate systems such as solid lipid nanoparticles (SLNs) and nanostructured lipid carriers (NLCs). Lipid nanoparticles such as SLNs may offer a viable alternative to liposomes for their improved stability, controlled drug release, and ease of scaling up.[5] NLCs, a new generation of lipid nanoparticles, are produced by mixing solid lipids with spatially-incompatible lipids leading to a lipid matrix with a less perfect crystalline structure. These NLCs overcome the limited drug loading capacity of SLNs and have been used for topical skin delivery.[6]

Polymeric nanoparticles have received increased attention as topical drug delivery carriers. Natural polymers such as chitosan and synthetic polymers such as poly(lactic-co-glycolic acid) (PLGA) have been shown to be non-toxic, biodegradable and capable of sustained drug release.[18] To further increase skin retention and alter their controlled release profile, researchers have developed physically or chemically cross-linked gel networks of polymeric nanoparticles.[6]

9.3.2   *Skin Diseases*

Skin diseases affect approximately 30 to 70% of the world's population and greatly impair one's quality of life.[19] It affects people regardless of age, from infants as young as three months old to the elderly. The causes of skin diseases are multi-factorial in nature and may be due to genetic, inflammatory or environmental causes, which make skin diseases challenging to treat despite advances in our understanding of dermatology. A survey conducted in 2010 revealed that fungal skin disease and acne are among the top ten most prevalent diseases worldwide, followed by eczema and psoriasis in the top 50 most prevalent diseases worldwide.[20] In addition, skin conditions have also been classified as the fourth leading cause of non-fatal disease burden.[20]

Dermatological disorders can range in severity from acute to chronic, very often starting with simply an itch or burn and developing into more severe conditions, some of which may cause permanent disfigurement and disability. Skin disorders may also lead to psychological, social and financial repercussions on patients, their families, as well as society.[21]

In many of these cases, although a cure may not be available, effective treatment is readily achieved with topical drugs. In this section, we discuss novel nanocarrier-based drug delivery systems for the topical delivery of drugs to treat four major skin diseases: acne vulgaris, atopic dermatitis, psoriasis and fungal infections.

9.3.2.1   *Acne vulgaris*

Acne vulgaris is a common dermatological disorder that affects approximately 95% of the population at some point in their lifetime, often causing physical and emotional distress.[22] Symptoms of the disease include the presence of papules, cysts and scarring, with a large number of sebaceous follicles in anatomical locations such as the face, back and chest. An elevated rate of sebum excretion and hypercolonization of the sebaceous follicles with *Propionibacterium acnes* is a major cause of inflammation.[23,24] Studies have shown that *P. acnes* triggers the release of lipases, free fatty acids and inflammatory cytokines, which leads to follicular hyperkeratinization and inflammatory lesion development.[25]

Treatment of acne depends on the severity of the disease, but first-in-line topical therapy includes the use of antibiotics, retinoids, benzoyl peroxide and anti-inflammatory drugs such as prednisone and dexamethasone. However, these medications can often lead to further irritation of the skin and resistance to antimicrobial agents when used in the long term.[26] As such, researchers have investigated alternative approaches to the standard-of-care, including drug-loaded vesicular systems such as liposomes, SLNs and polymeric microspheres.

In one such study, Honzak and colleagues designed liposomes encapsulating 1% clindamycin to improve the efficiency and safety of standard-of-care treatments.[27] They treated 73 acne vulgaris patients in a double-blind clinical study over six weeks. Results of the study showed that 33.3% of patients treated with the liposomal formulation had no open comedones, compared to 8.33% of patients treated with a clinically-available clindamycin formulation. No side effects were reported with the 1% clindamycin liposomal formulation. The researchers suggest that the mechanism of delivery is due to the liposomes disintegrating in the SC and releasing the entrapped drug locally.

As an alternative to liposomes (which use phospholipids), researchers have developed niosomes by using the self-assembly of non-ionic surface active agents. These novel carriers are capable of holding drugs of different solubility and have been tested for their ability to enhance the residence time of drugs in the SC and epidermis.[28] Goyal et al. designed benzoyl peroxide-loaded niosomes and showed that they significantly decreased the bacterial load compared to free drug and empty niosomes.[29] Inflammation studies were carried out in vivo by injecting P. acnes intradermally into the ears of albino mice. Different formulations were applied epicutaneously daily for a period of four days. Inflammation was inhibited by up to 54.16% when treated with the drug-loaded niosomal gel compared to 50.9% when treated with free benzoyl peroxide, suggesting that the presence of a carrier system enhances the transdermal penetration of the drug.

Recent research has shown that SLNs and NLCs have properties that make them useful for skin applications. These novel nanoparticles have an increased drug loading capacity and exhibit long-term physical and chemical stability.[30]

Ridolfi et al. designed tretinoin-loaded chitosan-solid lipid nanoparticles (SLN-chitosan-TRE) and evaluated their antimicrobial properties in vitro. SLN-chitosan-TRE inhibited the growth of P. acnes and S. aureus whereas SLN-TRE did not exhibit any antibacterial activity. This study shows that a combination of chitosan and SLN improves the carrier properties of SLN, which correspondingly improves the therapeutic efficiency of TRE.[31]

Stecova et al. designed various topical formulations of cyproterone acetate (CPA) using SLNs. Since CPA is poorly absorbed via the skin, it is usually administered intravenously, resulting in severe systemic side effects such as liver toxicity.[32] The CPA-SLN formulation was shown to increase skin penetration of CPA by four-fold compared to a cream formulation. Ex vivo experiments using human skin samples also indicated low levels of CPA present in the dermal layer, suggesting a lowered risk for systemic side effects.[33]

Retin-A Micro, a microsphere gel formulation containing 0.1 or 0.04% tretinoin by weight, is a commercially available topical treatment for acne vulgaris. The biodegradable polymer microspheres demonstrated good efficacy and tolerability in patients.[34]

Using only natural polymers, Friedman and group formulated chitosan-alginate nanoparticles to deliver chitosan for antimicrobial applications. Transmission electron microscopy images showed the ability of blank chitosan-alginate nanoparticles to osmotically disrupt *P. acnes* cell membranes, leading to decreased inflammation. Further studies also showed a similar antimicrobial effect against *S. aureus* and *E. coli*, highlighting the broad-spectrum effect of this platform against various skin inflammatory diseases. When treated with benzoyl peroxide-loaded nanoparticles, the production of *P. acnes*-induced inflammatory cytokines IL-12 and IL-6 was inhibited in human monocytes and keratinocytes, respectively.[35]

9.3.2.2   Atopic dermatitis

Atopic dermatitis (AD) is a chronic inflammatory skin condition that often results in dry, pruritic and inflamed skin. It commonly starts during early infancy and continues through into childhood, but a sudden onset can also occur in 1–3% of adults. The pathogenesis of AD involves a complex interplay of genetic, immunologic and environmental factors. The management of AD is expensive, as it requires long-term application of topical medications and dermatologist visits. In Canada, AD management amounts to an economic burden of approximately US$1.4 billion per annum.[36] At present, there is no cure for AD but topical treatments with ointments and creams are able to keep the symptoms under control.

The clinical hallmark of AD is extreme itching, which can cause patients to experience sleepless nights, anxiety and depression. Therefore, the treatment of AD should be directed at limiting the pruritus and skin inflammation that result from the itch-scratch cycle.[37] AD patients are often prescribed topical treatments comprising of emollients and anti-inflammatory corticosteroids or topical calcineurin inhibitors (TCIs), which commonly include betamethasone valerate, fluticasone propionate, hydrocortisone and tacrolimus.[36]

However, the prolonged use of topical corticosteroids and TCIs has numerous disadvantages, despite their excellent anti-inflammatory effects. First, most of these drugs induce the risk of adverse local side effects such as skin irritation, atrophy, striae and erythema. Second, systemic side effects such as Cushing's syndrome and hypothalamic-pituitary-adrenal axis (HPA) suppression have also been reported in rare cases.[38] Due to the potency of corticosteroids, dosing frequency must be limited to twice daily and only mild to moderately potent steroids can be used for treatment in children. Therefore, nanoparticle carriers have been investigated to improve drug efficiency and reduce drug dosage in the treatment of AD.

Using a solvent injection method, Zhang *et al.* synthesized two different types of SLN formulations for the prolonged and localized delivery of corticosteroids into the skin. The researchers carried out *in vitro* permeation studies on human epidermis samples in a Franz cell chamber comparing betamethasone 17-valerate (BMV)-loaded monostearin SLNs, BMV-loaded beeswax SLNs, a drug suspension, and a commercial lotion.[38] Monostearin SLNs showed a remarkably lower permeation rate than the commercial lotion. Skin treated with monostearin SLNs resulted in a higher BMV content in the epidermal layer, whereas treatment with the commercial lotion resulted in higher BMV content in the dermal layer. Hence, it can be seen that SLN formulations are useful for BMV delivery to inflammatory sites in AD, while also reducing systemic side effects due to the reduced penetration of BMV into the dermis.

In another study, Pople *et al.* designed tacrolimus-loaded lipid nanoparticles (T-LN) using a hot homogenization technique and compared its drug release properties, skin penetration capabilities and side effects with commercially-available ointment Protopic.[39] Tacrolimus is a primary immunosuppressant that reduces inflammation by inhibiting the activation of T lymphocytes. To achieve an optimal therapeutic effect, tacrolimus must target inflammatory cells in the dermis. *Ex vivo* studies on pig ear skin showed greater drug release and accumulation of T-LN in the skin compared to Protopic ointment. *In vivo* studies on albino rats with Protropic showed tacrolimus deposition mostly in the SC, with insignificant amounts reaching deeper into the epidermis and dermis. On the contrary, T-LN favored tacrolimus deposition in the deeper layers of the skin. In addition, the researchers also investigated potential *in vivo* side effects of T-LN and Protopic in rabbits. As the encapsulation of tacrolimus in lipid nanoparticles prevents direct contact between the drug and skin, local side effects such as burning sensations and itchiness should be reduced at the site of application. Rabbits treated with Protopic displayed moderate skin irritation while rabbits treated with T-LN showed no signs of erythema. These studies show the superior performance and safety of T-LN in treating AD, which may also help to improve patient compliance.

To further improve skin permeation, Shah *et al.* modified NLCs with polyarginine chains.[40] The researchers co-delivered two anti-inflammatory drugs, Spantide II (SP) and ketoprofen (KP), to achieve a synergistic anti-inflammatory effect. Comparison of NLC and NLC-R11 (modified) formulations on full-thickness rat skin showed greater penetration depths for NLC-R11, with a significantly higher amount of both drugs present in the SC, epidermis and dermis layers. To evaluate the therapeutic effect of NLC-R11, the swelling of mice ears was induced with dini-trofluorobenzene (DFNB). Measurement of ear thickness using vernier calipers indicated a two-fold decrease in the swell size for mice treated with NLC-R11 compared to NLC.

Polymeric nanoparticles have also been used in the treatment of AD. Hussain *et al.* designed chitosan nanoparticles for the co-delivery of hydroxytyrosol (HT), a potent antioxidant, and hydrocortisone (HC), an anti-inflammatory drug which is commonly used in a cream formulation for the treatment of AD.[41] *In vivo* studies were carried out in NC/Nga mice treated with DFNB to induce allergic contact dermatitis. Advanced digital light microscopy was used to observe microscopic changes in AD-like lesions on the mice skin after treatment. Mice treated with the HC-HT-NP formulation exhibited minimal signs of erythema, hemorrhage and swelling, indicating the potential of chitosan nanoparticles to be used for the co-delivery of HT and HC in AD treatments.

9.3.2.3 Psoriasis

Psoriasis is an autoimmune inflammatory skin disorder which is estimated to affect approximately 2–4% of the global population.[42] Skin cells undergo rapid proliferation in response to inflammatory signals, resulting in bright red patches of thickened plaques and scaling skin.[16] Recent studies have also indicated an increased risk of psoriatic patients developing diabetes, arthritis and heart disease.[3]

A chronic, lifelong disease, the mainstay of treatment for mild to moderate psoriasis is topical therapy. The variable course of psoriasis often results in periodic remission and worsening, requiring long-term pharmacotherapy plans. Several drugs are used in the treatment of psoriasis: dithranol being one of the most effective for topical treatment, methotrexate (MTX) as the gold standard of systemic treatment, as well as other widely used corticosteroids such as betamethasone dipropionate.[43,44] Despite their popularity, topical treatments for psoriasis generally give rise to low therapeutic efficacy. As a result of T cell-mediated inflammatory cascades, subsequent hyperkeratosis and anomalous keratinization may lead to the formation of thickened skin lesions. The thickened skin layer affects the absorption of topical drugs, while prolonged therapy results in toxicity. To improve the efficacy and safety of psoriasis treatments, researchers have developed nanoparticle carriers to increase drug penetration into thickened skin and decrease dose-related toxicities.

Liposomal formulations have been widely studied for antipsoriatic drug delivery due to their high encapsulation efficiency, biocompatibility, small size and elasticity. Bhatia *et al.* developed multilamellar liposomes using a thin film hydration method for the topical delivery of tamoxifen (TAM), which is commonly administered systemically for psoriasis treatment. TAM has low aqueous solubility and often results in side effects such as abdominal cramping and nausea. By encapsulating TAM in liposomes, the researchers showed that drug permeation flux in mice skin was significantly higher at 59.87 μg/cm^2/h for the liposomal gel compared to 24.55 μg/cm^2/h for the Carbopol gel. Furthermore, tamoxifen retention was six-fold higher when treated with the liposomal gel compared to the Carbopol gel.[45]

Aside from liposomes, ethanolic liposomes known as ethosomes are also attracting attention for transdermal delivery applications. Ethosomes are prepared using ethanol in high concentrations, and have a highly elastic structure that allows for improved skin penetration.[46] Using a mechanical dispersion method, Dubey *et al.* engineered MTX-loaded ethosomes that had a drug entrapment efficiency of 68.71% compared to 53.1% with conventional liposomes, because of a greater retention of the poorly-soluble MTX in the ethosomal core.[47] Confocal laser scanning microscopy studies showed that the ethosomes reached a permeation depth of 170 μm in nude rat skin models, compared to liposomes that were confined to a few micrometers deep. In addition, studies carried out by other groups have also indicated that ethosomal formulations are non-irritant and well-tolerated *in vivo* for topical delivery.[48]

Polymeric nanoparticles made from natural polymers such as chitosan have been studied extensively for their biocompatibility and biodegradability. Senyigit and colleagues designed self-assembled chitosan and lecithin nanoparticles loaded with clobetasol-17-propionate (CP) incorporated into a chitosan gel to show the superior performance of the carrier-based gel as compared to clinically-approved formulations.[49] CP is a high-potency topical corticosteroid used to treat moderate to severe psoriatic plaques. The supra-molecular nanoparticles were formed by interactions between the negatively-charged lipids and positively-charged polysaccharide chains. Although a tenth of the drug was loaded in the NP gel, *in vitro* studies showed that the amount of CP retained in pig ear skin was similar in both the NP gel and commercially-available CP cream formulations.

Another recent development is the use of novel tyrosine-derived nanospheres (TyroSpheres) for the delivery of paclitaxel (PTX), an anti-proliferative drug commonly used in the treatment of cancer. PTX has also been investigated for its potential to inhibit the hyperproliferation of keratinocytes associated with psoriasis, but the clinical use of PTX is often limited by its poor solubility. Here, the nanospheres developed by Kilfoyle *et al.* provide a promising platform for the topical delivery of PTX as their small diameters (70 nm) greatly favor permeation and accumulation in the skin.[50] Cytotoxicity and permeation ability of the formulation were tested on human immortalized keratinocytes (HaCaT) and human cadaver skin, respectively. Applied at the same drug concentration, PTX-TyroSpheres gave an IC_{50} value that was 45% lower than free drug. In addition, PTX-TyroSpheres showed selective accumulation in the epidermal layers of human cadaver skin, reducing the risk of adverse systemic responses.

Shah *et al.* designed a hydrogel embedded with PLGA and chitosan nanoparticles for the co-delivery of two anti-inflammatory drugs, SP and KP. Histological staining of samples from an imiquimod-induced psoriatic plaque rat model showed a reduction in epidermal thickness and leukocyte infiltration when treated with the SP-KP hydrogel in comparison to treatment with SP gel or KP gel alone. In addition, levels of the inflammatory cytokines IL-17 and IL-23 were significantly reduced with SP+KP treatment compared to SP and KP single-drug treatments. Increased percutaneous delivery of the drugs to the deeper layers of skin was also reported. It was also significant that the gel formulation helped to improve skin hydration.[51]

In one study testing nanocarrier-based microemulsion gels, Baboota and colleagues investigated combination therapy using a microemulsion formulation of betamethasone dipropionate and salicylic acid. The anti-inflammatory activity of the formulation was studied in a carrageenan-induced rat hind paw model. After a 24 h treatment period, it was observed that the microemulsion gel inhibited edema by 72.11% compared to 43.96% when treated with a commercially-available gel.[52] The increase in therapeutic action can be attributed to the enhanced drug permeation of the nanocarriers in the gel, which also contained penetration enhancers such as fatty acids and surfactants.[53]

9.3.2.4 *Fungal infections*

There has been a steady increase in the incidence of severe life-threatening fungal infections over the past few years.[54] Fungal infections are particularly prevalent in immunocompromised patients, and are a major cause of mortality and morbidity among that population. There are several clinically significant fungal infections in humans such as cryptococcosis, pneumocystis, mucormycosis, actinomycosis, aspergillosis and candidiasis, of which aspergillosis and candidiasis are the most common.[54] Even with the availability of several anti-fungal drugs, the unfortunate fact is that the mortality rate of invasive fungal infections exceeds 50%.[54] Various drug classes like polyenes, echinocandins and azoles are used commonly for treating such infections. However, they possess several major shortcomings such as non-specific toxicity, unwanted drug interactions, water insolubility, poor stability and development of resistance. For example, amphotericin B has poor water solubility[55] and fluconazole gel produces erythema and itching when topically applied.[56]

Researchers have used nanomedicine to overcome some of the current limitations in fungal infection treatments. Metallic nanoparticles, such as silver nanoparticles and zinc nanoparticles, have been explored for the treatment of fungal infections. Due to their high surface area-to-volume ratios, metallic nanoparticles exhibit antibacterial activity at the nanoscale by interacting directly with bacterial cell membranes.[57] In particular, when ionized, silver nanoparticles bind to tissue proteins and give rise to structural changes in the cell wall and nuclear membrane of bacteria, leading to cell distortion and death.[58]

Panacek et al. synthesized 25 nm silver nanoparticles using a modified Tollen's reaction.[59] The nanoparticles inhibited various Candida spp. at very low concentrations with the lowest minimum inhibitory concentration (MIC) at 0.21 mg/L. The silver nanoparticles were further modified with polyvinylpyrrolidone (PVP) to increase their stability and the stabilized nanoparticles had a lower MIC (up to 0.1 mg/L) in comparison to unmodified silver nanoparticles. In addition, the silver nanoparticles did not show any observable in vitro cytotoxicity when tested on human dermal fibroblasts in the therapeutic concentration range. This shows that the nanoparticles can selectively inhibit fungi without inducing non-specific toxicity.

An interesting study led by Gajbhiye and coworkers used fungi as bioreactors for synthesizing silver nanoparticles.[60] The nanoparticles were synthesized by an extracellular biosynthesis route using the fungus Alternaria alternata. When the fungal biomass was challenged with a silver nitrate solution, polydisperse and spherical silver nanoparticles in the size range of 20–60 nm were produced. The silver nanoparticles were shown to potentiate the effect of fluconazole. When tested on various strains like Phoma glomerata, Phoma herbarum, Fusarium semitectum, Trichoderma sp., and Candida albicans, the antifungal effect (computed using the area of zone of inhibition of disk-diffusion assay) of the combination of fluconazole and silver nanoparticles was significantly higher than fluconazole alone in the majority of the species tested.

Zinc oxide nanoparticles have also been found to have good antifungal properties and are commonly used in toothpastes, cosmetics, sunscreens and textiles. Lipovsky et al. synthesized ZnO nanoparticles using an ultrasonication method, which produced monodisperse particles with a diameter of 11.6 nm. The nanoparticles inhibited C. albicans in a concentration-dependent manner. Further, the nanoparticles produced free radicals upon excitation with blue light. The free radicals in turn induced a 60% reduction in cell number compared to control samples exposed only to ZnO nanoparticles or blue light.[61]

Haghighi *et al.* developed hybrid nanoparticles comprising of ZnO nanowires in the core and titanium dioxide (TiO_2) in the shell. ZnO nanowires with diameters of 50–100 nm were first synthesized and TiO_2 nanoparticles were deposited onto the wires by atmospheric pressure chemical vapor deposition.[62] When preformed biofilms of *C. albicans* were deposited onto these hybrid nanostructures, fungal growth was inhibited at a significantly higher level compared to when they were deposited on glass substrates or TiO_2 nanoparticles. Further, the nanostructures inhibited fungal films via photocatalytic degradation upon excitation with visible light.

Apart from metallic nanoparticles, non-metallic nanoparticles have also been used for anti-fungal therapy. Winnicka *et al.* synthesized hydrogels of drug-loaded dendrimers and tested it against various *Candida* spp. The anti-fungal drug keto-conazole was loaded into poly(amidoamine) or PAMAM dendrimers to enhance its stability. The minimum inhibitory concentration (MIC) and minimum fungicidal concentration (MFC) values indicate a 16-fold increase in the antifungal activity of ketoconazole when loaded into PAMAM dendrimers. The positive surface charge of the dendrimers enhanced binding to the negatively-charged fungal cell wall. To aid in topical application, the dendrimers were also formulated as a hydrogel using Carbopol 980.[63]

Polymeric nanoparticles have also been used in the treatment of fungal infections. Bachhav and coworkers formulated aqueous micelles using novel amphiphilic methoxy-poly(ethylene glycol)-hexyl substituted polylactide (MPEG-hexPLA) block copolymers encapsulating anti-fungal drugs such as clotrimazole, econazole nitrate and fluconazole.[64] The micelles were prepared by either a co-solvent evaporation method or co-solvent evaporation sonication method, and exhibited high drug loading efficiencies. The micelles were monodisperse with hydrodynamic diameters ranging from 70–165 nm. Using confocal laser scanning microscopy, the researchers showed that the micelles penetrated the skin through the follicles and hence may be useful for targeted follicular drug delivery. Skin deposition of micellar econazole nitrate was also shown to be ten-fold higher after a three-hour treatment period compared to treatment with Pevaryl (a commercially available liposomal formulation).

Ven and coworkers developed PLGA nanosuspensions encapsulating amphotericin B and compared it with traditional anti-fungal and anti-leishmanial treatments.[65] The nanoparticles were prepared by solvent-antisolvent precipitation and loaded with amphotericin B. The nanoparticles were tested on leishmania species such as *Leishmania infantum* promastigotes, intracellular amastigotes, and fungal species *C. albicans, Aspergillus fumigatus* and *Trichophyton rubrum*. The nanoparticles were compared with free drug (amphotericin B in solution), as well as the commercial preparations Ambisome, Abelcet and Fungizone. The drug-loaded nanoparticles performed significantly better than the other groups tested and also showed very low non-specific toxicity and hemolysis. Further, in an *A. fumigatus* mouse model, the nanoformulation showed the best *in vivo* therapeutic efficacy of all the formulations tested and required only half the active drug dose of Ambisome to achieve therapeutic efficacy.

9.4 Conclusion and Outlook

In recent decades, skin disorders have become a primary clinical concern due to their rising health and economic burden on society. Topical therapy has been identified as the main treatment route for most skin diseases. However, poor skin penetration and local side effects are significant drawbacks that limit the reliability and efficacy of topical drug delivery.

Instead of screening for and characterizing new medicinal drug compounds, a process that requires significant investment into research and development, many academic laboratories and pharmaceutical companies have instead diverted their focus into formulating nanoparticle-based treatments to alter the drug uptake and efficiency of FDA-approved small molecule drugs. Several recent studies suggest the potential for novel nanoparticle delivery systems to enhance site-specific drug delivery and reduce side effects when applied topically. These nanoparticle carriers not only enhance skin penetration into the deeper dermal layers of skin, but also control the rate of drug release and give rise to a drug reservoir in the skin by interacting with the lipid layer and hair follicles in the skin.

However, challenges lie ahead for the clinical approval and application of these nanoparticle systems. First, the synthesis of the various nanoparticle formulations involves multiple manufacturing steps, making the cost of production relatively higher compared to conventional creams and ointments. Second, there is insufficient clinical data regarding their efficiency and safety, making it necessary for more research to be carried out.

Nonetheless, as skin diseases continue to present both a terrible social stigma and growing economic cost, we are likely to find greater interest among nanomedicine researchers in the development of topical skin treatments. As the field matures, we expect to see a concomitant rise in the number of nanoparticle-based topical treatments in the clinic.

Acknowledgements

J.M.C. acknowledges financial support from a Nanyang Assistant Professorship start-up grant awarded by Nanyang Technological University.

References

1. Gupta, S. *et al.* Nanocarriers and nanoparticles for skin care and dermatological treatments. *Indian Dermatol Online J* 4(4), 267–272 (2013).
2. DeLouise, L.A. Applications of nanotechnology in dermatology. *J Invest Dermatol* 132(3), 964–975 (2012).
3. Rahman, M. *et al.* Nanomedicine-based drug targeting for psoriasis: Potentials and emerging trends in nanoscale pharmacotherapy. *Expert Opin Drug Deliv* 12(4), 635–652 (2015).
4. Desai, P., Patlolla, R.R. and Singh, M. Interaction of nanoparticles and cell-penetrating peptides with skin for transdermal drug delivery. *Molecular membrane biology* 27(7), 247–259 (2010).
5. Korting, H.C. and Schafer-Korting, M. Carriers in the topical treatment of skin disease. *Handb Exp Pharmacol* 197, 435–468 (2010).
6. Gupta, M., Agrawal, U. and Vyas, S.P. Nanocarrier-based topical drug delivery for the treatment of skin diseases. *Expert Opin Drug Deliv* 9(7), 783–804 (2012).
7. McGrath, J.A. and Uitto, J. Anatomy and organization of human skin. In *Rook's Textbook of Dermatology*, Wiley-Blackwell, 1–53 (2010).
8. Senyigit, T. and Ozer, O. *Corticosteroids for skin delivery: Challenges and new formulation opportunities.* In *Glucocorticoids — New Recognition of Our Familiar Friend*, Qian X. (Ed.), InTech (2012).
9. Paul, A.J., Kolarsick, M.A. and Goodwin, C. Anatomy and physiology of the skin. In *Site-Specific Cancer Series: Skin Cancer*, Oncology Nursing Society, 1–12 (2014).
10. Del Rosso, J.Q. and Levin, J. The clinical relevance of maintaining the functional integrity of the stratum corneum in both healthy and disease-affected skin. *The Journal of Clinical and Aesthetic Dermatology* 4(9), 22–42 (2011).
11. Hadgraft, J. Modulation of the barrier function of the skin. *Skin Pharmacol Appl Skin Physiol* 14(Suppl 1), 72–81 (2001).
12. Escobar-Chávez, J.J. R.D.-T., Rodríguez-Cruz, I.M. Domínguez-Delgado, C.L. Morales, R.S. Ángeles-Anguiano, E. and Melgoza-Contreras, L.M. Nanocarriers for transdermal drug delivery. *Research and Reports in Transdermal Drug Delivery* 1(1), 3–17 (2012).
13. Barry, B.W. Novel mechanisms and devices to enable successful transdermal drug delivery. *Eur J Pharm Sci* 14(2), 101–114 (2001).
14. Papakostas, D., *et al.* Nanoparticles in dermatology. *Arch Dermatol Res* 303(8), 533–550 (2011).

15. Hussain, Z., *et al.* Downregulation of immunological mediators in 2,4-dinitro-fluorobenzene-induced atopic dermatitis-like skin lesions by hydrocortisone-loaded chitosan nanoparticles. *Int J Nanomedicine* 9, 5143–5156 (2014).
16. Pradhan, M., Singh, D., and Singh, M.R. Novel colloidal carriers for psoriasis: Current issues, mechanistic insight and novel delivery approaches. *J Control Release* 170(3), 380–395 (2013).
17. Schafer-Korting, M., Mehnert, W., and Korting, H.C. Lipid nanoparticles for improved topical application of drugs for skin diseases. *Adv Drug Deliv Rev* 59(6), 427–443 (2007).
18. Zhang, Z., *et al.* Polymeric nanoparticles-based topical delivery systems for the treatment of dermatological diseases. *Wiley Interdiscip Rev Nanomed Nanobiotechnol* 5(3), 205–218 (2013).
19. Bickers, D.R., *et al.* The burden of skin diseases: 2004: A joint project of the American Academy of Dermatology Association and the Society for Investigative Dermatology. *Journal of the American Academy of Dermatology* 55(3), 490–500 (2006).
20. Hay, R.J., *et al.* The global burden of skin disease in 2010: An analysis of the prevalence and impact of skin conditions. *J Invest Dermatol* 134(6), 1527–1534 (2014).
21. Dehkharghani, S., *et al.* The economic burden of skin disease in the United States. *J Am Acad Dermatol* 48(4), 592–599 (2003).
22. Bergfeld, W.F. The evaluation and management of acne: Economic considerations. *Journal of the American Academy of Dermatology* 32(5), S52–S56.
23. Leyden, J.J., and Kligman, A.M. Acne vulgaris: New concepts in pathogenesis and treatment. *Drugs* 12(4), 292–300 (1976).
24. Leyden, J.J., McGinley, K.J., and Vowels, B. Propionibacterium acnes colonization in acne and nonacne. *Dermatology*, 196(1), 55–58 (1998).
25. Jeremy, A.H., *et al.* Inflammatory events are involved in acne lesion initiation. *J Invest Dermatol* 121(1), 20–77 (2003).
26. Vyas, A., Kumar Sonker, A., and Gidwani, B. Carrier-based drug delivery system for treatment of acne. *Scientific World Journal* 276260, (2014).
27. Honzak, L., and Sentjurc, M. Development of liposome encapsulated clindamycin for treatment of acne vulgaris. *Pflugers Arch* 440(5 Suppl), R44–5 (2000).
28. Kazi, K.M., *et al.* Niosome: A future of targeted drug delivery systems. *Journal of Advanced Pharmaceutical Technology & Research* 1(4), 374–380 (2010).
29. Goyal, G., *et al.* Development and characterization of niosomal gel for topical delivery of benzoyl peroxide. *Drug Deliv* (2013).
30. Müller, R.H., Radtke, M., and Wissing, S.A. Solid lipid nanoparticles (SLN) and nanostructured lipid carriers (NLC) in cosmetic and dermatological preparations. *Advanced Drug Delivery Reviews* 54(Supplement), S131–S155 (2002).
31. Ridolfi, D.M., *et al.* Chitosan-solid lipid nanoparticles as carriers for topical delivery of tretinoin. *Colloids Surf B Biointerfaces* 93, 36–40 (2012).
32. Katsambas, A., and Papakonstantinou, A. Acne: Systemic treatment. *Clinics in Dermatology* 22(5), 412–418 (2004).
33. Stecova, J., *et al.* Cyproterone acetate loading to lipid nanoparticles for topical acne treatment: Particle characterisation and skin uptake. *Pharm Res* 24(5), 991–1000 (2007).
34. Kircik, L.H. Evaluating tretinoin formulations in the treatment of acne. *J Drugs Dermatol* 13(4), 466–470 (2014).

35. Friedman, A.J., *et al.* Antimicrobial and anti-inflammatory activity of chitosan-alginate nanoparticles: a targeted therapy for cutaneous pathogens. *J Invest Dermatol* 133(5), 1231–1239 (2013).

36. Watson, W., and Kapur, S. Atopic dermatitis. *Allergy, Asthma & Clinical Immunology* 7(Suppl 1), S4, (2011).

37. Leung, D.Y., Atopic dermatitis: New insights and opportunities for therapeutic intervention. *J Allergy Clin Immunol* 105(5), 860–876 (2000).

38. Zhang, J., and Smith, E. Percutaneous permeation of betamethasone 17-valerate incorporated in lipid nanoparticles. *J Pharm Sci* 100(3), 896–903 (2011).

39. Pople, P.V., and Singh, K.K. Targeting tacrolimus to deeper layers of skin with improved safety for treatment of atopic dermatitis. *International Journal of Pharmaceutics* 398(1–2), 165–178 (2010).

40. Shah, P., P., *et al.* Enhanced skin permeation using polyarginine modified nanostructured lipid carriers. *Journal of Controlled Release* 161(3), 735–745 (2012).

41. Hussain, Z., *et al.* Self-assembled polymeric nanoparticles for percutaneous co-delivery of hydrocortisone/hydroxytyrosol: An ex vivo and in vivo study using an NC/Nga mouse model. *International Journal of Pharmaceutics* 444(1–2), 109–119 (2013).

42. Schmitt, J., and Ford, D.E. Psoriasis is independently associated with psychiatric morbidity and adverse cardiovascular risk factors, but not with cardiovascular events in a population-based sample. *Journal of the European Academy of Dermatology and Venereology* 24(8), 885–892 (2010).

43. Pinto, M.F., *et al.* A new topical formulation for psoriasis: Development of methotrexateloaded nanostructured lipid carriers. *International Journal of Pharmaceutics* 477(1–2), 519–526 2014 (2014).

44. Agarwal, R., *et al.* A novel liposomal formulation of dithranol for psoriasis: Preliminary results. *J Dermatol* 29(8), 529–532 (2002).

45. Bhatia, A., Kumar, R., and Katare, O.P. Tamoxifen in topical liposomes: Development, characterization and in-vitro evaluation. *J Pharm Pharm Sci* 7(2), 252–259 (2004).

46. Barupal, A.K., Gupta, V., and Ramteke, S. Preparation and characterization of ethosomes for topical delivery of aceclofenac. *Indian Journal of Pharmaceutical Sciences* 72(5), 582–586 (2010).

47. Dubey, V., *et al.* Dermal and transdermal delivery of an anti-psoriatic agent via ethanolic liposomes. *J Control Release* 123(2), 148–154 (2007).

48. Paolino, D., *et al.* Ethosomes for skin delivery of ammonium glycyrrhizinate: In vitro percutaneous permeation through human skin and in vivo anti-inflammatory activity on human volunteers. *J Control Release* 106(1–2), 99–110 (2005).

49. Senyigit, T., *et al.* Lecithin/chitosan nanoparticles of clobetasol-17-propionate capable of accumulation in pig skin. *J Control Release* 142(3), 368–373 (2010).

50. Kilfoyle, B.E., *et al.* Development of paclitaxel-TyroSpheres for topical skin treatment. *J Control Release* 163(1), 18–24 (2012).

51. Shah, P.P., *et al.* Skin permeating nanogel for the cutaneous co-delivery of two anti-inflammatory drugs. *Biomaterials* 33(5), 1607–1617 (2012).

52. Baboota, S., *et al.* Nanocarrier-based hydrogel of betamethasone dipropionate and salicylic acid for treatment of psoriasis. *International Journal of Pharmaceutical Investigation* 1(3), 139–147 (2011).

53. Som, I., Bhatia, K., and M. Yasir, Status of surfactants as penetration enhancers in transdermal drug delivery. *Journal of Pharmacy & Bioallied Sciences* 4(1), 2–9 (2012).

54. Brown, G.D., *et al.* Hidden killers: Human fungal infections. *Science translational medicine* 4(165), 165rv13–165rv13 (2012).

55. Mazerski, J., Grzybowska, J., and Borowski, E. Influence of net charge on the aggregation and solubility behaviour of amphotericin B and its derivatives in aqueous media. *European Biophysics Journal* 18(3), 159–164 (1990).

56. Sanjay, J.B., *et al.* Formulation, development and evaluation of Fluconazole gel in various polymer bases. *Asi J Pharm* 1, 63–68 (2007).

57. Sirelkhatim, A., *et al.* Review on zinc oxide nanoparticles: Antibacterial activity and toxicity mechanism. *Nano-Micro Letters* 7(3), 219–242 (2015).

58. Rai, M., Yadav, A., and Gade, A. Silver nanoparticles as a new generation of antimicrobials. *Biotechnology Advances* 27(1), 76–83, (2009).

59. Panáciek, A., *et al.* Antifungal activity of silver nanoparticles against Candida spp. *Biomaterials* 30(31), 6333–6340 (2009).

60. Gajbhiye, M., *et al.* Fungus-mediated synthesis of silver nanoparticles and their activity against pathogenic fungi in combination with fluconazole. *Nanomedicine: Nanotechnology, Biology and Medicine* 5(4), 382–386 (2009).

61. Lipovsky, A., *et al.* Antifungal activity of ZnO nanoparticles — the role of ROS mediated cell injury. *Nanotechnology* 22(10), 105101 (2011).

62. Haghighi, N., Abdi, Y., and Haghighi, F. Light-induced antifungal activity of TiO2 nanoparticles/ZnO nanowires. *Applied Surface Science* 257(23), 10096–10100 (2011).

63. Winnicka, K., *et al.* Hydrogel of ketoconazole and PAMAM dendrimers: Formulation and antifungal activity. *Molecules* 17(4), 4612–4624 (2012).

64. Bachhav, Y.G., *et al.* Novel micelle formulations to increase cutaneous bioavailability of azole antifungals. *Journal of Controlled Release* 153(2), 126–132 (2011).

65. Van de Ven, H., *et al.* PLGA nanoparticles and nanosuspensions with amphotericin B: Potent in vitro and in vivo alternatives to Fungizone and AmBisome. *Journal of Controlled Release* 161(3), 795–803 (2012).

Index

www.ingramcontent.com/pod-product-compliance
Lightning Source LLC
Chambersburg PA
CBHW050542190326
41458CB00007B/1874